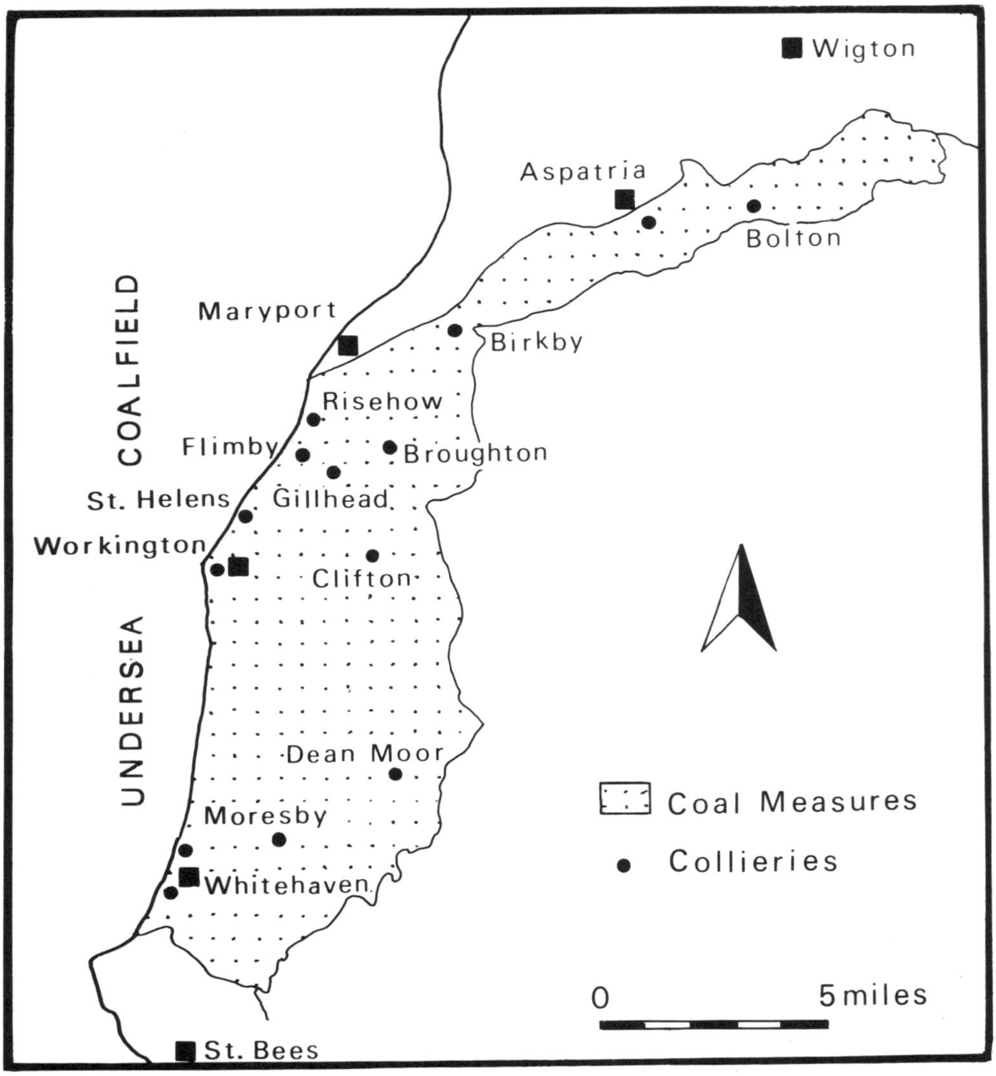

The Cumberland Coalfield

WEST CUMBERLAND COAL
1600-1982/3

by

OLIVER WOOD

CUMBERLAND & WESTMORLAND
ANTIQUARIAN & ARCHAEOLOGICAL SOCIETY
Extra Series XXIV

ISBN 0 9500779 5 X (Hardback)
ISBN 0 9500779 6 8 (Softback)
PRINTED BY TITUS WILSON & SON LTD., KENDAL
1988

TO MY WIFE

Acknowledgments

The Society gratefully acknowledges the generous financial assistance which they have received from the following: Cumbria County Council, Miss Aileen Armstrong for the estate of Mary Fair, The Trustees of the Lonsdale Estate and Copeland Borough Council. Without this help this book could not have been published.

Also Cumbrian Newspapers Ltd. for permission to publish the cover photograph, which was taken at the time of the closure of Haig Pit, Whitehaven and to Carlisle Museum for drawing our attention to it.

CONTENTS

SKETCH MAPS

PREFACE

Historians have devoted little attention to the industries of Cumberland largely because they were of relatively small importance and the area somewhat remote. The history of coal mining, for many years the chief industry of the county, has been neglected. No attempt has been made to trace the evolution of coal mining from early times since the publication of R. W. Moore's lengthy article in the *Victoria County History of Cumberland* eighty years ago. The brief accounts of the industry in the older histories of the county are repetitive and restricted in scope, little consideration being given to such matters as costs, profits and labour relations and, above all, to the position of the miners themselves. In addition, and this is understandable in view of the limitations imposed by the nature of their task, the authors of these works made little use of primary sources.

A new history of coal mining in West Cumberland from early times to the present day is long overdue and this book is an attempt to fill in this gap. It too is long overdue for work on it began over thirty years ago when, as a native Cumbrian, I was first drawn to investigate this subject. It is based not only upon printed sources but also upon a detailed study of original documents, many of which have never been examined before. As a result it has been possible to correct in places the work of older historians and of some very recent authors. To supplement the text, and because they are not quickly accessible elsewhere, fairly copious statistics are provided in the appendices.

In writing a book of this kind a mountain of debt has accumulated, a debt which it is a pleasure to acknowledge. I am particularly grateful to the Earl of Lonsdale for allowing me unrestricted access to the coal mining records in his vast collection of family papers. I owe a special debt to Mr B. C. Jones who for many years, with unfailing courtesy and enormous patience, has guided me through the large and varied collections of documents in the Carlisle Record Office. The successive Area Agents of the Cumberland Miners' Association, Mr M. Rowe, Mr W. S. Proud and Mr H. C. Hanlon, made their records freely available and from their personal knowledge and experience of the industry were able to shed a great deal of light upon it. Mr H. Fancy, curator of the Whitehaven Museum, kindly provided information about the location of many early pits in the Whitehaven area. I learned much about mining methods from Mr J. Graham, Mr B. Lowther and Mr S. Lowther who for a large part of their lives worked in different capacities in the Cumberland coalfield. I am also indebted to Miss Beatrice Warren and Mr Ian Caruana who drew the sketch maps and to Dr J. D. Marshall and Mr R. L. Wyllie for their kindly encouragement and valuable advice over the years.

It would be impossible to write a history of post-war coal mining in Cumberland without the help of the National Coal Board. Much of the post-1946 information in this book has not been published and I owe a large debt to many officials of the Board. Mr L. R. Hoten, Area Statistician of the Western Area and his successor, Mr G. W. Wood, have made considerable efforts to provide much-needed statistics. In addition, the last chapter of this book owes much to their informed criticism. Their colleagues, Mr K.

Hutchinson and Mr R. Crossley of the Public Relations Department, have been very helpful in keeping me abreast of developments in the coalfield since the industry was nationalized. Since there is little printed information about opencast mining in Cumberland the brief section on it in this book could equally not have been written without the willing and ready help of officials of the Opencast Executive – Mr J. McK. Guyan, Mr E. G. Robson, Mr J. C. Graham and Mr S. G. Ross.

I should like to thank the British Academy for a generous grant towards the expenses of research.

Since the last should be first I have to record my gratitude to my wife for accepting cheerfully the demands made upon her patience and understanding for far too many years whilst this work was in progress.

For all sins of omission and commission I am responsible.

CHAPTER I

1600-1705

After more than three centuries of exploitation the small, isolated coalfield of West Cumberland now has a very uncertain future. Out of the hundreds of pits sunk there only one still survives and for years it has failed to show a profit. Possibly the least valuable of all the National Coal Board's assets this coalfield once had national importance, particularly in the eighteenth century when Whitehaven Colliery was the biggest colliery in Great Britain and its mines, the deepest in the country, were a source of wonder to visitors and an object lesson in mining practice. As coal mining grew in West Cumberland it brought enormous changes to the economic life of the region. The coal mining industry was important not only in itself, but also because it made possible the development of coal-based industries, especially the manufacture of iron and steel in the nineteenth century, and of related industries such as iron ore mining and shipbuilding. As a result much of the county's population became concentrated in the coalfield, particularly in the coastal strip between Whitehaven and Maryport. Without its coal there would have been little industrial growth in the county. Yet for over a hundred and fifty years the bulk of Cumberland coal was mined, not for local use, but for export.

Coal mining in Cumberland on a significant scale dates from the latter half of the seventeenth century when geology and geography combined to make it both practicable and profitable. Few records exist to illuminate the history of the industry before the reign of Elizabeth I. Until that time coal mining in the county was minimal, a little coal being worked from outcrops at small expense to meet local domestic needs and to provide fuel for salt pans along the coast. A small stimulus to the industry was supplied by German immigrants who settled near Keswick where they found and worked valuable deposits of copper. They used both charcoal and coal in smelting the copper and in 1565, and again in 1568, they were complaining that their operations were seriously handicapped by a lack of fuel. Their need proved helpful to colliery owners. Coal was delivered to the copper works by pack horses from Bolton Colliery and from mines belonging to Henry Curwen, lord of the manor of Workington. Curwen, apart from having a customer for his coal, was able, in 1574, to borrow iron tools from the company and also men to work in his coal mines.[1]

Coal shipments were impossible until output exceeded local demand and harbour facilities were greatly improved. A report (28 April 1566) by commissioners appointed to survey the 'Ports, Crekes and Havens in Cumberland' shows that there was little accommodation for shipping and that trade was confined to a few pickerdes (small vessels of seven to ten tons) which carried herrings and codfish to Chester and Liverpool from Whitehaven, Parton, Workington and Ravenglass.[2] The origins of the coal trade are obscure but the first recorded coal shipments from Cumberland were made in 1604-5 when Workington exported 1,366 chaldrons of coal to Ireland and overseas. Similar shipments from Workington later increased to 1,908 chaldrons in 1615-16 and to 2,288 chaldrons in 1625-6.[3] So long as the Curwens wanted large quantities of coal for themselves there could be no substantial export of coal from the Workington pits. Sir Patricius Curwen, who succeeded to the family estate in 1623, in addition to using coal in his home and at his salt pans also used it to burn large amounts of lime on his demesne, leaving little coal for export.[4] After 1600 there was a rapid increase in coal consumption at towns all along the Irish coast, particularly Dublin. This stimulated the Workington coal trade but between 1600 and 1640 Workington coal does not appear to have been as successful as that of Flintshire and Lancashire in meeting the growing demand in Dublin.[5] Many small freeholders were then mining coal at Clifton, near Workington. A petition to the Privy Council in 1635 from Dorothy Salkeld, the owner of a mine at Clifton, shows that coal there was conveyed to Workington "and from thence transported and conveyed for Irelande, diverse parts both of Englande and Scotlande". Her petition alleged that Curwen was hindering the movement of coal from Clifton to Workington along roads which he claimed were his and not common highways. The Privy Council, however, supported Curwen's counter-petition that the roads were his and that he was entitled to compensation for damage done to his property by the carriage of coal through his grounds.[6] Jealousy among the Clifton freeholders and the excessive wayleave rents charged by Curwen prevented systematic exploitation and thereby hampered the growth of the coal trade.[7] Competition from Whitehaven also injured Workington's prospects and Curwen was anxious to thwart such competition. In a letter, 17 November 1662, he stated, "I shall also desire to farm the coal custom. It is only 8d. a ton, and all that it has yielded is £60 a year; but as it may be used to divert trade from Whitehaven to Workington, I shall offer £50 a year for it."[8] Mining at Workington declined in the latter half of the century; in 1676 Workington was described as "a fair haven but not so much frequented now with ships, the coleyery being decayed thereabout". Four years later Curwen's colliery and salt pans were valued at only £20 per annum,[9] in 1660 they had been valued at £120 per annum.[10]

The Fletcher family, lords of the manors of Distington and Moresby, were responsible for mining development in the parish of Distington, some three miles south of Workington. In 1614 they leased, and in 1624 bought for £20, the coal royalties in Distington belonging to John Fearon. George Fletcher added to these in 1633 when he leased coal royalties in Distington owned by William Harrison and Christopher Pattison.[11] According to Nef,[12] the Fletchers in 1620, when attempting to extend and consolidate their mineral resources, found their plans being hindered by the rights of freeholders. Six of these husbandmen were eventually deceived, not only into giving away their mineral rights in perpetuity, but also into paying £20 when Henry Fletcher engaged himself to free them from their "lyveryes and wardshippes then claymed" by the King. As Fletcher was careful, in taking over the mines, to keep the only copy of the agreement, the men were left without means of holding him to its terms. In 1675 various Distington royalties belonging to Henry Fletcher were sold for £53 to Sir John Lowther who, in the same year, also bought the adjacent coal royalties of Anthony Walker for £13. 6s. 8d. and Stubscales Colliery from John Banke for £30. Additional coal royalties in Distington owned by Edward Woodhall and others were bought by Lowther in 1699 for £60.[13]

Mines in this parish were also being worked in the last decade of the century by William Christian and Thomas Lamplugh.[14]

As Workington declined the way was open for the rise of Whitehaven as the most important mining centre in the county. This was dependent upon the vigorous exploitation by the Lowther family of the rich coal deposits in the lands once owned by the priory of St Bees. The monks had done little to raise coal and in 1553 came a change of ownership when the priory of St Bees with the manor and rectory were granted by Edward VI to Sir Thomas Chaloner. He, although aware of the existence of coal in his new possessions, did nothing to profit by it; in his time there was little mining development in the Whitehaven area. He made an agreement, 25 October 1560, with ninety of his tenants at St Bees whereby, in consideration of £88 and the payment and performance of customary dues, they were granted a lease of their lands for fifty years and allowed to supply themselves with coal.[15] To the governors of St Bees School in 1586 he granted liberty "to take forty loads of coal at his coal pits in the parish of St Bees, for the use of the school".[16]

The history of the manor of St Bees after 1553 has given rise to confusing and conflicting accounts of it by different writers. An abstract of title to the estate[17] explains concisely what happened:

16 June 7 Ed. 6. Grant to Sr Tho Chalenor Kn^{t}
Sr T. Chal^{r} Kn^{t}, his Son, & W. Fleetwood & Hen. Fleetwood Esq^{rs}
Grant the same to Tho. Wybergh of St Bees Gent 30 Novr 42^{d} Eliz.
Tho. Wybergh Grants to Geo. Lowther of Graystock Gent 11 Feb 42^{d} Eliz.
Geo. Lowther Grants to Ger^{d} Lowther of Grays Inn & Thos Wybergh 14 Feb. 42^{d} Eliz.
10th Mar. 1 Jac. Divided the Cell & Rect^{ry} &c
Gerrards p^{t}

St. Bees Walton Hens^{m} Hindhouse	Tyth	Mortg^{d}. to R^{d} Fletcher of Cockth

31 Aug. 1620 18 Jac. Sr Ger^{d} to his Bro^{r} Wm Lowther of Ingleton in York.
Rich^{d} (son of W^{m} of Ingleton) Esqr, marrys Isabella D^{r} of S^{r}
R^{d} Fletcher, who conveys to him the Tyths of St. Bees &c in Mortg to ye s^{d} Sr R. F. 15 June 1 Car. 1625.
W. Lowther of Ingleton Grants St. Bees to his Son Richd 15 Ap. 6 Car.
W^{m} had Lancelot a 2^{d} son parson at Kirbythore.
16 Ap. 6 Car. W. Low^{r} of Ingleton, his Son & heir Richd Low^{r} & his 2^{d} Son Lanc^{t} joyn in Conveying the Estate at St. B. to Sr J. Lowther of Lowther. Consid^{n} 2450£.
21 Oct 7 Car. S^{r} JL & Mr Wybergh came to a partition of ye Dem^{n} & Saltpans &c.
5 Jan. 14 Car. S^{r} JL. of L^{r}. Bar^{t}, Grants to his Bro^{r}. Chr Lowther.

This document makes it clear that, contrary to the view hitherto held, the St Bees estate was granted to Christopher Lowther by his elder brother Sir John Lowther, Bart. and was not bequested to him by his father Sir John Lowther, Kt.* Not until 1670 did the Lowthers take over all the St Bees estate when Thomas Wybergh sold his part of it to Sir John Lowther of Whitehaven for £850.[18]

* The original grant deed of 5 January 1639 by Sir John Lowther Bart. to his brother Christopher is at the Carlisle Record Office. See D/Lons/W St. Bees Deeds 1/41.

Dr Nef[19] has suggested that the substitution of the lay landlord for the ecclesiastical landlord helped the rise of the coal industry. The layman was more likely to develop his coal resources, to invest heavily in his mines, and lay holdings could be purchased, unlike ecclesiastical holdings which were rarely for sale. In fact there was little mining development at St Bees for a century after it was acquired by laymen. The possession of coal-bearing land by laymen was not enough in itself to ensure the rise of the coal industry. More than this was required. The landowner who wished to profit from the coal lying beneath his land needed abundant capital to finance costly mining operations, a willingness to undertake risks in a highly speculative industry and a market for his product. Wealth was needed to bore for coal, sink shafts, drive levels, install machinery, hire workmen and bear losses caused by flooding or explosions. Waggon-ways added to costs and often expensive harbour works had to be constructed before coal could be exported. Apart from the vagaries of the market, nature often thwarted the mining speculator. Geological faults could bring mining operations to a halt. Exploratory work, at times to no purpose, could be very expensive and mines could be destroyed by fire, floods or explosions, wiping out the work of many years and invested capital amounting to thousands of pounds. It is not surprising that some landowners were never able to make their collieries pay and others, more cautious, avoided the risk of investment losses by leasing their coal royalities to speculators and becoming rentiers. A handful of landowners was responsible for the growth of the Cumberland coal industry. Owning the minerals and with influence to obtain Acts of Parliament favourable to their plans, these landowners – the Lowthers, Curwens and Senhouses – had the capital necessary to exploit the coal deposits on the estates they acquired by inheritance or purchase. Not until the nineteenth century did iron-masters provide capital in large measure for the coal mines, although some owners of furnaces and prosperous merchants did contribute to colliery development in the eighteenth century.

Coal mining on a large and profitable scale was impossible without a substantial and accessible market and one did not exist in Cumberland. This mountainous and sparsely populated county could consume little of its coal. Its remoteness from populous districts and the high cost of transport, before the age of the canal and the railway, restricted the sale of coal to places within a few miles of each pit and there were few buyers. Nor were there industries capable of using coal in large quantitites. Fortunately for the landowners of Cumberland there was an important and growing market in Ireland which provided the outlet they needed for their coal. Since much of the coalfield is coastal the difficult and expensive movement of coal by land was avoided and after a short sea voyage most of the output sold readily in Ireland. Down to the middle of the nineteenth century Cumberland coal was mined primarily to provide fuel for Ireland; it was not until after 1850 that the local iron industry retained a large proportion of the coal raised from neighbouring pits. The rise of the ports of Whitehaven, Workington, Harrington and Maryport was based upon colliery development and more particularly upon the coal trade with Ireland.

A shortage of timber accounted for the Irish demand for coal. Deforestation in Ireland began with the systematic destruction of forests in the Elizabethan period and in the early seventeenth century large amounts of timber were required for buildings and for domestic fuel. Timber was also needed for a growing iron industry and to provide staves for casks, then a very profitable export. These factors, together with an increasing population, created a demand for coal.[20] Coal had been little used; in 1600 Ireland imported a little over 5,000 tons but after that date there was a rapid increase in coal consumption in towns along the Irish coast, above all in Dublin where two-thirds of all the coal imported was burned.[21] By 1636 coal was said to be "almost the only material used for firing along this coast all the winter from Knockfergus (Carrickfergus) to Youghal".[22]

It was the Lowthers who made the biggest contribution to the growth of the coal mining industry in Cumberland. For over two centuries this family worked continuously and profitably the coal under their lands in and about Whitehaven and their success stimulated other Cumbrian landowners to follow their example. The Lowther connection with Whitehaven began in 1630 when Sir John Lowther (1581-1637) of Lowther bought the manor of St Bees for £2,450 and soon afterwards, together with his second son, Christopher (1611-1644), he began mining there. Not much is known about the scale of coal production at Whitehaven in his time, but at least 2,400 tons of coal were sold in 1636. Lowther and his son were partners in the sale of salt and coal to Dublin and as soon as he took over the manor Sir John began to keep a record of payments for coal and salt sold at St Bees. Receipts from the sale of coal indicate that the development of coal mining did not proceed very far:[23]

Year	*Coal Receipts*
	£
1631	—
1632	10
1633	30
1634	30
1635	50
1636	60
1637	40

Coal exports from Whitehaven were hindered by the lack of good port facilities. A letter from the King, 17 May 1633, to the Lord Deputy in Dublin describes Whitehaven as commodiously situated for Ireland and suitable for supplying it with coal. "It is, however, decayed and dangerous, and has been the cause of the wreck of many good ships. A pier would cost £600, and as Dublin will profit largely by it, Dublin must pay for it by paying a small impost on coal in addition to that already paid. The duty shall stop when the sum is raised."[24] After his arrival at Whitehaven Christopher Lowther (who was created a baronet in 1642) soon saw that a pier was necessary to convert the creek there into a harbour capable of dealing with increased traffic and he was largely responsible for the building of a new pier between 1633 and 1634.

Sir Christopher laid the foundations of the family fortunes at Whitehaven. He initiated the policy of buying up land in and near the town, a policy followed by his successors which was to give the Lowthers a monopoly of land ownership in that area. As a younger son he spent much of his brief life in trade. In partnership with his father, his uncle Robert Lowther and Rowland Jackson he was a merchant in Dublin trading in coal, salt, wheat, herring, hides, tallow and other commodities. Coal mining at Whitehaven in his time was on a small scale and was not his first consideration; for him it was a useful adjunct to his salt pans. His pits at Flattfield and Davis Field provided coal for his three salt pans, for the few inhabitants of Whitehaven and for the Dublin market. In 1632 100 tons of coal put on board at Whitehaven at 2s. 4d. per ton and sold for £49 in Dublin, yielded, after the deduction of freight charges, port and customs duties, a net profit of £6. 6s. Whilst in Dublin Lowther bought, among other things, candles for sale to the colliers in his Whitehaven pits. He realized that Dublin could prove valuable as an outlet for his coal but the coal trade was then severely restricted by the small capacity of the vessels engaged in it. He informed his uncle, Robert Lowther, 7 October 1632, "coles this year hath bene at 16s. and 14s. a ton ever since August, soe coles I thinck would be our best trade for as much stock as they would require if we had Flemish bottoms to carry them, but our owne country barks are soe little they carry nothing at a time". A letter of his in 1635 deals with the same problem, "For coles we must thinke of some way to get them carried by great shipping and stored up at Dublin against Winter". Although statistics are not available it is certain that very little coal in those days was exported to Ireland from Whitehaven.[25]

A notebook (dated 1705) sheds some light on what was done at Whitehaven in Sir Christopher's time,

> "The Coal Trade at W(hitehaven) which is now become the only support of the place & also of ye adjoyning Country, has been more or less time out of mind, but the first rise thereof is owing to Sr. Ch. Lowther, who at his own charge in the year 1637*, built a Peer & some ships at Whitehaven (then a very Inconsiderable place & of no Trade) whereby he made some advances towards Introducing a Trade by Sea in Cumberland, & by the Encourag^{t} he gave to Strangers, invited Ships from Liverpool & other Places to trade hither for Salt & Coals, but Salt was then the principal comodety, & of no small advantage to Sr. C. L.
>
> Before the Peer was built the Coal Trade was very Inconsiderable & only suppl'd Sir C. L's salt pans, the County thereabout's & some Open Boats that went coastways, which being all the Consumption there was for Coals, it cannot be imagined the Collieries should be much wrought into in those times when (besides the smallness of their vent) people were very ignorant in the manner of working their Coals and only got some small Quantities in the outburn of such of their collieries as were most easy to come at, which were all wrought at Bearmouths & left Roof full of Water by which they are put beyond an ordinary freeholder recovering owing to ye want of skill, as not knowing the Use of Pitts whereby they might wth more Facility have taken out their Coals, nor the use of Levells to Drain their Collieries, but contented themselves wth bearing out their Coals on men's backs at Bearmouths; where they also discharged ye water by the help of Dams made of earth."[26]

Clearly coal mining was somewhat primitive and not very productive.

Sir Christopher Lowther died in 1644, his son Sir John Lowther (1642-1706) succeeding to the Whitehaven estate when he was an infant. During the latter's minority nothing was done to increase the output of coal from his land; between 1648 and 1659 only two hewers were employed at Greenbank Colliery.[27] Thereafter Sir John, man of considerable energy and foresight, quickly altered the situation by acquiring coal-bearing land which he worked systematically and skilfully. Because of his frequent and prolonged absences from Whitehaven he had to rely heavily upon Thomas Tickell, his principal steward, who served him loyally and dilligently from 1666 to 1692. Tickell, whose annual salary

* The date given here is wrong. It should be 1633-4.

of £20 was raised to £40 in 1679,[28] in addition to carrying out the normal duties of an estate agent also kept a close eye on the management of the collieries and the coal trade. Sir John embarked upon a policy of territorial expansion, first in the Whitehaven area and then in Moresby and Distington, designed to give him control of the collieries in those districts. The ascendancy established by the Lowthers in the Cumberland coal industry did not come easily or quickly; it followed, over many years, the planned elimination of rival coal owners who were either bought out or forced out of business. In the latter part of the century Sir John was not the only one working coal in the Whitehaven district. Among others were William Christian (who between 1678 and 1690 leased and worked coal at Hensingham, Whingill, Corkickle, Moresby and Distington, and in his own land at Moresby),[29] Thomas Addison and Thomas Patrickson. To make his mining operations as efficient as possible and to reduce competition Sir John bought coal-bearing land whenever he could. He spent over £11,000 in buying property – houses, land and coal royalties – but his main concern was to acquire land which contained coal or which helped to provide unhampered access to the harbour. Near Whitehaven he bought tracts of land at St Bees, Corkickle, Hensingham and Akebank.[30] The advantages resulting from the concentration of land in the hands of one man are clearly stated in James Lowther's "Advice about Parton",

> "It is impossible the Coal Trade can be manag'd for the good of any Country or Place unless considerable Collieries are united in one hand. People can neither dry their Coals, nor gett them nor lead them without raising the Price of Coals & that tends to drive away the Trade unless they have such a spread of Coals that they can dry their own Coals without drying their neighbors Coals. Where people have but the Coals of a smal compass, they can work it to little or no advantage the Colliery can but be wrought in part, the best of the Coals are drown'd & lost."[31]

The amalgamation of adjoining collieries would facilitate coal extraction, simplify drainage, and ventilation and enable coal to be moved freely above ground, thus avoiding the payment of extortionate wayleave rents.

According to a Lowther publication[32] Sir John was the first in Cumberland to drain his collieries by means of levels and to employ engines to draw water from coal seams below level. He began by working coal from the outcrops of the Yard, Burnt and Prior bands along the western side of St Bees valley between Greenbank and the Ginns in the locality called Howgill. Outcrop working of this kind was naturally impeded by surface water and to drain the workings a level or watercourse was driven westwards from Pow Beck until it intersected the Burnt (or Bannock) Band and continued along that seam. Begun about 1663 the Bannock Band surface level was eventually extended to a total distance of 1,800 yards and was used for draining the outcrop water into Pow Beck. In 1669 he obtained from the king the grant of 150 acres of land between high and low water marks near Whitehaven which enabled him to control the loading of ships. Between 1668 and 1696 he worked Bransty colliery (formerly worked by Robert Bigland) on the Duke of Somerset's estate, situated on the north of the harbour, to supply coal to the salt pans at Bransty.[33]

Thomas Jackson was appointed manager of Lowther's Whitehaven collieries 20 January 1666 at a salary of £15 per annum and he remained there until April 1675.[34] Other managers[35] followed him in quick succession until the appointment of John Gale in 1682 as "colliery steward". He held office for over twenty years. Few details of the working of these collieries in this period have survived but the account books of the Greenbank and Three Quarters Band Collieries, both dating from 1675, provide a little information. Working costs at Greenbank for the week ending 7 April 1675 were:

Tons	*Bur**		*Days*	*£*	*s.*	*d.*
15	6	John Gilliat & Ro. Milburn 8½d.	3		4	3
15	6	Henry Wilson & Tho. Tyson	3		4	3
5	2	Richard Hunter	2		1	5
36	6	Bearers 14 days work 7½d.			8	9
3 Wattermen 6½d. per day & candles 8d.			5		8	9½
In ye dike 10 men at 8d. per day					6	8
Bankman 3 days					2	0
Pd. Leadings 5 Ton Coales					5	0
		Sum pd.		2	1	1½

* A ton consisted of 8 burthens

In the quarter ending 30 June 1675 Greenbank Colliery produced 876 tons 5½ burthens of coal at a cost of £107. 7s. 10d. Receipts from the sale of coal amounted to £205. 15s. 9½d., leaving a profit of £98. 7s. 11½d. In the same period the Three Quarters Band Colliery raised 444 tons of coal; costs amounted to £34. 13s. 2½d. and receipts to £56. 9s. 3d. leaving a profit of £21. 16s. 0½d. Coal was then sold to the ships at 3s. per ton and to the country at 2s. per ton. These were not large collieries, only five hewers were employed at Greenbank and four at the Three Quarters Band Colliery, each man being paid 8½d. per day.[36]

Sir John Lowther had to meet competition from the colliery owners of Moresby whose coal, shipped from Parton, a village about a mile to the north of Whitehaven, vied with his coal in Dublin. Among his competitors were the Fletcher family, Thomas Addison, Richard Sanderson, William Christian and William Gilliatt. After buying a coal royalty at Croft Pit in 1674, Addison, for a time, was associated with William Fletcher in working coal at Moresby. They jointly purchased land and coal at Lowyeat belonging to Edward Benson in 1676 for £140 and two years later made an agreement with Richard Sanderson to drain his coal at Latters by constructing a level from Lowca to Lattera. Sanderson was to pay five-eighths of the cost of the watercourse and the other two the remaining three-eighths. The three of them were then to join together in mining coal at Lattera, Sanderson bearing five-eighths of the costs and receiving five-eighths of the profits. This plan came to nothing because Lowther in 1679 bought Sanderson's coals at Lattera for £300. In 1680 Addison bought half of Fletcher's coal royality at Howgate and worked a colliery there himself until 1691. He also purchased other coal royalties in Moresby belonging to Aldred Skelton, Matthew Lawrence and John Straughton and in 1684 bought half of Anthony Benn's coal in the manor of Hensingham for £217. 13s. 9d. The other half of Benn's coal royalty was sold to Sir John Lowther at the same time and at the same price.[37]

Sir John disposed of Addison's competition by buying his "collieries in and about Whitehaven" for £500 in 1691. Other competitors in Moresby were also bought out, e.g. William Gilliatt and William Christian, both of whom had pits at work there. In addition

Sir John bought coal royalties belonging to William Fletcher at Lowholme, William Lawrence's coal at Goosegreen and coal at Blearbank belonging to William Asbrigg.[38] These purchases enabled Lowther in 1692 to develop his own Lattera Colliery at Moresby; by 1699 he had five pits there. His acquisition of property was aided by the financial difficulties of his competitors. Colliery owners with limited coal resources and insufficient capital were unable to compete sucessfully and were obliged to sell out to him, many small freeholders having to sell their land as well as their coal. William Fletcher, one of Lowther's chief competitors, was compelled to sell much of his estate and died a debtor in Carlisle gaol.[39]

Lowther's purchases did not end the Moresby threat to his interests for the possibility arose of Parton rivalling Whitehaven as a harbour. The growth of trade in the seventeenth century had been accompanied by little improvement in port facilities and the merchants and ship-owners of Whitehaven, alleging that accommodation there was inadequate, decided to improve the haven at Parton by erecting a pier which would rival Lowther's at Whitehaven. The site chosen was a piece of foreshore between high and low water marks near Moresby. Lowther realized that a flourishing harbour at Parton would not only assist the colliery owners in that area but also diminish the fees payable to him by merchants who would otherwise be using his pier at Whitehaven. He thwarted this scheme by petitioning the king to grant him the Moresby lands between high and low water marks, i.e. the land on which the proposed pier was to be built, so that he could refuse consent to the erection of a pier. In 1678 the whole of the 150 acres was granted to him by fealty and the rent of £1 per annum. Two years later William Fletcher, lord of the manor of Moresby, attempted to build a pier at Parton on Sir John Lowther's land but had to abandon the attempt when Lowther obtained a perpetual injunction against it in the Court of the Exchequer. Lowther followed this up by petitioning the king to appoint Whitehaven a member-port under the head-port of Carlisle, the port of Whitehaven to include the coastline at Moresby. In this too he was successful. In 1681 within the port of Whitehaven legal quays were appointed at Whitehaven, Workington and Milnthorpe, but Parton was not so appointed and the terms were such that further development at Parton was impossible. As a result the merchants had to continue to use Whitehaven harbour and pay Sir John fees for the use of his pier.[40]

The struggle to provide a harbour at Parton was renewed in 1695, when Thomas Fletcher (son of William) and Thomas Lamplugh, who was then working coal at Distington, began rebuilding the pier at Parton. Again Lowther obtained an injunction from the Court of Exchequer prohibiting such work. However, Lowther since 1693 had been mining coal at Lattera Colliery and some improvement at Parton would facilitate the shipment of this coal. He therefore allowed Fletcher and Lamplugh to repair "a little old pier" at Parton which "made a harbor capable of holding about 7 or 8 Ships big enough indeed to carry on the Coal Trade sufficient for that place".[41] The growing output of coal in the neighbourhood made better harbour accommodation necessary and both Fletcher and Lamplugh invited Sir John to join them in obtaining an Act of Parliament to enlarge the harbour, especially because, in their view, he had benefited more than most from the improvements they had already made. They wrote to him, 23 June 1705,[42]

"Notwithstanding the Expence we have been at to make ye Harbour at Parton Sufficient for our Coal Trade yet we find that by the Advantage that you and others have taken to Ship your Coals there, as well as we, that were at the Sole charge of it; the place is not Capable of containing Ships sufficient for Venting the Coales of the several Coal Owners, but yours being nearest the Market, you gett most of the Trade from us who have been at all the Charge. The propperest method therefore whereby to redress this grievance, as we think, would be to gett an Act of Parliament to enlarge the Harbour, and hopeing that you will concurr with us in a thing of so publick as well as private advantage, we thought fitt to Communicate to you our Intentions therein, and if you please to joyn with us we shall look upon it as a favour to ye publick and alsoe to

Sr Yo^r most humble Servt^s
Tho. Lamplugh
Tho. Fletcher"

Sir John refused to co-operate with his rivals who carried on their campaign without him and against him. Their pertinacity was rewarded. Despite the influence of the Lowthers an Act was passed in 1706 empowering trustees to raise £1,600 to pay for the improvement of the harbour by means of a levy of 4d. per ton on coal shipped from Parton over the next eleven years. Sir James Lowther has left an account of this affair[43] which shows the strength of the opposition to his father and explains how this Act came to be passed,

"The Masters also all along encourag'd the building another Harbor at Parton in opposition to him . . . this they did as wel out of Spite to him, as to play the Coale Owners one against another & impose upon them all . . . in the beginning of the winter 1705 Mr Lamplugh & Fletcher gott a Petition sign'd by the Masters of Ships using the Parton Trade, & by the cheif Merchants of Whitehaven out of spite to my Father, wherein they sett forth the great advantages that w^ld arise to the Publick as well as their Neighbourhood by enlarging the Harbor at Parton. This Petition was presented to the House of Commons in Decr 1705 . . . Mr Lamplugh serving for Cockermouth was assisted by the two Knights of the Shire Sir R^d Musgrave & Coll Fletcher out of Spite they envying the improvement of our Estate. He was also assisted by Coll Stanwix out of Spite because he & I had formerly stood against one another in two Elections at Carlisle . . . it was also supported by false witnesses who gave evidence that Parton was more comodious for a Harbor than Whitehaven & my Father dying in the midst of the Proceedings Jan. 1705 I was forc't to leave that affair at London to go to Whitehaven to take possession of my Estate. The Bill was in a short time after carried thro both Houses & past into an Act."

Sir John had also to contend with opposition in Dublin, the Lord Mayor and aldermen of that city in 1680 complaining "That Sr. John Lowther by ingrossing ye Trade to himself occasions Whitehaven Coals to be sold at much dearer rates than formerly". It was further alleged "That the Old Mines next ye peere are poore and almost spent" and "The peere at Whitehaven is much decayed and dangerous". Sir John was also accused of serving the coal traders with writs, cutting their sacks, hindering them from staithing and delaying the loading of ships. Lowther denied these accusations and declared that he could supply Dublin with coal for generations, that the pier at Whitehaven had never been in better condition and was to be enlarged and that he had in no way hampered the coal traders. As to the main grievance, "The price of Coals at Dublyn is so far from being inhanced, that whereas they were formerly Twenty-four Shillings a Chaldron in Winter and Twenty Shillings in Summer, since Whitehaven gott the Trade, they are now about Fifteen or Sixteen Shillings a Chaldron in Winter and Twelve or Thirteen in Summer . . . and that they will be yet Cheaper rather than dearer".[44]

It is not surprising that about this time Sir John felt like giving up his coal mines. The high cost of colliery development, competition with neighbouring and distant coal owners, the struggle over Parton harbour and the dissatisfaction expressed in Dublin must have combined to discourage him from continuing his work. Some of his difficulties and his escape from them are described in an account written by John Spedding c. 1705:[45]

> "When Mr Gale first entered Sr John's service in ye year 168* his Interest in the Coal Trade was very low compared with what it was afterwards & his Circumstances were such that he offer'd his Collieries to Mr Gale for £300 per Ann; he had then several Competitors in the Coal Trade wch equall'd if not outdid him in their vent; for the Country people by keeping their horses upon the Comons which cost them little were alwayes at hand to serve the Country Coal owners which gave them an advantage over Sr John, for his Leaders were forced to be at greater Charge in maintaining their Horses, in and about this Town . . . .
>
> Sr John at that time having lost his Drift Colliery at Flattfield & having nothing discovered but the B(annock) Band & other small Bands at Howgill & Gr(een)bank, was as to ye State of his Collieries reduced very Low, but discovering the Prior Band at Howgill his affairs gained new Life."

After the discovery of the Prior or Main Band, the thickest of all the coal seams in the Whitehaven area, Lowther now had an abundant supply of coal available to him and what he needed was a quicker and cheaper method of moving it to the harbour. His colliery manager, John Gale, suggested the use of carts instead of pack horses and on Lowther's instructions put his suggestion into effect by

> "making a Causey . . . wch when finished he presently Sett on 14 Carts (5 wherof were his own) each 4 bags apeece & for their more ready dispatch had Slings fastened to them so that they had nothing to doe but let down the Tackle & hoyse them up Imediately . . . . This gave Sr John the Comand of the Coal Trade here & brought him so great proffit that the Collieries wch he would have formerly lett for £300 a year now made him £1,700 in one year."

The "causey" or "causeway" recommended by Gale in 1682 was constructed in the following year from Woodagreen Pit to the harbour and consisted of a road bounded on each side with wooden baulks on which the cart wheels would run. According, to Thomas Tickell, "The Carts carry generally 4 loads apeece, some of ym one more, some one lesse and some 8 carriages ye day & some six according to their abilityes".[46] This was much more economic than moving coal by pack horses although pack horses continued to be used in Cumberland long after 1683. It may be noted that Spedding's account disproves the statement made by many local writers that this method of transporting coal "would obviate the use of sacks in loading vessels in the harbour". Among the regulations for the loading of ships it was laid down in 1714, "That if any sack of Coals want Measure, it is to be given back to the Leader & no more of the Sacks that are remaining in his Cart to be taken by any ship in ye Harbour, till he has return'd with them & made up the Coals that wanted".[47]

By this time Lowther had begun "staithing" or storing his coal in bulk near the harbour at Whitehaven. When the coal trade was slack the pits could be kept going by staithing the coal and when trade revived the accumulated supplies could act as a buffer stock to stabilize prices. He wrote to Tickell, 28 August 1680, "As to staithing my

* The complete date is not given in the original, but it should be 1682.

Intentions (as I writ you) are not to staith a Coal when ye ships do take, but my single design is y^{t} Dublin may not Complain of Dearness of coals in Winter or Spring as they did last year, for w^{t} betwixt ye Imployment of Ships in ye cattle Trade, and a stop ther was of Moresby coals, they were 30s ye ton at Dublin".[48]

In the latter half of the seventeenth century coal at Whitehaven could only be loaded at low tide. It was "brought in sacks on Horseback to ye sides of ye Ships at Low water & so put on board . . . but if it be more than half Tyde & consequently ye Ships afloat, ye Coals are either set down upon ye Shoar, or in ye streets of ye Town, or carryed back a qrter of a Mile where each Coal owner has his Winter's provision of Coals & ther left til ye return of ye Tyde".[49] Trouble arose at Whitehaven over colliers being loaded out of turn which not only irritated the masters and owners but also, they complained, helped to raise coal prices in Dublin. Lord Essex, Lord Lieutenant of Ireland, passed on these complaints to the Privy Council. On being questioned Sir John Lowther claimed that the high price of coal was caused by the closing of three or four pits at Whitehaven and that an employee he had dismissed was largely responsible for the complaints. The Privy Council in a letter to Lord Essex, 4 August 1675, decided that "since Coales may be had from other parts of this Kingdome at as cheap rates sufficient to Supply & furnish ye said citty (i.e. Dublin)" to order him if further complaints should be made to "to empower ye Citty by an Act of State, or otherwise as your Lo^{p} shall judge most proper to prohibite ye Importation & buying of any Coales coming from Whitehaven".[50] Fortunately for Lowther and for Whitehaven such an embargo was not imposed. Discontent over loading turns was reduced when fifty-two merchants and shipmasters trading out of Whitehaven bound themselves by an agreement, 28 April 1679, "to loade Coales . . . as wee first come within the Souther head of St. Bees to shut it up within the North Head as formerly; nor refuse any coales in any of our respective turnes of loading that are Merchantible and measure". Any breach of the agreement would cost the offender £40.[51]

Records giving annual production figures at Lowther's collieries are not available before 1695. Between 1695 and 1705 the output of coal from his pits at Whitehaven and Lattera amounted to:–[52]

Year	*Tons*	*Loads*	*Year*	*Tons*	*Loads*
1695	18,905	0	1701	34,410	5
1696	16,686	4	1702	33,866	3
1697	21,939	4	1703	40,869	5
1698	20,153	1	1704	33,275	0
1699	20,364	7	1705	33,426	2
1700	27,286	3			

The success of the New Gin Pit (sunk in 1699) in working coal from the Prior Band led to the sinking of neighbouring pits to the same seam – in 1701 Stone Pit was producing coal and three other pits, Tysons, Millers and Double Shaft Pits, were being sunk nearby. Unfortunately, 7 June 1702, "the Main Dam in the Lowest Endgill at Howgill Broke & drown'd all the New Gin Workings" and the miners there had to be transferred to Knockmurton Pit.[53]

To sell his coal in Ireland Lowther had to compete with the colliery owners of

Ayrshire, Flintshire, Pembrokeshire and Lancashire. In all these counties coal mines could be worked within six miles of the sea and freight charges to Ireland were roughly equal. As a result Ayrshire coal could compete with Whitehaven coal in Ulster and south Wales coal could compete successfully in southern Ireland. Competition for the Dublin market between the mine owners of Cumberland, Flintshire and Lancashire proceeded on fairly even terms in the first half of the seventeenth century. According to Nef Sir John Lowther by 1675 "had driven practically all his rivals out of the Dublin market".[54] This statement antedates the Lowther control of the Dublin market because twenty years later Whitehaven coal had still to compete with coal from Workington, Scotland and Wales. William Gilpin, Lowther's principal steward, in a letter to his master dated 9 January 1695 says,

> "The latest acct I had of ye price of Whitehaven coals at Dublyn was 18s. per Tun Dublin measure (wch is near double of ors). Workington Coals (I know not how it happens) sell for 12d. per Tun more, Scotld sells much cheaper, but are much worse, and vends but few, Liverpool none, Milford affords only a Smithy Coal, Swansey &c afford not so good a coal as yors and sells cheaper, and carry most to Cork &c . . . . 'Tis comonly reckon'd yt Whitehaven supplies Dublin wth one half of wt they use. There were exported of yors this last year above 21,000 Tun."[55]

In Gilpin's opinion Scottish competition was most to be feared, particularly coal from Saltcoats which sold for 12d. per ton less than Whitehaven coal.[56] By 1700 Lowther had gained almost a monopoly of the coal trade with Dublin which his family were to retain for a century.

For the trade between Dublin and Cumberland some merchants between 1660 and 1690 built a new fleet of sixty colliers, ranging in burthen from seventy to 150 tons, whereas the average coal cargo shipped at this time from west coast ports other than Whitehaven and Workington did not exceed thirty or forty tons. Of the 180 vessels said to have been employed in the export of coal to Ireland in 1685 sixty belonged to citizens of Whitehaven and Workington and the remainder to inhabitants of other west coast ports between Plymouth and Cumberland.[57]

The Whitehaven traders when selling their coal in Dublin were harassed by the gabard owners who exacted "double rates upon them under the Notion of Demurrage if they discharge not within so many days, by wch means their Sales being limited to a certain time the Inhabitants there take Advantage thereof and have very much brought down their prices".[58] Two years later they were complaining that they were "forc'd to have them (i.e. gabards) from Eight dayes to Eight dayes & cannot have them for less time, so yt sld a ship sell all her coals in 3 dayes they must nevertheless pay for Eight, or if in Nine they must pay for sixteen, wch seems not to be proper turn of Gabberd, but a penalty enforcing them to sell".[59]

Relations between Sir John and the masters and owners of local vessels were far from amicable, the latter fearing a Lowther monopoly of the coal industry and with it the power to dictate coal prices. The masters were not prepared to accept any price that he might care to fix; they realised that he needed to sell his coal to them just as they needed to sell it in Ireland. In 1699 Lowther had to report that "all ye Ships here combined not to take Coals but at an abatement of price, less than ever was paid before. I submitted rather than loose ye Works, ye consequence was yt much of ye other Pitts, some not under a Water Charge were laid in & so continue, expecting a time for a reasonable Market".[60] For him the situation was made worse by the delusion he then harboured that his collieries were approaching exhaustion. He informed Lord Lonsdale, "The Case is this, ye Collieries are worn out to y^{t} degree ye Trade must remove, not to Workington where they are as scarce as in this place, but to Elnefoot, where ye main of ye Collieries are now left".[61] His forebodings were groundless and Whitehaven continued to produce coal until the present day. Nor were the masters diposed to conduct the coal trade in accordance with his views. As output from his collieries increased Sir John was anxious that the masters should seek new markets for his coal, but they paid little attention to his wishes. Since Whitehaven coal sold readily in Dublin they were unwilling to look for new customers in other ports. In Lowther's words, "We are so used to Dublin y^{t} when Coals bear a price there as they now doe, there is no prevailing wth any of our Ships to goe elsewhere".[62]

Trade with Ireland suffered during the many years of war in the late seventeenth and early eighteenth centuries, numerous ships being sunk or captured by the French. Contemporary records[63] show that frequently Cumbrian masters redeemed their vessels by paying a ransom to their captors and on regaining their freedom resumed their trading activities. Coal exports were further interrupted by a shortage of seamen when they refused to sail without a naval escort or who deserted to avoid impressment. Such difficulties are reflected in the fluctuations in profit at Lowther's collieries and in the volume of coal exports; in 1689 his collieries showed a loss.

Coal Shipments from Whitehaven and Lattera Collieries[64]

Year	*Tons*	*Loads*	*Year*	*Tons*	*Loads*
1695	17,640	0	1701	32,655	4
1696	17,595	0	1702	30,183	5
1697	24,292	4	1703	35,498	0
1698	20,354	7	1704	27,105	2
1699	22,488	4	1705	28,546	4
1700	26,366	3			

In the last years of his life, despite wartime impediments, there was an upward trend in coal shipments from Sir John's pits. Compared with production figures these figures also make clear how much his collieries depended upon the Irish trade. Out of 18,905 tons of coal raised in 1695, 17,640 tons or 93·3 per cent were exported; in 1703 35,498 tons or 86·8 per cent of the 40,869 tons produced left the country.

To increase output from existing pits and to work new ones it was essential to build up the labour force. In Sir John Lowther's time most miners were native Cumbrians, many of them small farmers and labourers who at busy times in the farming year left the mines to return to their husbandry and in doing so seriously injured coal production. Sir John encouraged the immigration of all who could contribute to the economic growth of Whitehaven and skilled miners were especially welcome. To accommodate his colliery employees twelve stone cottages were built in Hodgson's Croft in 1692 and let at £1 per annum; six brick cottages were also built for the same purpose and let at 30s. per annum.[65] A liberal outlook in religion probably helped to attract workmen. When the Protestant dissenters of Whitehaven, 6 July 1694, requested the grant of a small piece of ground on which to build a chapel of their own he readily acquiesced.[66] William Christian who came from Beamish, some eight or nine miles from Newcastle, is said to have employed four Newcastle hewers at two of his Whitehaven pits in 1685.[67] As well as men Newcastle supplied lessons in mining technique as Sir John acknowledged, "Not being able to visit New Castle myself as I have formerly done to my gr^t satisfaction I send ye bearer one of the Stewards of my Collieries to see w^t methods are used there different to what we practice here".[68] There was however, no sudden flood of immigrants into the Cumberland coalfield; large scale immigration, particularly of Scotsmen and Irishmen, did not take place until the latter years of the eighteenth century.

According to the account books of the Greenbank and Three Quarters Band collieries a total of thirty-one persons were employed there in 1675. As Lowther extended his mining operations the number of his mineworkers increased, from ninety-two in 1699 to 113 in 1701. In view of the large number of women and girls employed in the Lowther pits later in the eighteenth century it may be pointed out that all his colliery workers in these years were males. Two lists of employees,[69] which give the names and occupations of the men employed in and about the Lowther collieries, make it possible to analyse the distribution of the labour force.

Employees in the Lowther Collieries

	*1699**		*1701*		
	Howgill	*Greenbank*	*Howgill*	*Greenbank*	*Lattera*
Haggers	10	2	17	5	—
Odmen	10	—	19	—	3
Trailers	22	2	17	3	—
Winders	4	3	7	3	3
Bankers	5	—	4	—	—
Bankman	—	1	1	—	—
Slatepickers	2	—	1	—	—
Level men	13	—	3	—	—
Endgill	—	1	—	—	—
Tubmen	2	—	3	—	—
Dikemen	3	2	—	—	—
Quarrymen	2	—	—	—	—
Sinkers	—	—	11	—	2
Attending ye borers	—	—	—	—	1
Workmen at new causeys	—	—	3	—	—
Highwaymen	3	—	2	—	—
Highway cart	—	—	1	—	—
Carter	1	—	1	—	—
Corvemaker	1	—	1	—	—
Overseers	3	—	2	—	—
Total	81	11	93	11	9

* There are no figures available of the number of employees at Lattera Colliery in 1699.

A recent publication† contains a list purporting to show the number of workmen

† J. V. Beckett, *Coal & Tobacco* (Cambridge, 1981), 65.

employed at Sir John Lowther's collieries in 1705.[70] It gives the total employed at the Howgill, Lattera and Greenbank collieries respectively as 347, 39 and 48. i.e. about four times the total employed in 1701. Such an increase in so short a time is most unlikely; it is highly improbable, for example, that at Howgill the number of haggers would increase from 17 to 82, the odmen from 19 to 82 and the trailers from 17 to 108. The author, unfortunately, appears to have misread the original documents and confused the number of shifts worked with the number of men employed.

Costs of working varied from one colliery to another, much depending upon the depth and thickness of the coal seam and upon the distance from the face to the bearmouth or to the pithead. In the quarter ending 30 June 1675 the Greenbank Colliery's output of 876 tons cost £107. 7s. 10d. or 2s. 5⅜d. per ton; in the same quarter coal from the Three Quarters Bank Colliery cost 1s. 6¾d. per ton. Costs of leading coal to the harbour at Whitehaven from these collieries were, respectively, 1s. and 7d. per ton.[71] Between 1695 and 1705 costs at Howgill and Lattera came to:–[72]

Year	*Output*	*Cost*			*Cost per Ton*	
	Tons	£	*s.*	*d.*	*s.*	*d.*
1695	18,905	2,059	5	5	2	2·14
1696	16,686	2,048	18	5½	2	5·47
1697	21,939	2,577	5	2	2	4·19
1698	20,153	2,424	14	7½	2	4·87
1699	20,364	2,931	12	7	2	10·55
1700	27,286	3,056	8	10½	2	2·88
1701	34,410	3,870	14	0	2	3·00
1702	33,866	3,477	4	9	2	0·64
1703	40,869	3,801	1	10¾	1	10·32
1704	33,275	3,450	9	6	2	0·89
1705	33,426	3,300	8	3	1	11·70

Details of the costs and profits of working coal at the Lowther pits are given in "An Estimate of the Charge and Produce of Severall Pitts Compar'd June 24th 1702":–[73]

Pit	*Price Per Ton*		*Profit Per Ton*		*Cost Per Ton*	
	s.	*d.*	*s.*	*d.*	*s.*	*d.*
Knockmurton B.B.	2	8	1	3½	1	4½
Ditto Pr. B.	2	8	1	4	1	4
Thickett Pr. Band	2	8	1	5¼	1	2¾
Ditto Rd. Coal	2	8		11	1	9
Yard Band	2	8	1	2¾	1	5¼
New Gin	2	8	1	3¾	1	4¼
Gill-close Pit	2	8	1	4	1	4
Greenbank	2	8		11	1	9
Lattera	2	8	1	0½	1	7½

These costs are not total costs, they are only the wage costs involved in cutting the coal, bringing it to the surface and leading it to the harbour. Leading costs varied with distances from the harbour, 1s. per ton from Greenbank and Lattera and 7d. to 9d. from the other pits. Haggers and odmen were paid 10d. per day, trailers 8d. and bankers and winders 7d.-8d.

Total profits from Sir John Lowther's collieries for the years 1666-1705 inclusive amounted to £26,184. 17s. 0¾d., an average of £654. 12s. 5d. per annum. Only once during this period did his collieries fail to show an annual profit, in the year ending 4 September 1690 when there was a loss of £170. 17s. 8¾d. Annual profits ranged from £239 to £1,531, varying with the volume of coal shipments.[74]

Sir John died in 1706 and was succeeded to the Whitehaven estates by his second son James (1673-1755), the elder son, Christopher, having been disinherited because of his drinking and gambling habits and his aversion from work. During his lifetime Sir John had seen output from his mines reach 40,000 tons and coal shipments 35,000 tons in a single year. Not only did he lay the foundations of a flourishing coal industry and provide guiding principles for its later development but he also had the satisfaction of seeing Whitehaven change from "a Town of two or three fishing Boats, and it may be a smal Bark or two"[75] to a town with a rapidly growing population busily engaged in producing and exporting coal. In 1660 Whitehaven had a population of some 250 inhabitants which increased with the growth of coal mining, to 1,089 in 1685, 2,281 in 1696 and to 2,977 in 1702. Carlisle in 1698 had a population, "exclusive of ye soldiers" of 210 families numbering 996 persons.[76]

Coal mining in Cumberland during the seventeenth century, outside the Whitehaven and Workington districts, existed on only a very small scale. Mining at this time was of little importance in inland areas; even if rich deposits of coal had been available transport costs to the coast would have made extensive development impossible. At a colliery in Westward coal was being won in 1603 and in the following year five shallow pits were sunk there. Four hewers were employed who were each paid 5d. per day. The few surviving records provide no information about this colliery's output but with such a small number of men to cut coal it cannot have been very large.[77] There was some activity in the manor of Aspatria in the latter half of the century. In October 1657 Robert Highmore and William Baine applied to "the Rt. Worshipfull Mr Potter & the rest of the Commissioners" for permission to bore for coal there and offered to pay an annual rent of £4 for twenty-one years should coal be found.[78] Nothing is known of what transpired but there was a colliery at Oughterside in 1681 which belonged to William Orfeur of Plumbland.[79] North of this colliery the Duke of Somerset in 1701 had two pits near the River Ellen.[80] Letters written by Sir John Lowther indicate that in 1699 mining was going on near Ellenfoot (i.e. Maryport) "wher Ships begin to goe already".[81] This marks the beginning of the coal trade at Maryport and, although this is not stated, it seems likely that these vessels loaded with coal from the colliery at Broughton.

Notes and References

1 *CSPD 1547-1580*, 258, 319. M. B. Donald, *Elizabethan Copper: The History of the Company of Mines Royal* (1955), 155. W. G. Collingwood, *Elizabethan Keswick* (Kendal, 1912), 31, 155, 188.

2 P. H. Fox, "Cumberland Ports and Shipping in the reign of Elizabeth", CW2, xxi, 74-7.

3 J. U. Nef, *The Rise of The British Coal Industry* (1932), Appendix D, iii.

4 C. Phillips, "The Gentry in Cumberland and Westmorland 1600-1665". (Lancaster University Ph.D. thesis, 1973), 201.

[5] Nef, *op. cit.*, I, 70.
[6] *PCR*, vol. xlv, 248, 487-8.
[7] Nef, *op. cit.*, I, 70.
[8] *CSPI 1660-2*, 627.
[9] I. Fletcher, "The Archaeology of the West Cumberland Coal Trade", CW1, iii, 297.
[10] Phillips, *op. cit.*, 201.
[11] D/Lons/W, An Abstract of Deeds and Conveyances.
[12] Nef, *op. cit.*, I, 308.
[13] D/Lons/W, An Abstract of Deeds and Conveyances.
[14] D/Lons/W, Collieries, List 2, 31-46. VCH, 371.
[15] J. P. Steel, "Genealogical Gleanings Relating to Cumberland", CW2, xxiii, 61.
[16] *VCH*, 359.
[17] D/Lons/W, Misc, Estate Paper.
[18] D/Lons/W, An Abstract of Deeds and Conveyances.
[19] Nef, *op. cit.*, I, 135, 143-4.
[20] J. C. Beckett, *The Making of Modern Ireland 1603-1923* (1966), 28-9. G. O'Brien, *The Economic History of Ireland in the Seventeenth Century* (Dublin, 1919), 43-4.
[21] Nef, *op. cit.*, I, 90-1, 109-10.
[22] *CSPI 1633-47*, 130 – Lord Deputy and Council to the English Privy Council, 30 May 1636.
[23] Phillips, *op. cit.*, 203-4.
[24] *CSPI 1633-47*, 12.
[25] D. R. Hainsworth, *Commercial Papers of Sir Christopher Lowther 1611-44*. Surtees Society, vol. clxxxix. (1977).
[26] D/Lons/W, John Spedding's notebook.
[27] D/Lons/W, Accounts 8.
[28] D/Lons/W, Sir John Lowther to Thomas Tickell, 1 Apr. 1679.
[29] D/Lons/W, Collieries, List 2, 31-46.
[30] D/Lons/W, An Abstract of Deeds and Conveyances. J. V. Beckett, *Coal & Tobacco: The Lowthers and the Economic Development of West Cumberland, 1660-1760* (Cambridge, 1981), 220-1.
[31] D/Lons/W, Estate Memo Books.
[32] *The Case of Sir Jon Lowther Bart.* 1705.
[33] *VCH*, 359. Fletcher, *op. cit.*, 274-5.
[34] D/Lons/W, Estate Accounts 1666-85; Estate papers, List 2, No. 23.
[35] These were: Henry Addison 2 Apr. 1675-18 Aug. 1675; Richard Bettesworth 18 Aug. 1675-8 Aug. 1677; Laurence Branthwaite 8 Aug. 1677-13 Sept. 1682.
[36] D/Lons/W, Account Books of Greenbank and Three Quarters Band Collieries.
[37] D/Lons/W, List of royalties acquired by Addison; An Abstract of Deeds and Conveyances.
[38] D/Lons/W, An Abstract of Deeds and Conveyances.
[39] D/Lons/W, Estate Memo Books, James Lowther's "Observations about the Collierys".
[40] *CTB 1676-9*, 924. R. C. Jarvis, "The Appointment of Ports in Cumberland, Westmorland and Lancashire", CW2, xlvii, 125-65. *The Case of Sir John Lowther Bart.*
[41] D/Lons/W, Estate Memo Books, Lowther's "Observations about the Collierys". *The Case of Sir John Lowther Bart.*
[42] D/Lons/W, Sir John Lowther's letter books.
[43] D/Lons/W, Estate Memo Books, James Lowther's "Observations about the Collierys".
[44] Fletcher, *op. cit.*, 311-13.
[45] D/Lons/W, John Spedding's notebook.
[46] D/Lons/W, Tickell to Sir John Lowther, 13 Oct. 1683.
[47] Spedding, 3 Oct. 1714.
[48] *VCH*, 360.
[49] D/Lons/W, Undated statement among Tickell correspondence.
[50] D/Lons/W, Sir John Lowther to Tickell, 12 May 1674. *PCR*, vol. lxiv, 480.
[51] D/Lons/W, Whitehaven 6.
[52] D/Lons/W, Collieries 1; Colliery Abstracts.
[53] D/Lons/W, Collieries 42; Colliery Abstracts.

[54] Nef, *op. cit.* I, 109-10.
[55] D/Lons/W, Gilpin to Sir John Lowther.
[56] D/Lons/W, Gilpin to Sir John Lowther, 13 Jan. 1695.
[57] Nef, *op. cit.*, I, 392, II, 30.
[58] D/Lons/W, Gilpin to Sir John Lowther, 2 Mar. 1698.
[59] D/Lons/W, Letter books of Sir John Lowther, Lowther to Mr Carleton, 5 Sept. 1700.
[60] Ibid., Lowther to Carleton, May 1699.
[61] Ibid., Lowther to Lord Lonsdale, 3 July 1699.
[62] Ibid., Lowther to Sir Richard Southwell, 10 Apr. 1701.
[63] D/Lons/W, Lowther-Gilpin correspondence; Colliery Abstracts.
[64] D/Lons/W, Collieries 1; Colliery Abstracts.
[65] D/Lons/W, Estate Account Book 1693-99.
[66] Historical Manuscripts Commission, Thirteenth Report, Appendix, Part VII: *MSS. of the Earl of Lonsdale.* (1893), 246.
[67] *VCH*, 360.
[68] D/Lons/W, Letter books of Sir John Lowther, Lowther to Sir W. Blacket, 15 May 1701.
[69] D/Lons/W, Colliery papers; Collieries 42.
[70] J. V. Beckett, *op. cit.*, 65. He gives as his source Carlisle Record Office, D/Lons/W, St. Bees, Box 13, list 1.
[71] D/Lons/W, Account Books of Greenbank and Three Quarters Band Collieries.
[72] D/Lons/W, Collieries 1.
[73] D/Lons/W, Colliery Papers.
[74] D/Lons/W, Estate Accounts; Colliery Abstracts.
[75] D/Lons/W, W. Gilpin to Sir John Lowther, 2 Nov. 1694.
[76] D/Lons/W, Gilpin-Lowther correspondence.
[77] D/Lec 119, Westward Colliery papers.
[78] D/Lec 41, Oughterside Colliery papers.
[79] R. S. Ferguson, "The Colliery, Harbour, Lime and Iron Tokens of West Cumberland', CW1, xv, 408.
[80] D/Lons/W, Plan of Estates marking coal pits, 1701.
[81] D/Lons/W, Letter books of Sir John Lowther, Lowther to Lord Lonsdale, June 1699, 3 July 1699.

CHAPTER II

1706-1755

WHITEHAVEN DISTRICT

By 1700 a beginning had been made to exploit the coal seams of Cumberland, but it was little more than a beginning. Pits were shallow, methods primitive and mining science in its infancy. Large areas of coal remained untapped. Mining throughout the coalfield on a comparatively large scale came later, when its geology became better known and when other wealthy landowners, besides the Lowthers, were prepared to risk their capital in colliery enterprises.

At this point a brief description of the geology of the coalfield may be desirable. The Cumberland coalfield extends along the coast from Whitehaven to Maryport, a distance of fourteen miles, and is from four to six miles in width. It then turns inland in a narrow strip, two miles wide, to the north-east for a distance of twelve miles, almost to Wigton. This field has an area of ninety square miles; in addition an area of undersea coal from St Bees to Maryport has been worked for two centuries, the present workings at Whitehaven stretching about four miles from the coast. A further six square miles of coal have been worked under the Permian strata of St Bees Head, bringing the total area of the coalfield to 108 square miles. There are about twenty coal seams, of which eight have been worked, varying in thickness from two feet to sometimes fifteen feet in the Main Band at Whitehaven. Mining operations are hampered by frequent faults and variations in the thickness of the seams. Along the coast between Flimby and Whitehaven the coal measures dip seawards, the dip gradually increasing southward, a geological feature which makes undersea mining possible. North of Flimby the coal rises upwards towards the sea floor and the limited amount of cover (fifty fathoms of cover being the minimum requirement) makes undersea winning of coal impossible.[1]

In the eighteenth, as in the seventeenth century, it was the Lowther family who contributed most to the development of coal mining in the county and James Lowther (1673-1755), who succeeded to the baronetcy on the death of his elder brother Christopher in 1731, was the greatest entrepreneur of them all. With skill and foresight he carried on his father's work to make Whitehaven Colliery one of the leading collieries in the kingdom. A shrewd, clear-minded business man continually in pursuit of money, he used twentieth century methods against eighteenth century competitors. His inherited wealth became the basis of a great fortune since it was employed to acquire more lands and to exploit their coal deposits. Endowed with an extraordinary talent for making money, he not only invested in mines but also in other industries and in shipping, and supplemented his income by lending money at interest to local merchants and to impoverished landowners on the security of their estates. Known as 'Farthing Jemmy' because of his notorious parsimony he died one of the richest men in the country. His success at Whitehaven is all the more remarkable when it is remembered that he was an M.P. for over fifty years who spent long periods in London. During his many protracted absences from Cumberland he was in constant touch with his senior employees at Whitehaven, insisting upon frequent and detailed reports on all matters relating to his collieries and there was very little that could be done without his consent. Fortunately for him his subordinates proved both loyal and competent, particularly William Gilpin (1657-1724) who succeeded Thomas Tickell as principal steward in 1693 and the two Spedding brothers, John and Carlisle. The Spedding family gave almost a century of devoted service to the Lowthers and their holding of key positions for many years was a major factor in the successful development of the Lowther collieries.

A contemporary writer expressed the view that, "The 3d curiosity at Whitehaven or rather the greatest, is Sr.Ja.Louder himself who in the midest of great riches lives but in a poor way. He is an indolent old man and knows nothing about coal works but in order to grow rich carried them on by the best advice and seems indeed to be very well served."[2] That Sir James Lowther lived frugally for a man of his wealth is beyond question; his passion for economy was obsessive. All expenditure was closely controlled and strict accounts had to be kept. But no one who has read his enormous correspondence would describe him as "indolent". A lazy man could not have built up such a vast fortune as he bequeathed; it was the product of an active and skilful money-maker who made the best use of his own inheritance. Throughout his long career it was he who made the decisions about his collieries and other business interests. Nothing that could in any way benefit his coal mines was overlooked – the elimination of competition, the purchase of land, timber, iron and cordage, the hiring of labour, the adoption of improved mining techniques and the management of the harbour all received Lowther's personal attention. The allegation that he was ignorant of coal mining is baseless. He was not a colliery manager – he hired colliery managers. Nevertheless, he was well informed about coal mining thanks to the innumerable reports sent to him from Whitehaven and, since a large part of his income came from collieries, he had the sense to try to find out all he could about the latest developments in mining methods.

John Spedding (1685-1758) was the eldest son of Edward Spedding[3] and entered the service of the Lowthers in 1700 at a salary of £10 per annum. For over fifty years he laboured diligently to promote their interests and his ability was soon recognized. Sir John Lowther's will, 8 October 1705, states: "I bequeath to John Spedding and William Cuppage, each of them, two years' wages, and for that they are fully apprized of all the particulars of my Estates, especially my collieries, I recommend them both in a particular manner to my son, James Lowther, to be imployed in controuling the steward's accounts, or otherwise, as he shall think fit."[4] An important part of his work during the early years of his service was to keep a close eye upon the activities of John Gale and to report upon them to his master.[5] After Gale's dismissal in 1707 Spedding managed the collieries and in 1722, when William Gilpin retired, he joined the latter's son, Richard, in the management of the Lowther estates. Richard Gilpin neglected his work and on his removal from office in 1730 John Spedding became principal steward, his brother Carlisle then taking over the direction of the collieries. He kept detailed accounts of mining costs, coal sales, profits and losses, collected money from shipmasters and maintained a voluminous correspondence with James Lowther to whom he wroted literally thousands of letters, many of them accompanied by colliery accounts. He was constantly seeking coal bearing land or other land that might be useful to his employer; indeed he would have welcomed a Lowther monopoly of coal in Cumberland. Lowther's many land purchases were made on Spedding's advice and it was Spedding who usually conducted the negotiations prior to purchase. Similarly it was he who recommended which vessels should be bought by his master.

Throughout his career his devotion to the Lowther cause was complete and he displayed an exceptional capacity for work combined with a considerable talent for business. But he had less desirable traits. Although convinced that his own conduct reached the highest standard of moral rectitude he could, to gain his ends, descend to deceit and dissimulation and at times show a large measure of ruthlessness. When he had decided that a colliery was worth buying it was often after securing all the information he could from employees of the colliery who were then rewarded with jobs in the Lowther mines. Sometimes, to lower the price or to make the purchase possible at all, the colliery was bought in Spedding's name, or in that of some other supporter of the Lowther interest, and subsequently conveyed to Lowther. Other tasks, unconnected with mining, fell to the lot of this indefatigable worker. At election times he did his utmost to ensure Lowther's return to Parliament, contacting voters, and, when opportunity offered, buying burgages. In addition he saw to it that the Lowther mansion at Whitehaven, together with its furniture, plate and gardens were kept in good order, that sufficient ale was brewed before his employer returned home from London and he had, on occasion to settle the squabbles of the domestic staff.

In the first half of the eighteenth century coal output and coal shipments from Whitehaven Colliery were more than trebled. Such progress resulted from the purchase of more land, the sinking of many new and deeper pits to richer seams and the adoption of more efficient mining methods. Early in the century it was found necessary to dismiss the colliery manager, John Gale. In his time a great deal of slackness had grown up in the pits and his inordinate delay in handing over money received from the sale of coal aroused suspicion that he was using it for his own purposes. Following a series of complaints from Spedding, Gale was dismissed in March 1707 after handing over his accounts and what money he had in his possession. Investigation showed that Gale had embezzled a large sum of money which he was compelled to pay back and this required the help of members of his family. Lowther's comment on Gale's "knavish" actions was, "It will be his Interest to lay open the whole Truth & to pay me w^t^ he can & Secure the rest for tho' with his Jugleing and Jesuitical Evasions he may Shuffle on a little longer yet it will fall the heavier upon him at last."[6]

It was not until 1710 that the greatest of all the Whitehaven mining engineers entered the service of the Lowthers. This was Spedding's youngest brother Carlisle (1695-1755), then a boy of fifteen who might quite easily have become a sailor instead. His introduction to mining is described by John Spedding, "As to what Mr Gilpin was pleased to mention about my young Brother being Imployed about ye Collieryes; I intended to have put him to Sea wth some of Mr. Gale's sons, but none of them being willing to doe it for this Year . . . . I shall make him assist me in ye business, which if he takes well to I hope will be to yor advantage, if not 'twill be no more charge for me & better for himself to keep him to business than otherwise."[7] Carlisle's fortuitous change of career was to bring great benefit to the Lowthers and to the mining industry in general. The boy learned quickly and after his appointment as manager of the Lowther collieries in 1730 his skill and inventiveness earned him a national reputation. According to Hutchinson,[8] Carlisle Spedding spent some time in the Newcastle mines as a hewer, working under an assumed name, in order to learn all he could about coal mining and colliery management. Suspicion was aroused when he was injured in a pit explosion, for this apparently obscure collier received the best medical treatment then available. There is no evidence to confirm this story. Since the Lowthers had close contact with Newcastle colliery owners and Whitehaven coal did not compete with Newcastle coal it seems unlikely that James Lowther would have to descend to industrial espionage to learn details of Newcastle mining methods.

The adoption of the Newcomen steam pumping engine made possible the sinking of Saltom Pit, perhaps the most notable of Spedding's mining achievements. A bore on the seashore at Saltom had shown Main Band coal at 480 feet and Spedding, showing enterprise unusual at that time, sank a pit close to the high water mark to work not only the land coal but also the coal under the sea, relying upon the steam engine to drain the workings.[9] Sinking began 4 March 1730 and coal getting 9 February 1732, an event celebrated by the expenditure of £1. 6s. on "Ale at Coaling the Pit". Detailed accounts show that by the middle of 1733 £4,521. 13s. 3¼d. had been spent on this pit. In addition £950 was spent in "Driving the Air Race in the Ban^{ck} Band at Saltom begun Octob^{r} 12 1732".[10] The Saltom shaft, oval in shape, was between eight and ten feet in diameter, "divided in the middle all the way down and wainscotted all over. In one half of this sink (i.e. shaft) the pumps draw, as will likewise the pumps of the new engine built on the side. At the other half the coals are drawn up by a horse engine".[11] This pit, one of the first in the country to raise undersea coal, was highly productive and had a working life of over a century.

Since the pit was close to the shore it was decided to export its coal from Saltom Bay in order to avoid the cost of transporting it to Whitehaven harbour for shipment. A small staith and quay were built there in 1732 but the harbour did little business for it was visited by only a few small vessels. In 1733 861 tons of coal were shipped from Saltom and 1,607 tons in the following year; in the next two years coal exports slumped to twenty-two and ten tons respectively. To dispose of Saltom coal it therefore became necessary to construct a waggon-way to the staith at Whitehaven and this was completed in 1735. The coal was raised to shore level, conveyed along a short drift to the bottom of Ravenhill Pit and there raised by a second gin to the waggon-way at the top of the cliff, a height of twenty-seven fathoms. In addition to Saltom Pit, Spedding sank Thwaite, King, Duke, Kells, Ravenhill, Fish, Fox, Arrowthwaite, Parker, Newton, Country, Moss and Hind Pits on the Howgill (i.e. western) side of Whitehaven.[12]

Following his purchase of the Fletcher coal royalty at Priestgill in 1710 James Lowther decided to re-open Whingill Colliery where several shallow pits had been worked by William Christian in the last twenty years of the seventeenth century. Coal production was resumed in 1714 and this marked the beginning of the Whingill or eastern division of Whitehaven Colliery. Many new pits were sunk and among those for which Carlisle Spedding was responsible were Carr, Pearson, Pedlar, Taylor, Fox, Daniel, Jackson, Hunter, Watson, Harras and Green pits.[13] To drain the colliery a level had to be driven about one and a half miles. A letter written by him in 1750 describes this difficult task,

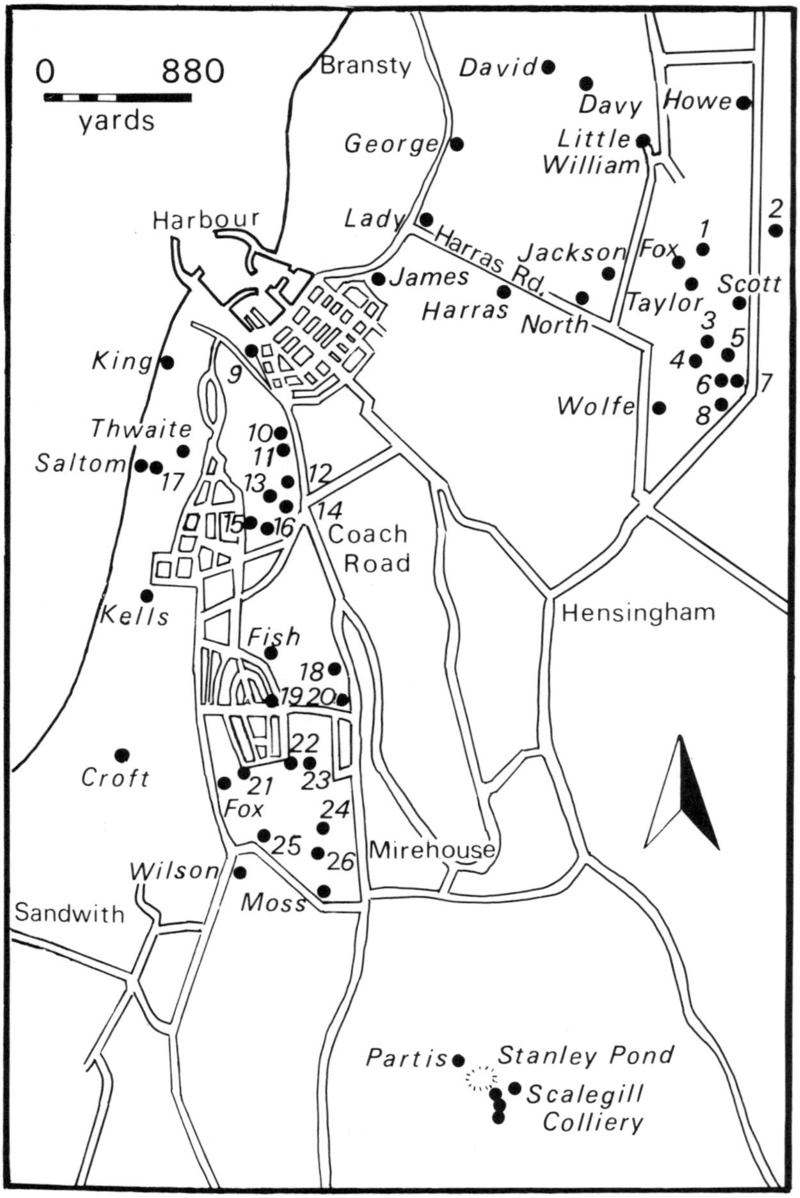

FIG. 2. – Whitehaven Pits in the 18th Century

• **Numbered Pits**

1 Hunter	8 Daniel	15 Arrowthwaite	22 Bank
2 Bateman	9 Duke	16 Swinburn	23 Baxter
3 Pedlar	10 Newtown	17 Ravenhill	24 Thicket
4 Carr	11 Pedlar	18 Knockmurton	25 Watson
5 Watson	12 Harris	19 Hind	26 Country
6 Pearson	13 Clay	20 Greenbank	
7 Cauliflower	14 Engine	21 Gameriggs	

'We have just Thirld our level to Whingill Colliery which we have been 12 years in driving. Our first pitt is near 100 yards from the Levell mouth, our 2nd pitt near 700 yards from the first and then beyond the 2nd pitt 'tis 700 yards to the Coale and from that place where we Coaled the Levell we have wrought 200 yards upbank . . . and by Continueing the Levell thro' Severall Large downcast Dykes to the rise shall I hope make a Dry Colliery for Ages."[14]

With the sinking of new pits there was a marked increase in production at the Whitehaven Colliery and its progress is shown in the following output figures:

Year	*Howgill*		*Whingill*		*Total*	
	Tons	*Lo.*	*Tons*	*Lo.*	*Tons*	*Lo.*
1701	31,980	5			31,980	5
1711	35,968	1			35,968	1
1721	41,883	5	9,193	1	51,076	6
1731	69,776	0	17,674	3	87,450	3
1741	82,524	5	7,328	0	89,852	5
1750	108,209	7	17,918	2	126,128	1

During this period the annual output ranged from a minimum of 26,414 tons in 1706 to a maximum of 126,128 tons in 1750. For the first time in its history the colliery produced over 100,000 tons of coal in a single year, in 1736 when 101,905 tons were raised. Between 1730 and 1744 the annual output exceeded 80,000 tons and in the next six years over 100,000 tons of coal were mined each year. Whingill was much less productive than Howgill. Only once, in 1730, did it yield more than 20,000 tons and in the years 1713-15, 1720-25 and 1735-46 annual production was less than 10,000 tons.[15]

Throughout his life Sir James Lowther, like his father, bought land systematically to consolidate his coal empire; whenever he sold land he invariably reserved the coal rights for himself. Between 1709 and 1754 he spent over £46,000 in buying land and coal royalties in Cumberland as part of a carefully planned scheme designed to destroy his competitors. Land purchases were supplemented by leases of land which might contain coal or provide wayleave. Although the most acquisitive member of a highly acquisitive family and most anxious to establish a local monopoly of coal, this was not mere greed on his part, but sound economics. Such purchases reduced competition, eliminated the need of surface wayleaves and made possible an efficient drainage system, a system quite impossible where small pieces of land were held in separate ownership by men who could not afford expensive levels. More than anything else it was the power of the purse which was responsible for the dominance of the Lowthers in the Cumberland coalfield. Among his many purchases in the vicinity of Whitehaven were estates at Corkickle, Weddicar, Wreah, Stockhow, Arlecdon and Frizington, smaller pieces of land in St Bees, Arrowthwaite and Woodhouse and coal royalties in Corkickle, Whillimoor and Bigrigg. In 1753 he also leased Lord Egremont's coal royalties in Corkickle and Whillimoor for twenty-six years at an annual rent of £20.[16]

Some of Lowther's land transactions depended upon his deliberate exercise of financial pressure upon landowners much poorer than himself. To acquire properties he wanted he was quite ready to lend money to landowners on the security of their estates in the hope that they would be unable to repay the loans and he would then be able to foreclose. This policy is summed up in his own words, "I shal be willing to lend Wm Beeby money on a Mortgage of his Estate at Flimby as it is the likelyest way to gett the Purchase of it."[17] As it happened he did not obtain this estate although he did lend money to Beeby. His devious tactics, however, were successful when dealing with other landowners. One of these was Thomas Patrickson who owned an estate at Stockhow and a colliery at Scalegill which had been worked since 1680. In 1706 he was said to send "more coals to W(hitehave)n than all ye Country Coal owners" but by 1720 he was in financial difficulty and had to mortgage his estate to James Lowther for £2,000. His delays in paying interest on the debt led to threats of foreclosure and in 1726 he had to surrender Stockhow to Lowther for £2,252. 4s. Four years later he sold his collieries at Scalegill and Linethwaits to Lowther for £500. Scalegill Colliery, about two miles to the south-east of Whitehaven, was worked by the Lowthers from 1732 to 1776 when it was flooded. Until about 1759 its output was exported from Whitehaven to which the colliery was connected by a branch waggon-way completed in 1739; afterwards the colliery was worked for inland sale.[18]

Lowther, with a complete lack of scruple, abused his position as a school governor to obtain a lease of the St Bees School coal royalties in 1742 and in this he had the willing and active support of John Spedding who was another governor of the school. As early as 1726 Spedding informed his master of the possibility of a long lease of these royalties on very moderate terms, but this affair moved very slowly. In 1741 Spedding announced his intention of securing the lease with the support of three other school governors; the four of them together would form a majority of the school's seven governors. Eventually Spedding was able to report to Lowther, 29 October 1742, "We had a Court at St Bees today – Mr Ritson, Mr Patrickson and I were all that attended . . . The lease of ye Collierys in the School land to you for £3. 10. 0. per ann. was Executed by the 3 Go^{rs} present by affixing the common seal to it in the presence of the Master." A manifest injustice was done to the school. Not only was the rent absurdly low, the term of the lease, 867 years, was also excessively long. Nor was there any tonnage rent payable. According to Matthias Dunn, "The fair term in 1742 ought to have been about 3½d. per ton, a lease not exceeding 42 years and a Certain Rent of from £100 to £150 per annum." Lowther himself seemed to have some doubts about this lease for two years later Spedding wrote to him, "As to the Lease of the Royaltys in the School land it is settled by a Lease under the common seal wth the approb^{n} of Mr Ritson, Mr Ponsonby, Mr Patrickson and myself who are a majority of ye Gov^{rs} ye 29th Oct. 1742 at the Rent of £3. 10. 0. & I have pd 2 yrs Rent."[19] How very odd!

Sir James continued his father's policy of acquiring collieries in the Moresby district in order to remove competition with Whitehaven coal. He bought collieries in Moresby which were too small to function successfully on their own and whose efficiency depended upon consolidation. In June 1718 he wrote, "To recover such Collierys in Moresby & make them usefull to the Neighborhood my Father & I at a mighty expence have bought in & laid together a great many of these ruin'd Collierys. While we were buying the People of Whitehaven made great outcrys of a monopoly tho' the only way to work them to any advantage to the Country."[20] In addition to small parcels of land he bought the Priestgill, Blearbank and Goosegreen estates in Moresby and coal royalties belonging to Thomas Fletcher, Richard Sanderson, John Sanderson and others. One competitor who had to sell out to Lowther was Thomas Lamplugh who in 1723 was said to command two-thirds of the trade at Parton. Another was John Brougham. Although he was 'a stranger to ye working of Collierys' Brougham, after buying the manors of Moresby and Distington for £4,100 from Thomas Fletcher in 1722, believed he could mine coal profitably in Moresby. He invested heavily in collieries and worked with Lowther to restore the harbour at Parton from which his coal would be exported. By 1726 he was regretting his mining ventures although it was not until 1737 that he sold his manors at Moresby and Distington to Lowther for £6,600.[21] With these purchases Sir James now controlled mining in Moresby as well as Whitehaven and both the harbours at Parton and Whitehaven were firmly in his grasp.

Between 1695 and 1708 output at Lowther's Lattera Colliery ranged from 855 tons to 4,599 tons per annum and during these years the colliery showed an annual loss. Sir James therefore decided to close the colliery and production ceased in December 1708. In 1714 he bought Thomas Lamplugh's coal royalties in Distington for £2,400 and in the same year re-opened Lattera which he worked together with the Gunerdine Colliery at Distington, the coal from both being shipped from Parton. Lowther had two more collieries in the Parton district, Birketts which was in production 1718-23 and Quarter-banks 1725-31. Although the Parton pits made losses in nine of the twenty-four years between 1714 and 1737 they still cleared a profit of over £2,000.[22] After 1737 very little mining took place. The new Hall Colliery at Moresby began in 1738 but its one pit, Hutton Pit, ceased to draw coal in the following year. At Lattera coal was worked intermittently. In 1741 the Lister and Fisher Pits closed and coal work was not resumed at Lattera until 1742 when Lamb Hills Pit was opened out and the Metal Band worked until 1750. Blearbank Pit was sunk in 1751 but stopped soon after 1776 and coal mining in Moresby ceased for the next fifty years.[23]

South of Moresby Gilfrid Lawson between 1710 and 1730 was raising coal from a number of pits in Hensingham. His efforts were not very successful; output was usually less than ten and never above twenty tons per day and excessive water in the pits led to high drainage costs. The methods employed to extract coal seem to have been remarkably inefficient. At one of his pits where there was a seam of coal twelve feet thick the daily output did not exceed twenty-seven corves. Despite its limitations Lawson's mining activity was of some concern to John Spedding who kept a close eye upon all competitors. Lawson was able to attract some of Lowther's colliery workers and coal leaders to work for him and also to take over some of the country sale from the Howgill pits. Spedding responded by hiring Lawson's men whenever possible and, to prevent Lawson from exporting coal, fined shipmasters who bought the latter's coal and threatened to deny them coal from the Lowther collieries. Gilfrid Lawson, however, was at no time a serious competitor. Although he boasted in 1722 that he was making a profit of £500 per annum at Hensingham, in 1728 he was ready to let his collieries for £100 a year and two years later his mining operations were at a standstill.[24]

At Frizington, some four miles east of Whitehaven, coal was first worked about 1718. Between 1728 and 1730 William Wood mined coal at Howth Gill Colliery, using it in conjunction with local iron ore in a vain and costly attempt to make bar iron. Coroners' reports show that a pit was being wrought at Frizington in 1752 and another at Arlecdon in 1735.[25]

WORKINGTON DISTRICT

Little information has survived concerning coal mining in the Workington area in the first half of the eighteenth century. Henry Curwen, who succeeded to the family estate in 1673, owned collieries in Harrington, Lowca, Seaton and Clifton but seemed indifferent to both coal mining and the coal trade and was content to lease his collieries to various tenants. The collieries at Lowca and Harrington, together with the salt pans, were let at £130 per annum, Seaton Colliery brought in £140 a year and Clifton Colliery which was "confin'd to a small compass of ground" yielded an annual rent of £45. These lessees apparently worked their collieries successfully. In 1720 the collector at Whitehaven reported to the Commissioners of Customs, "a great increase of ye exportation of Coals at Workington, by the Ld. of the Mnr. there letting his collieries to undertakers that have been at a great expence, and thereby dry'd large quantities of coals wch. have occasioned very great resort of shipping there."[26]

James Lowther was most anxious to add the Curwen collieries to his own and an opportunity arose in 1725 when Henry Curwen died and was succeeded to the entailed estates by another Henry Curwen (1680-1727) of Sella Park, a member of a collateral branch of the family. He ordered John Spedding to try to obtain a lease of these mines, but Spedding realized that some cunning would be necessary; it would be fatal to the project if it immediately appeared that he was seeking to lease the collieries for his master's benefit. Accordingly he set out to ingratiate himself with the new owner and offered him abundant advice about the management of his estate, particularly his collieries. It was on his advice that Curwen refused to renew the lease of the tenants of Harrington Colliery and decided to work it himself. Spedding wrote that Curwen, "desired me to send a Man over to view it (i.e. Harrington Colliery) wch I accordingly did & instructed him w^{t} sort of a Report to make, wch has Confirm'd him in takeing the Colliery into his own manag^{t} . . . .This is the only way I could take to keep him from Leasing it to some or other, from whom it c^{ld} not be had again . . . I am pretty much in his Esteem & in time things may wind about so as to get from him w^{t} is for yor Service." Shortly after taking over the family estate Curwen was not only managing Walton Wood Colliery at Harrington but also a small pit near Workington Hall where the coal seam was only twenty-two inches thick. The lessees of Clifton were allowed to continue mining at the old rent of £45 per annum but Curwen's inheritance did not include Seaton Colliery which, together with the manors of Seaton and Stainburn, had been left by his predecessor to Charles Pelham of Brocklesby in Lincolnshire.[27]

Curwen soon tired of colliery management and within weeks professed himself willing to lease all his collieries to Spedding for £340 per annum. Spedding, concealing the fact that he was acting on Lowther's behalf, tried unsuccessfully to obtain a reduction in the rent. Given a free hand to inspect Curwen's collieries he determined to exploit this freedom for his master's advantage. "I have liberty to view them as often as I please upon pretence of adviseing his managers (who are very Ignorant in those affairs) & I shall make use of it either to draw them into Works that will be Expensive, or manage them in some other way as I see Occasion, & I hope that matter will be work^{t} about to yor Sattisfaction." Eventually Curwen offered his collieries to Lowther himself but they could not agree upon the terms of a lease, Lowther being unwilling to pay a rent of more than £300 per annum. Curwen, therefore, had to continue with his collieries and according to the Lowther-Spedding correspondence enjoyed some success at Walton Wood Colliery which he claimed in 1726 would bring him a profit of £300.[28]

He died in 1727 and his brother Eldred (1692-1745) succeeded to the Curwen estates. Again Spedding engaged in devious manoeuvres to help his employer to take over the Curwen collieries. Realizing that Curwen would not lease his collieries directly to Lowther, Spedding's plan was to persuade Curwen to lease them to Walter Lutwidge, Thomas Benn and himself and subsequently to assign them to Lowther. In 1728 Curwen agreed to let all his collieries (except Clifton Colliery already leased to John Porter), his salt pans at Walton Wood and all his rights over Workington harbour for thirty-one years at an annual rent of £300 and signed a memorandum to that effect. He soon regretted the bargain and tried to extricate himself from it on the grounds that it was "made in a Drunken bout (wch was not realy the Case, tho' he did fall into that Condition after it was Concluded & sign'd.") and offered Spedding and Lutwidge £100 to have the agreement cancelled. Despite considerable pressure Curwen refused to provide the lease he had promised and retained the collieries in his own hands. By 1729 both Lutwidge and Benn had lost interest in the plan to lease the Curwen collieries; the growing costs of their Seaton Colliery made them reluctant to saddle themselves with an additional rent of £300 per annum. Thus Spedding's guile had achieved nothing and Lowther had lost his last chance of combining the Whitehaven and Workington collieries.[29]

Although Eldred Curwen seemed to find little pleasure in managing his collieries he continued with them and embarked upon new mining ventures. Between March 1734 and July 1738 35,540 tons of coal were sold from his Walton Wood and Ennon collieries. When Porter's lease of Clifton Colliery expired Curwen took it over himself. In 1736 he opened out an old colliery near Clifton chapel which had a daily output of about twelve tons and at the same time worked Pasture Pit which had a similar output. The old colliery proved expensive; its reconstruction cost Curwen a great deal of money, drainage costs had to be met and the charge for leading the coal to Workington amounted to 1s. 10d. per ton. Curwen also joined in the coal trade, buying three vessels outright at a total cost of £2,150 and shares in other vessels. As a colliery owner and as a shipowner he was concerned about the condition of the harbour at Workington and did what he could to improve its facilities.[30]

Eldred Curwen died in 1745 and was succeeded by his son Henry (1728-1778) who was responsible for sinking more pits near Workington. The coal accounts show that in addition to Ennon Colliery, Moorbank, Scows, Moorclose, Hunday and School House pits were at work in 1750. Coal was then led to the harbour at a cost of 10d. per ton and sold to the ships at 3s. 4d. per ton. At Harrington, Park Pit was producing coal between 1749 and 1757. Since the coal was transported to Workington for export leading costs were high, 1s. 6d. per ton in 1749 rising to 1s. 8d. in 1754. Between 1750 and 1755 he spent £2,128 in re-opening the old Curwen colliery at Lowca which had been worked thirty years earlier by lessees. During these years he had two more pits at work in Clifton, Pasture Pit and Brow Top Pit. When he died in 1778 his collieries were producing a profit of £5,500 per annum.[31]

Two Yorkshire miners, James and William Gorton, played a prominent part in the early development of coal mining in the manor of Seaton. Together with Robert Greer they leased Seaton Colliery from Henry Curwen for thirty-eight years from 1 May 1722 at a yearly rent of £140. When the manor passed by bequest to Charles Pelham he renewed the lease of the colliery to the Gortons on the same terms as the Curwen lease. Spedding realized that this colliery could be of value to his employer and as early as October 1722 he was making overtures to James Gorton in an attempt to persuade him to surrender it to James Lowther. From the beginning the Gorton-Greer partnership was an unhappy one and quarrels about the management of the colliery in 1723 seemed likely to end in the dissolution of the partnership. James Gorton was willing in 1727 to assign his two-thirds share of the colliery to Spedding and Walter Lutwidge provided his rent arrears of £120 and debts of £200 could be paid off and he could remain as manager of the colliery. The Gorton brothers sold their interest in Seaton Colliery, 24 April 1728, to Spedding and Lutwidge for £400 and they were soon joined by a third and equal partner, Thomas Benn. Throughout, Spedding's intention was to get hold of the colliery and then assign it to Lowther. In 1731 Lowther bought out Spedding for £300 and Benn for £700, but Lutwidge, much to Spedding's indignation, refused to give up his share of the colliery and insisted on remaining a partner in the enterprise until the lease expired.[32]

Seaton Colliery consisted of two groups of pits, one at St. Helens and Moorhouse Guards and the other in Seaton Banks, on the north bank of the Derwent. James Gorton in 1727 calculated the cost of working Seaton coal at 2s. per ton: cutting 3½d., trailers 1d., drawing to bank 1½d., leading to the ships 1s. 4d., bankman, leading horses, sinking shafts etc. 2d. Since the coal was sold at 3s. per ton there was a clear profit of 1s. per ton. These calculations, produced for Spedding's benefit, were coloured by optimism and self-interest for the colliery in the next five years showed an annual loss:

Year	*Output* *Tons*	*Shipped* *Tons*	*Country Sale* *Tons*	*Loss* £	*s*	*d*
1728	7,015	5,629	485	208	11	1½
1729	8,291	7,990	408	205	6	3½
1730	7,818	6,428	367	608	3	11
1731	4,836	4,005	249	483	5	0½
1732	11,700	9,866	513	1,474	2	2½

There was a marked improvement thereafter; in the next twenty years coal output trebled and the colliery made profits instead of losses. In 1741 it produced 34,566 tons, 32,188 tons in 1748 and 39,328 tons in 1753. Profits amounted to £719. 7s. 11½d. in 1736 and to £1,236. 1s. 6½d. in 1743. At the four working pits in 1745 fifty-seven hewers and thirty-five trailers were employed.[33]

Lutwidge was largely responsible for the decision to cut the high cost of leading coal by constructing a waggon-way from the colliery to the harbour. By December 1732 the waggon-way was completed and it became possible to carry the coal in waggons directly to the ships, either from the pits or from a staith built on the north side of the Derwent. Branch lines were built in later years so that by 1740 all the Seaton pits were connected to waggon-ways. To improve the conduct of the colliery new regulations were drawn up in 1735, their main purpose being to define clearly the duties of the senior staff. Under James Gorton, the general manager, John Medcalf was put in charge of underground working, John Moor was to supervise business at the harbour and Thomas Atkinson was to direct work at the coal banks. After 1739, following a long period of close and friendly relations, the Lowther-Lutwidge partnership turned sour. Lutwidge was convinced that the colliery was mismanaged by old, idle, ignorant and dishonest men, that wages were excessive and that too high prices were paid for the ropes and timber supplied to the colliery. Lutwidge's repeated protests to Lowther about these matters went unheeded for the most part, but they both made an agreement in 1741 with John Westray who became responsible for repairing the Seaton waggon-way; for this he was to be paid 13d. for each waggon of coal shipped. Lutwidge, a man usually *fortiter in re* but rarely *sauviter in modo*, addressed Gorton in forthright terms,

> "If what Sev^ll people informs me this Day be thrue and I believe it is your more culpable than any man that has been hanged att Carlisle these forty years. Ye charge is no less than that you directed the (men) . . . not to Lead down Coals nor suffer others to Lead accordingly few or none is Led. Your charged with Sinking pitts to the rize and Trealing from ye Dib. Paying 4½ per ton when you may have them Trealed at 2 or 2½. You pay Twenty five shillings per ton for w^t was formerly fifteen. In short that you now are more barefaced than before and Stick at nothing to Sacrafize the work."

For years Lutwidge complained about the high costs, overmanning and underworking of the colliery and he laid much of the blame on Spedding who, he alleged, did not want Seaton Colliery to flourish since it might lead to the diminution of profits at the Whitehaven Colliery and thereby injure the latter's standing with Sir James Lowther. There was much ill-feeling between Lutwidge and Spedding and the colliery suffered in consequence. They had both been concerned in a ropery at Whitehaven and when the partnership was dissolved a long dispute followed over the financial arrangements, Lutwidge finally bringing and losing an action against Spedding in the court of Chancery. Lutwidge believed Spedding overcharged for the ropes he supplied to the colliery and disregarded any suggestions to reduce costs and increase efficiency at Seaton. In a letter to Spedding in 1746 he put several blunt queries,

> "Q^r were you concerned in reality for the good of those works, would you neither go nor send above twice a year. Q^r would you take mo^thly pay Bills in as you do & pass over such Numbers of unreasonable Charges as you do. Q^r would you suffer such Eregular Charges to be maid. Q^r would you suffer Perkin to have his Smidie Coals for 14 or 15 years for nothing . . . Q^r would you suffer the ways to be laid with Ash & Elm in lew of Oake. Would you suffer the way to be so bad that no cariges can go upon them by which wee lost the leading of 3 or 400 waggons last Month . . . would you as you have done all along discouraged me from Indeavouring to redress these Grievences."

It seems likely that there was a clash of authority and Spedding had the full support of the owner of two-thirds of the colliery.[34]

There is little direct evidence to substantiate Lutwidge's allegations of inefficiency at Seaton, but his continuing complaints over many years, even allowing for his animosity towards Spedding, suggest that there was some truth in them. It is unlikely that he would have addressed to Spedding the accusations quoted above if they were without foundation. That Spedding supplemented his salary by overcharging for the ropes and timber he sold to the colliery is a distinct possibility. Apparently John Westray failed to keep the Seaton waggon-way in good repair for according to Lutwidge both the waggons and the waggon-way were badly neglected. Nevertheless, between March 1741 and March 1748 77,760 waggons of coal were shipped from the colliery which, at 13d. per waggon, entitled Westray to a payment of £4,212, or on average, £600 per annum. This certainly seems excessive. If the colliery was mismanaged and its costs were unnecessarily high it is difficult to understand why no effort was made to reproduce the high level of efficiency which existed at Whitehaven. Lowther was wealthy enough to supply whatever capital the colliery required and a prosperous colliery at Seaton would have added to his profits provided that it did not in any way compete with his Whitehaven Colliery. When Spedding in 1728 recommended that Lowther should take over Seaton Colliery it was because he regarded it "as the one most likely to stand in your way of any upon ye whole Coast" and that Lowther therefore should take steps to prevent the colliery being worked by others. If Lutwidge's allegations were true, and this is far from certain in view of the colliery's growth in output and profits after 1740, it may be that Lowther, who was quite prepared to buy or lease collieries in order to deny them to possible competitors, was not too concerned with maximum efficiency at Seaton if his main consideration was control of, rather than exploitation of the colliery. Lutwidge, despite his bitterness towards Lowther and Spedding, retained his one-third share of the colliery until 1760 when the lease expired and the colliery was then worked by Sir James Lowther, later Earl of Lonsdale.[35]

After acquiring a new colliery it was normal practice for Lowther to buy or lease land in its vicinity to enable coal to be moved from it free of wayleave rent. In 1732 he bought Richard Richmond's tenement at Seaton for £100 and in the same year leased four more pieces of land at Seaton at a total annual rent of £64. 10s. In the following year he and Lutwidge leased Richard Cook's coal royalty in the adjacent manor of Camerton for thirty-one years at £20 per annum if the coal remained unworked; if they worked the coal the rent would be £40 per annum. Another coal royalty in Tysons Fields, Camerton was leased by Lowther in 1745 for seventeen years at an annual rent of ten guineas. It is not known whether Sir James worked coal in Camerton. The terms of the lease from Cook do not indicate a strong determination to do so; it is possible that his Camerton leases were merely designed to thwart potential rivals.[36]

In this period coal mining continued in the manor of Clifton. John Cookson & Co. established a furnace in 1723 at Little Clifton on the banks of the River Marron, which supplied the necessary water power; ironstone was obtained from Branthwaite and coal from their own mines at Clifton and Greysouthen. James Gorton worked several pits at Clifton but there is little information available about his activities there, only a few brief items in the Spedding-Lowther correspondence. In 1721 it was said that "Gorton has already w^{rt} some years in the Coll^{ry} at Clifton".[37]

In the early years of the eighteenth century Thomas Salkeld was mining coal in his own land at Great Clifton, but like so many colliery owners he found it a costly business. In 1725 he tried to borrow £600 on the security of his estates, "including ye mannor & Collieryes at Clifton". Apparently he failed to raise the money and shortly afterwards he leased his colliery in February 1726 to John Porter of Workington for nine years at an annual rent of £190. Porter in the following year also leased Henry Curwen's collieries in Great and Little Clifton. On the expiry of the Salkeld lease in 1735 the colliery was then taken over by Messrs. Cookson and Williams, owners of the Clifton furnace, for twenty-one years at a yearly rent of £250. Before they took possession of the colliery

Williams made an agreement, 20 May 1734, with Sir James Lowther and Walter Lutwidge to supply them annually throughout the term of the lease with 7,000 tons of coal which he would deliver to Workington Harbour. He also agreed not to sell coal to any other person for export at Workington and should the colliery's annual output exceed these 7,000 tons the surplus coal would be for country sale only. Lowther and Lutwidge promised not to work any of their coal in Clifton or Greysouthen throughout the period of twenty-one years. This agreement was designed to increase Lowther's share of the coal trade at Workington and according to Spedding, it "has brake the heart of ye Coal own^{rs} at Work^{n}, and must put the Trade of that place intirely into your hands". Spedding was careful to safeguard Seaton Colliery from Clifton competition. When there were large stocks of coal at Clifton he preferred to store the coal in neighbouring fields "for as to Leading them to Workington it w^{d} but just hinder so much of the Sale from Seaton". Williams, presumably, was satisfied with his share of the bargain; he had a sure customer for much of the outut from Clifton Colliery and greater control of the land sale in that area.[38]

Lowther extended his domain to Distington where, as in other districts, he disposed of competition by purchase. He bought coal royalties in Distington belonging to John Fearon, Joseph Steel, Anthony Dixon and Jeremy Cowman and in 1714 he bought all the coal royalties in Distington belonging to Thomas Lamplugh, the leading colliery owner in that area. Lamplugh had worked Gunerdine Colliery since 1693 and Lowther continued to work it until 1718 and again for short periods in 1728 and 1736. John Brougham had no more success with his mining ventures in Distington than he had in Moresby and after failing with pits in Barngill and Distington Outgang he sold his manors to Lowther in 1737. Richard Piper, who had a colliery in his land at Barngill, mortgaged his property to Lowther in 1741 and two years later had to sell it to him for £500. With these acquisitions Lowther added control of mining in Distington to his control of mining in the Whitehaven and Moresby districts.[39]

MARYPORT DISTRICT

It was coal mining that accounted for the birth of Maryport, known before 1750 as Ellenfoot or Elnefoot. Once an important Roman fort its history ceased with the departure of the Romans. The first modern house, now the site of the "Golden Lion", was not built until 1718. Pennant, writing in 1772, described Maryport as "another new creation, the property of Humphrey Senhouse, Esq., and so named by him in honour of his lady: the second house was built only in 1750. Now there are above one hundred, peopled by thirteen hundred souls, all collected together by the opening of a coal trade on this estate".[40] Little was done in the hinterland of Maryport in the first half of the eighteenth century. Large scale mining development was impossible since the land was divided among many different owners; the small and shallow mines which existed were worked independently of each other and could achieve little. Their owners, lacking in both capital and vision, could not produce, and did not think of producing, a single, co-ordinated plan of development for the whole area; the Lowther approach, in fact, was out of the question. The Senhouses, lords of the manor of Ellenborough held a crucial position. Not only did they own coal mines but they also controlled an important route to the mouth of the Ellen from which coal might be exported. They hindered the growth of mining in their neighbourhood by their delay in creating a harbour at Ellenfoot. With insufficient means to buy nearby estates themselves, the Senhouses, doubtless jealous of the success of the Lowther family, were unwilling to co-operate with Sir James Lowther's efforts to expand in that area. Their limited abilities and narrow horizons may be contrasted with the technique of the Lowthers in their development of Whitehaven and its hinterland.

Between 1721 and 1727 Humphrey Senhouse I (*d.*1738) made sustained but unavailing attempts at Ellenborough to mine coal for his salt pans and for export. However, he continued his efforts and by 1730 he was shipping coal. At the same time he was hindering Lowther's shipments of Dearham coal by refusing to allow it to be staithed near the harbour. Senhouse's colliery cannot have been very successful for by an agreement with Sir James Lowther, 7 September 1736, he undertook not to work any of his coal in Netherhall and Ellenborough for the next twenty-one years. By this agreement he also allowed Sir James the use of an old level in Ellenborough Common for draining his coal in Dearham, wayleave for coal through any land on the south side of the River Ellen and a piece of ground, forty yards square, near the harbour at Ellenfoot, for the purpose of staithing coal. In addition, Senhouse covenanted not to grant wayleave or staithroom to any other person during the twenty-one years which the agreement had to run. For all this Lowther was to pay him an annual rent of £5.[41]

In the adjoining manor of Ewanrigg (Unerigg) Ewan Christian (1718-1752) was working coal in a small way in 1750 and selling it to the ships at Ellenfoot at 3s. per ton. He was succeeded to the estate by his brother John (1719-1767) who continued the colliery which he later worked in conjunction with the adjacent colliery at Broughton. For years Lowther manoeuvred in vain to secure the Duke of Wharton's estate and colliery at Broughton. As a preliminary the Duke's bankman at Broughton was given a similar job at Lowther's Dearham Colliery in order that he could provide full information about his former place of employment. Lowther in 1723 was ready to buy the estate provided the price was reasonable and rumour-mongering, a favourite tactic of his, might help in that direction. "We must now talk of the ruine of the Coale Trade & the great number of Estates to be sold in our County in order to gett Broughton at a moderate price wch otherwise wil be run up extravagantly out of spite, but I won't bid in my own name." Nothing came of this or of his attempt to buy the Wharton estate in 1727 which was eventually sold to the Duke of Somerset in 1738.[42]

Writing in 1723 Spedding expressed his belief that the Broughton estate was "one of the best places for great plenty of Coal that ever I mett with in this Country" but it had never been thoroughly surveyed. Without full knowledge of the depth and extent of the various seams it was impossible to work the colliery efficiently. Mining there had been confined for the most part to working outcrop coal and no attempt had been made to win coal below fourteen fathoms. Between 1722 and 1738 Broughton Colliery was worked by the Duke of Wharton to provide coal both for export and local domestic use. Accounts show that in 1722 the colliery employed twelve haggers, five trailers and five banksmen who were each paid respectively 10d., 7d. and 7d.-8d. per day. High leading costs reduced profit margins; it cost 18d. to lead a ton of coal to Ellenfoot, 16d. to Flimby and 2s. to Workington. After buying the estate the Duke of Somerset carried on the colliery for a short time; between 1 November and 28 March 1739 it earned him a profit of £28. 5s. 3½d. By 1742 the colliery was in the hands of Ewan Christian who found little pleasure or profit in managing it and who would have liked Sir James Lowther to take it over. In 1749, on an output of 7,680 tons, the colliery made a profit of only £41. 1s. 5d. or 1¼d. per ton.[43]

In 1722 Lowther bought Richard Lamplugh's collieries and estates in Ribton and Dearham. The purchase price involved the settlement of Lamplugh's debts, which amounted to £5,000, the payment to him of an annuity of £200 during his lifetime and finally, in 1725, the payment of £1,000 to his widow. Lowther added to these properties by buying several pieces of land in Ribton at a total cost of £1,080. In Dearham he bought a coal royalty belonging to Fletcher Partis for £52. 10s. in 1737 and ten years later he leased all the coal in the parish of Dearham belonging to the Dean and Chapter of Carlisle for twenty-one years at an annual rent of £7. From 1722 until 1757 the Lowthers worked Dearham Colliery continuously, during which time over a score of pits were sunk there. In 1723 the colliery produced between eighteen and twenty tons per day; its output of coal rose to 100 tons per week in 1750 and to 150 tons per week in 1752, most of it being led to Ellenfoot for shipment. Except for three years when small losses were made, the colliery until 1756 made an annual profit ranging from £2 to £268.[44]

A small colliery at Crosby belonging to John Sanderson was being worked in 1722. Having neither the knowledge nor the capital to make a success of it he leased his colliery to others who were equally unsuccessful and had to cease mining in 1726. This colliery was conveniently situated for leading coal to Ellenfoot and Lowther was always interested in collieries near the coast. After some consideration he decided against buying the colliery but he did buy a coal royalty at Crosby from John Brisco in 1739 for £150.[45]

As part of his expansion policy Sir James had designs upon the land and coal in the neighbourhood of Flimby. The Gorton brothers and Greer, partners in Seaton Colliery, also shared a lease from William Blennerhasset of a colliery in Flimby. There is no information available about the working of this colliery, but its prospects must have been favourable because James Gorton in 1722 was said to be "for forwarding Flimby harbour" if he could persuade Greer to withdraw from the partnership. Lowther wished to take over the lease from Gorton or, if the price was right, he was willing to buy the colliery. He failed to lease or buy it. His next effort was to lend money to Blennerhasset and William Beeby on the security of their estates at Flimby; if, as he hoped, they failed to repay the loans he would then be able to take over their lands. He lent Beeby £600 in 1737 and Blennerhasset £300 in 1738 and both of them further sums in later years, Blennerhasset in 1753 owing Lowther £1,180. Presumably the two debtors remained solvent, regrettable though this might have been from Lowther's point of view, for he did not acquire either of their estates.[46]

In the first half of the eighteenth century lessees were raising a little coal for local consumption at the Duke of Somerset's inland collieries at Oughterside and Bolton. The few contemporary records suggest that production at Oughterside was intermittent; the pits there were closed in the winter of 1726 and in the early months of 1740. Oughterside Colliery (sometimes called Aspatria Westmoor Colliery) "formerly occupied by William Fletcher of Moresby", was leased in 1728 to Joseph Brisco, William Langcake and Robert Kirkhaugh for twenty-one years at a yearly rent of £10. Nearby Sir Wilfrid Lawson worked coal in his own land at Oughterside, losing a great deal of money in this venture. Oughterside Colliery passed from the Duke of Somerset to his nephew the Earl of Egremont from whom it was leased in 1753 by Lawson's coheiresses, Elizabeth and Charlotte Lawson, for seventeen years at an annual rent of £15. They did not, however, make an attempt to mine coal there. Information about Bolton Colliery is scarce although it seems to have been working throughout this period. Bishop Nicolson noted in his diary his two visits to Mr Robinson's colliery at Bolton in 1707. It competed with the collieries at Oughterside for the local market and when the price of Bolton coal was reduced in 1734 it made profits at Lawson's Oughterside Colliery impossible. In 1753 Lord Egremont's collieries in the manors of Bolton and Westward were leased for twenty-six years at a yearly rent of £60.[47]

These inland collieries had no chance of making large profits. Since they were situated in rural areas the local demand for coal was small and high transport costs to the coast debarred them from the Irish market.

The lack of statistical evidence makes it impossible to determine either the total ouptut of coal in West Cumberland or its distribution. Cumberland coal found a ready market in Ireland but only a small proportion of it was consumed locally. The smallness of the population restricted the demand for household coal and there were few industries capable of consuming large quantities of coal. An inadequate labour supply, poor communications and the absence of men able and willing to invest in industrial enterprises militated against industrial development. Salt pans dispersed along the coast between Maryport and Whitehaven made use of local coal but how much they consumed is unknown. Humphrey Senhouse between 1721 and 1727 tried to raise coal at Ellenborough in order to provide coal for his salt pans. He was unsuccessful and had to buy coal from Broughton Colliery. This colliery in 1749, out of a total output of 7,680 tons, sold 369 tons of coal to salt pans, 3,713 tons to the country and 3,560 tons to the ships. The Curwens and the Lowthers were able to supply coal to their salt pans at Walton Wood and Whitehaven respectively from their own collieries.[48]

A small outlet for Cumberland coal was found in the ironworks established in the county during the first half of the eighteenth century. At the Clifton furnace Messrs. Cookson & Co. in 1723 proposed "to use pitt Coal in Smelting the Iron Stone" but four years later they were using charcoal for this purpose. After 1735 they were able to produce pig iron and a variety of castings made with coal obtained from their own colliery at Clifton. According to Mr Williams, a partner in the firm, they used "nothing but Pitt coal coak'd" and could "make no such mettle with wood charcoal". William Wood's ironworks at Frizington was supplied with coal from Howth Gill Colliery between 1728 and 1730 and from 1755 the Maryport furnace obtained its coal supplies from Broughton Colliery. Four of the seven partners in the Maryport furnace also shared, with two others, the lease of Broughton Colliery granted by Lord Egremont in 1755 and one clause in the lease stipulated that coal sold to the furnace should be "at the same price as other coal got out of the pits".[49]

Sir James Lowther was well aware that coal-based industries in Cumberland would provide an additional market for his coal but he did little to promote them. His industrial investment was limited to his salt pans, a small share in the Whitehaven glasshouse and a loan of £5,500 at three and three-quarters per cent to the owners of the Low Mill ironworks. At Whitehaven coal from Lowther's pits was used in the production of copperas (i.e. green vitriol made from the marcasites or iron pyrites found in the coal mines), glass bottles, bricks and pottery. The amount of coal sold to these industries can be compared with the total sold locally from Lowther's Whitehaven collieries:[50]

Year	*Copperas Tons*	*Glasshouse Tons*	*Bricks Tons*	*Pottery Tons*	*Total Tons*	*Total Local Sales Tons*	*Lo.*
1735*	207	1,587			1,794	6,502	3
1736	267	2,406	15		2,688	10,543	0
1737	336	2,343			2,679	11,136	1
1738	381	621			1,002	11,235	4
1739	453	9			462	10,398	0
1740	375	834			1,209	11,459	7
1741	333	1,377		78	1,788	12,588	4
1742	360		39	138	537	11,146	6
1743	465		36	63	564	11,118	5
1744	705		48	3	756	13,480	3
1745	534		54		588	13,302	3
1746	591		51		642	12,953	7
1747	567		63		630	14,629	0
1748	531		135		666	16,174	0
1749	669		165		834	16,457	3
1750	459		135		594	17,894	3

* Half-year only.

Because of the scarcity of records it is not possible to provide statistics of coal consumed in Cumberland, but the mountain of Lonsdale MSS. does give that information about the Lowther collieries. The local demand for coal from Lowther's Whitehaven pits rose from 4,560 tons in 1705 to 17,894 tons in 1750; from 1736 it exceeded 10,000 tons per annum.

Output & Local Sales from Lowther's Whitehaven Collieries.[51]

Year	*Output Tons*	*Lo.*	*Local Sales Tons*	*Lo.*	*Local Sales as % of Output*
1705	30,595	6	4,560	6	14·90
1710	40,232	4	4,233	5	10·52
1715	52,550	1	4,584	1	8·72
1720	51,577	4	6,721	2	13·03
1725	64,759	1	2,898	5	4·47
1730	81,886	2	4,988	1	6·09
1735	93,348	5	8,356	2	8·92
1740	97,220	4	11,459	7	11·78
1745	102,390	5	13,302	3	12·99
1750	126,128	1	17,894	3	14·18

MINING METHODS

Miners approached their working places via bearmouths or day-holes (i.e. openings in the side of the hill) and descended along sloping roads cut through the rock. This method of reaching the coal face was described by a visitor to Whitehaven in 1755 in these words,

"The reader may suppose that he hath entered the mines by an opening at the bottom of a hill, and hath already passed through a long adit hewn in the rock, and arch'd over with brick, which is the principal road into them for men, and for horses; and which, by a steep descent leads down to the lowest vein of coal. Being arrived at the coal, he may suppose himself still to descend, by ways less steep, till after a journey of *a mile and a half*, he arrives at the profoundest parts of the mine. The greatest part of this descent is through spacious galleries which continually intersect other galleries."[52]

Until the nineteenth century the bord and pillar system of extracting coal was exclusively employed and the coal was removed by picks and wedges. At intervals along the face hewers dug out the coal (the areas from which the coal was excavated were the "bords") leaving the coal between the bords untouched to serve as pillars to support the roof. This system was wasteful since large quantities of coal were left unworked. When pits were shallow the coal pillars were small, sometimes only four or five yards square, others were fifteen yards long by two yards wide. As the pits became deeper the pillars increased in size; in 1765 the pillars were fifteen yards square. For many years the Whitehaven mines were the deepest in the country. Carlisle Spedding in 1750 was sinking King Pit which he intended to reach a depth of 123½ fathoms; at that time the deepest borehole in the Newcastle district reached 83 fathoms. By 1755 pits at Whitehaven had been sunk to a depth of 130 fathoms.[53] Whitehaven Colliery was one of the first collieries in Great Britain to use gunpowder for blasting; it was used there as early as 1712.[54]

The Greenbank Colliery account book for 1675 shows that coal was carried out by bearers. According to R. W. Moore, "the coal was probably borne out, in the earliest days, in baskets carried out by women and girls on their backs". This is mistaken, the coal was carried out to the bearmouth by men.[55] At the Three Quarters Band Colliery the corves of coal were placed on ashen runners, resembling sledges, and pushed or pulled by trailers along the corf-way, constructed of two parallel lines of wooden rails, to the bottom of the shaft. From there the corves were wound to the surface. These wooden roads for the conveyance of coal remained in use until the end of the eighteenth century when they were replaced by cast iron tram rails. The corf was made of wooden boards shaped into "an oblong square 32 inches long, 22 deep and 18 broad by measure taken in the inside, so that the contents of each is 12672 cubical inches". A writer described the construction of these corves in 1756, "The Boarded Corfs have their Feet & Posts of Alderwood which is not liable to splitt, the Barrs & Balks are made of Ash, the Boards for the sides, ends & bottoms of the Corfs, are made of Barrel, or Pipe Staves brought from Virginia, which being splitt out of clear green Oak Timber, don't break like Boards that are Sawn". The capacity of these corves varied from one-sixth to one-twelfth of a ton.[56] Moore's statement that the Newcastle corf, a circular basket made of hazel rods, was introduced into Cumberland about 1675 has been generally accepted and is incorrect. Until 1732 the corf used in Cumberland was made of wooden boards and not until then was the Newcastle type of corf adopted as a letter from John Spedding shows, "We have this day got a man from NCastle to make Corves after their fashion of a Larger size that we have used, or can make in our way with Boards, theirs are made Round with Wands like Baskets, & Contain above 70 Gallons & are drawn upon Trames." These corves were made of young hazel rods "of 4, 5 or 6 Years growth, cost about 8d. a Bundle containing fourscore Rods, the Ribs of the Baskets are made of largest sort of these Rods, round which the small ones are Twisted". An iron bow at the top of the corf enabled it to be attached to the hook at the end of the winding-rope. Originally these corves carried two-and-a-half cwts. but in later years corves with a capacity of twelve cwts. were employed in the thick seams of some of the Main Band pits. Having decided to use these baskets it was essential to import corvers from Newcastle who knew how to make and repair them. Thus we find Carlisle Spedding, when seeking help in finding a Newcastle corver to replace one who had left Whitehaven Colliery, writing, "The best we have has 10s. per week and the Second 9s. per week both which we had from NCastle. I desire you'l agree with one of the Best you can for 9s. per week and I shall allow 7s. 6d. for his Traveling hither." Boarded corves, however, continued to be used, particularly in the thinner seams, for many years after the introduction of the Newcastle corf.[57]

When pits were first sunk the coal was brought to the surface by jack-rolls, i.e. winches worked by manual labour, but by 1676 horse gins were being employed in some pits to cope with increasing pit depths.[58] In 1723 a horse gin was erected underground at Corporal Pit to draw the coal out of the dip workings. Since horses could not descend the narrow pit shafts they reached the coal by means of a sloping road driven from the coal seam to the surface. As the face advanced the cost of drawing coal to the shaft increased and horses were found to be cheaper than men, boys being employed to drive the horses as they pulled the trams or sledges. Horses and baskets (i.e. corves of the Newcastle type), could only be used where there was sufficient headroom for horses; in thinner seams men continued to trail the coal in boarded corves.[59]

By 1755 coal from the deepest mines at Whitehaven was raised in successive stages from one seam to another by means of underground horse gins and blind pits, or staples.[60] When a corf reached the surface a banker pulled it towards him and placed it on a tram which he then drew to the pit heap or bank where it was emptied; the tram and corf were then drawn back to the mouth of the pit in readiness for the next corf to come up. At the pit bank boys were employed as slatepickers to remove the slate and other refuse from the coal. Work on the surface was supervised by a banksman who kept account of the corves sent up, checked that they were full and controlled the loading of waggons.[61]

The high cost of buying, hiring and keeping horses made an alternative method of winding coal desirable. It is strange that after the adoption of a successful steam pumping engine more than seventy years were to elapse before a steam winding engine was installed in a Cumberland colliery. George Richardson, a colliery engineer at Whitehaven, tried to win support from colliery owners between 1734 and 1735 for his method of applying the steam engine to coal winding. In a broadsheet entitled, "A Proposal for the Advantage and Service of the Hon. Company of Coal Owners in England, &c." he claimed that his invention would "draw Coals out of the Pits, at any Depth, at one Third less Expence than they can be drawn by Horses". Nothing came of his efforts. His journey to London in 1734 proved fruitless; the Newcastle coal owners showed some interest but would not commit themselves to backing his invention, nor could he obtain a patent for it.[62]

The sinking of new pits in the early years of the eighteenth century led to difficulties that could have caused the cessation of large scale mining at Whitehaven. As the pits went deeper the drainage problem grew to such proportions that they were in danger of being drowned. In the early days drainage had been simple since outcrop coal was worked to the rise or along the level from the bearmouth, the water flowing naturally from the workings, When pits were sunk horizontal drifts or adits were cut through the strata from the lower grounds to drain the workings. Where gravitation drainage was impossible horse gins were employed to draw the water and these were in use at Whitehaven as early as 1676; the Ginns, a district of Whitehaven, takes its name from these gins. Sets of horses were used in relays to raise the water, but as pits increased in depth the efficiency of the horse gins diminished. Spedding's letters show that the situation was deteriorating and it became evident that more powerful machinery was needed to drain the mines. In 1712 he wrote, "Capt. Savory's invention is the only thing of that kind that can do us any Service".[63]

James Lowther made many enquiries to discover some means of solving the drainage problem and finally he turned to Newcomen's invention. By an agreement dated 10 November 1715 he hired a Newcomen engine with a 17 inch cylinder for £182 per annum which was erected at Stone Pit, Howgill and this was the first steam pumping engine in Cumberland. Construction of the engine began in 1716 and was completed in the following year; the engine house was built at a cost of £86. 9s. 9d. After some initial difficulties with leaking joints the engine was successful and coming at a time when it was vitally needed it saved the Lowther mines; it was now possible to win coal at greater depths and to the dip of the adit level. In February 1727 a new agreement was drawn up by which Lowther paid £200 for the materials of the existing engine and on payment of a further £350 was licensed to erect another Newcomen engine in its stead whose cylinder should not exceed 22 inches in diameter. It was ability of this second Newcomen engine to pump out water which led directly to the sinking of Saltom Pit.[64]

Further agreements were made for the installation of additional Newcomen engines. Lowther was empowered, 20 November 1730, to erect a pumping engine with a 17 inch cylinder on paying £105; should this engine be inadequate he could, if he wished, on paying an extra £52. 10s., erect a larger engine provided its cylinder did not exceed 36 inches in diameter or 9 feet in length. The 17 inch cylinder engine was set up at Saltom Pit in 1731 whilst sinking was in progress but was soon replaced by a more powerful engine. A receipt shows that this change of engines was effected on lower terms than those in the original agreement, "Rec^{d} this 6 June 1732 of the within nam'd James Lowther Esq., now Sr James Lowther Bar^{t} the Sum of Twenty six Pounds five shill^{gs} for liberty to sett up a 36 Inch Cylinder & in discharge of his covenant within written for paym^{t} of Fifty two Pounds ten shil^{gs}." Other engines followed – a 38 inch cylinder engine at the Ginns in 1736 and a 42 inch cylinder engine at Saltom in the following year. A second 42 inch cylinder engine was set up at Saltom in 1739 to supplement the work of the first one. In 1756 five men were employed to look after the two Saltom engines which were then "kept pretty constantly to Work"; at the Ginns it was seldom that more than one of the two engines was working and only two enginemen were employed. Water was drawn out of Saltom Pit by four pumps in four lifts or stages, each pump raising the water about twenty fathoms to a cistern, until it was eventually brought to the surface.[65]

Throughout their history pits in Cumberland, particularly in the Whitehaven area, were plagued by "fire-damp" and as greater depths were reached the problem of ventilation became critical. Accumulations of the gas together with the use of naked lights precipitated explosions which killed or maimed miners and seriously damaged the underground workings. One of the earliest of these explosions, at Priestgill in 1682, killed one man and injured six more. To the employer the damage done to the mines was more important that the loss of life. John Gale reported to Sir John Lowther, in 1697, "Wee have (of late) had severall workmen Burnt in the Pitts, as well at G^{n}bank as Howgill, one dyed of his hurt. I only acquaint you herewith because wee are frequently putt to charges either in the Care or burial of such". One man was killed in 1727 and two more at Whitehaven in 1733 by explosions of fire-damp. An explosion at Corporal Pit in 1730, caused by a miner carrying a lighted candle into some old workings, burned seven men, three of them very badly and led to a fire throughout the entire pit. To persuade the terrified miners to return to the pit after this explosion Carlisle Spedding had constantly to show himself in those areas of the pit where they most feared danger.[66]

In the struggle against fire-damp Sir James Lowther enlisted the help of the Royal Society. Samples of "damp air", i.e. fire-damp, were collected at Saltom Pit and sent in bladders to London in 1733 for testing by the Royal Society. At the same time Spedding sent Lowther "An acco^{t} of the Damp Air at Saltom Pitt" which was later published in the *Philosophical Transactions* of the Royal Society. Further experiments with fire-damp were carried out in 1736 but the Fellows of the Royal Society were unable to provide any means of overcoming the danger from fire-damp.[67]

It was essential to find some means of illumination whereby miners could work safely in areas of a pit where the presence of explosive gas made the use of candles highly dangerous. Carlisle Spedding, after conducting a number of experiments and finding that fire-damp was not liable to be ignited by a red hot iron or by the sparks from flint and steel, invented the steel mill about 1730. This consisted of a steel wheel which was made to revolve rapidly by means of a small toothed wheel while a flint was pressed against it and the resulting stream of sparks gave a light to enable the men to work in comparative safety. The steel mill was not completely safe for the sparks could ignite gas and to operate it was very hard work, two persons being needed to relieve each other. Spedding's invention, which was used during the sinking of Saltom Pit, was adopted in other coalfields and was not superseded until it was replaced by the Davy safety lamp after 1815.[68]

Where gas was abundant the area was dammed off and the gas piped out of the pit to the surface where it was ignited. It was said that the gas could burn night and day for months and years and that the flame could easily be seen at the Isle of Man. Sometimes explosions were followed by underground fires which lasted for months and which could only be subdued by flooding the mine; in 1743 the coal at Thwaite, Hind and Gameriggs pits was set on fire by gas explosions. Where there was gravitation drainage it was sometimes found efficacious to allow a small quantity of water to run constantly down one side of a divided shaft, "wch Spreading over the pitt has a surprising Effect in carrying the fresh air down the pitt allong with it, & also keeps a constant course down that pit, thro' the Works that is so full of foul air, wch the fresh air drives out".[69]

Carlisle Spedding* invented the system of ventilation known as "coursing the air" by which the current of air was directed up certain workings and down others until it ventilated the whole waste. It seems to have been first employed at Saltom Pit where an air race, driven between 1732 and 1733, required two partitions, each about 600 yards along. The Saltom shaft was divided by a timber partition "carried down in the midst

* Writers are still confused about the inventor of this system. Some ascribe it to Carlisle Spedding in the 1760s, but he died in 1755. Others, presumably relying upon John Buddle's evidence in the *Report of the Select Committee on Accidents in Mines (1835)*, 134, mistakenly credit James Spedding, Carlisle's son, with this invention. James Spedding was not born until 1720.

of it, plank'd & Caulk'd very Close, so that in Effect it became two pitts . . . The Air was also Contriv'd to go down one, and come up the other side of the Pitt". This system depended upon numerous partitions, doors and stoppings to compel the fresh air current to sweep through every part of the mine between its entry at the downcast side and its exit at the upcast side of the shaft. John Wesley noted this method of ventilation when he visited Whitehaven in 1759, "When a coal-pit runs far under the ground, it is customary here to build a partition wall, nearly from the shaft to within three or four yards of the end, in order to make the air circulate, which then moves down one side of the wall, turns at the end and then moves briskly up the other side". Where a pit had two shafts the air current entered the pit by the downcast shaft, passed through the workings and left via the upcast shaft, taking with it much of the foul air. Boys, usually the sons of miners and themselves destined to spend most of their lives in the pits, were employed as trappers to attend the air doors constantly "to open & shut them imediately after People have pass'd thro', lest the Current of Air should take a Wrong Course and leave the foul air to Stagnate & gather into a Body". In some pits the circulation of the air was assisted by a large underground furnace which was kept constantly burning. Spedding's system of ventilation was copied in most coalfields; it was in use in the Newcastle area by 1754.[70]

Beneficial though it was coursing the air could not eliminate the presence of gas, it could merely alleviate it. Fire-damp remained a perennial hazard and Whitehaven Colliery one of the most dangerous collieries in the country. In 1737 an explosion of fire-damp at Corporal Pit killed twenty-two persons and three horses. This disaster, according to John Spedding, was caused by miners who, having had too much to drink, ventured into dangerous workings which they had been forbidden to enter. In reply to Spedding's letters describing this accident Lowther said that he was "willing to be at any charges to find out and encourage proper persons to be knowing and skilfull in all matters about the Fire Damps" and that he would "give £100 to be disposed of among the sufferers as you think proper". Explosions of firedamp killed more men in the Whitehaven pits, two in 1740, five in 1753 and three more in 1754. Carlisle Spedding himself was killed by such an explosion in 1755.[71]

In his attempts to master the problems raised by firedamp Carlisle Spedding was joined by his friend William Brownrigg (1711-1800) an eminent physician and scientist who was elected a Fellow of the Royal Society in 1741. Dr Brownrigg devoted a great deal of time and effort to experiments with fire-damp conveyed by a pipe to his laboratory from a nearby pit. His research, valuable though it was, did not prevent further pit explosions and in 1744 John Spedding had to report, somewhat despondently, "The Observations Dr B. has made upon the Fire Damp are of Singular Use to my B^{r} in the Conducting those affairs & they have many Conferences about it. We have long been in hopes those matters w^{d} mend, but the works grow worse and worse".[72]

After the construction of the cartway from Howgill in 1683 pack horses continued to be used to move coal from Lattera and from Whingill Colliery when it began working. In 1716 coal from Gunerdine and Lattera was still being led to Parton by pack horses and pack horses at Whingill were said to be "very numerous". This form of transport was expensive and in bad weather extremely difficult. It was more economic to employ carts but leading costs remained high, particularly for pits at some distance from a harbour, and an even cheaper method of transport became necessary. This led to the construction of "railways" in Cumberland, the first of these, which was completed in 1732, running from Seaton Colliery to the harbour at Workington. Carlisle Spedding built similar railways at Whitehaven. Swimburn and Corporal pits were connected to a newly-built staith on the quayside at Whitehaven by a railway or waggon-way constructed between 1732 and 1733 and Saltom Pit by a railway begun in 1734 and completed in the following year. The railway consisted of parallel wooden rails along which waggons, mounted on cast iron wheels, ran from Howgill to the staith. As the waggon-way had a descent to the harbour the waggons moved down under their own weight and were braked by means of convoys (i.e. levers) on the front wheels; when empty they were drawn back to the pits by horses. On reaching the staith, a large wooden structure thirty-seven feet above the level of the quay and projecting over the vessels waiting to be loaded, the coal was discharged into the ships by opening the bottom of the waggons, the coal running down spouts into the hold. This was a much quicker and cheaper method of loading ships than the manhandling and emptying of sacks of coal. If there were no vessels waiting to be loaded the coal was dropped through holes into the bottom of the staith where there was storage space for 3,000 waggons of coal.[73]

Coal from the Whingill pits continued to be brought to the harbour in sacks until the construction of the Whingill waggon-way. Work on this waggon-way began in 1754 under Carlisle Spedding's direction. He wrote then, "I am very Thronk (i.e. busy) with our new waggon-way from Whingill and shall begin the Laying of Timber in a week's time. We are covering the way about 8 or 9 inches thick with flatt Stones and Rubbish for 'tis all been clay and w^{d} not Support the Waggons". By the following summer the waggon-way was in use.

> "We have been leading Down our New Waggon Way from Whingill this 5 weeks and Manage the Waggons pritty well in the Run with our Two Convoys on the fore Wheels. When the Railes are wett we lay Small free stones about the bigness of hen eggs which crushes to Sand upon the Railes and makes the Convoy take more hold of the Mettle Wheels so that we can do pritty well in wet weather . . . but to prevent great Mischeifs by the Running of Waggons I intend at the Bottom of the Run to have a Branch laid and a Switch Rail to be Shutt by an Old Man Siting in a Cabbin and with an Iron Rod thro' a Conduit fixt to a Small leaver (when he sees a Waggon got Loose) he may Shut the Switch Rail and turn the Waggon into the Branch which Riseing till it Runs against a Battery of Earth and Turf will take no harm and will fall back again gently. And I'll have another Switch Rail which shall Shutt of itself and turn the Waggon (in its coming back) into another Branch where it will Rest till bro^{t} into the Main Way below the first mention'd Branch by which means if another Waggon or more should^{d} follow they will escape breaking one another."

The untimely death of Carlisle Spedding in 1755 prevented the completion of the Whingill waggon-way. Until 1803 Whingill coal was conveyed along the waggon-way to a large staith near the quay and had then to be carted to the ships. To eliminate the cost of carting the waggon-way was extended to the shipping berths by carrying it over Bransty Arch, a bridge specially built for this purpose, which was opened 9 August 1803.[74]

THE MINERS

For the Whitehaven Colliery the first half of the eighteenth century proved to be a period of increasing output and profits. These would have been impossible without an adequate supply of labour and this was not always available. For many years the Lowthers had to reply upon small farmers and labourers to work their mines but they frequently left the collieries to resume farming and this sometimes led to the temporary closure of pits. In 1710 John Spedding had to report, "We sett forwards the Yard band & Corporal pitt yesterday but cannot yet gett a compleat Sett made up at ye latter for want of Haggers, some of which begin to leave us & fall to their husbandry, as they usually doe at this time of ye year". Six months later he renewed his complaint. "Ye Gettings now fall a great deal short of what we can Lead by Reason a great many of ye workmen goe off to looke after their Hay so that we must leave off some of ye pitts to make up ye workmen for the Rest."[75]

There was competition for labour not only with the other collieries along the coast but also with the industries then developing in the vicinity of Whitehaven. Nor was it easy to persuade men who had had no experience of mining to undergo the rigours and dangers that characterized life in the pits. In the winter months sailors whose ships were laid up were sometimes employed, but when the general demand for labour was high even these were not to be had. "The same scarcity of Labourers is observ'd in most business's, & we find it also about the Collierys, where in the winter time we had about 15 sailors that constantly work'd, some Hag^{rs} others Traylers &c but the beginning of last month the same scarcity was found amongst the Sailers." Spedding realized that it would be beneficial if the miners could be housed in Lowther property for then pressure could be exerted to compel them to continue working in the mines. "It is certainly much for yor Interest to have Colliers take up little pieces of ground for Houses & gardens on the Sides of the Comons about Ribton, Harras & Akebank, & when they are all fixt there you are sure of them & may Employ them either at Howgill, Whingill, or Morresby." In the same way when letting farms he took care to obtain tenants who could serve the colliery by providing horses and carts for leading coal to the harbour. Tenant farmers who forgot their obligations to the colliery were summarily evicted, a process made possible by compelling the coal leaders to enter into "Bonds payable weekly that they may be turn'd out at any time if they go off to any other work".[76]

Experienced pitmen were welcome at Whitehaven; in an expansionist and competitive period additional manpower was valuable and if other colliery owners were thereby deprived of necessary labour so much the better. A Newcastle colliery manager wrote to Carlisle Spedding, "There is 9 or 10 more Run away from our place and I hear some of them is with you". Spedding in his reply to this letter, doubtless for good reasons, carefully omitted to comment upon the matter of the absconding Newcastle miners. Colliers who had fled across the border to escape from the servitude of Scottish mines were readily employed at Whitehaven.[77]

The earliest information about miners' wages at Whitehaven is contained in a letter from Sir John Lowther to his agent in 1666, ordering that a man about to be employed as a sinker or hewer should be paid 9d. per day. Four years later Lowther was trying to reduce hewers' wages to 8d. per day and to make them provide their own candles. In 1675 Greenbank Colliery haggers were paid 8½d. and bearers 7½d. per day; at the Three Quarters Bank Colliery haggers and trailers were paid 8½d. and 7½d. respectively per day. By 1702 haggers were being paid 10d. and trailers 8d. per day. Piece rates began to replace day rates in 1716, a change which makes it difficult to determine the precise earnings of the hewers since piece rates varied not only from one pit to another, but also in different seams in the same pit, higher rates being paid to those working in harder seams. In 1716 hewers at Andrew and Baxter pits were paid 6d. per ton, but at the other five pits hewers continued to receive 10d. per day; corvers were paid 1s., trailers and bankers 8d. and slatepickers 2d.-3d. per day. From the colliery account books it is evident new wage rates for hewers were agreed at the beginning of each quarter and that some of them, having tried piece rates, opted for a return to day rates. Thus hewers at Corporal and Swinburn pits who had been paid 4d. per ton in 1719 were being paid 10d. per day between 1721 and 1723. However, by 1728 the system of paying hewers by the ton was in general use.[78] When hewers were paid by the day they were expected to cut a certain amount of coal each shift before they left the pit; this quota was known as the "darg". A dispute over the amount of the darg led to a brief and unsuccessful strike at Bell Pit in 1710. Spedding refused to consider lowering the darg from thirty-two to twenty-eight corves and the hewers had to return to work on his terms.[79]

Miners were liable to fines if they committed "misdemeanours", a word which covered a variety of offences – stealing coal, rope or old iron, claiming payment for insufficiently filled corves, stripping coal pillars, sending up "dirt" as coal and many more. The fines, which usually ranged from 1s. to 5s., were deducted from the offender's wages and paid into a separate fund from which payments were made to sick workmen and colliery widows or to help to pay funeral expenses. In 1716, for example, one sick miner received 2s. 6d., another 5s. after being burned by an explosion of fire-damp and the widow of a man killed at Andrew Pit was paid 10s. In an attempt to eliminate dishonest practices John Spedding resorted to hiring informers who, for small sums, were prepared to denounce their colleagues. It was typical of Spedding that the money to reward informers came from the fines imposed on offenders. Unpleasant though it was the system of delation certainly brought to light many petty abuses.[80]

The miners were never subservient and soon learned the value of collective action against their employers. In 1712 Spedding was again having trouble with the hewers over the amount of the darg at Baxter and Darby pits and this time he had to accept their restriction of output. The coal leaders, i.e. carters, also proved awkward and they were in a position to do much harm by blocking coal supplies to the ships. A prolonged strike by the coal leaders would compel the shipmasters to obtain their coal at Workington or some other port. In addition, as coal piled up on the pit banks at Whitehaven, miners would have to be paid off and these would be quickly re-employed in neighbouring collieries and in this way a valuable supply of manpower would be lost to Lowther's competitors. An attempt in 1710 to lower the price of leading from 10d. to 8d. per ton precipitated a strike and once more Spedding had to admit defeat. "We have had a very hard contest with ye Coal Leaders . . . but their Combination was so strong that all we could possibly do or think of could not break it at this time; we held out ten days without Leading any but with one Single Cart . . . we were forced to Comply with ye Leaders for ye present." Apart from organized strikes the carters could cause difficulties by neglecting their work and it was not easy to compel them to attend to their duties regularly throughout the year; the strict labour discipline engendered by extreme poverty did not yet apply to them. In 1727 Spedding complained, "I was never so perplex'd about any thing as about keeping up the Carts. They are not above 35 when they are all on & they are grown so horribly Careless & Extravag^{t} that they care not whether they work or not or wch end goes foremost. If the weather is any thing wett or dirty it is next to impossible to get them driven out to work & all this last 2 weeks we have many days had not above half of 'em". However, the situation changed in 1729 when Spedding embarked upon the policy of letting land to coal leaders on condition that they continued to lead Lowther's coal to the harbour; if they failed to do so they could be evicted from their farms at a week's notice. Faced with the possibility of losing their farms the coal leaders became much more amenable.[81]

By the middle of the century there was some improvement in the wages of the Whitehaven miners. Hewers were paid 1d. per corf in the Main Band and 1½d-2d. per corf in the Bannock Band. In some pits the hewers both cut and trailed the coal to the shaft bottom or hired men to do the trailing; in the latter case hewers and trailers together received a certain price for hewing, filling and trailing, the price varying with the nature of the seam and the distance to trail. Other workmen were paid as follows:

Slatepickers	3d.	per day	Underground waymen	5s.-9s.	per week
Trappers	2d.-6d.	,, ,,	Level men	4s.-5s.	,, ,,
Fillers	10d.-16d.	,, ,,	Enginemen	6s.-9s.	,, ,,
Bankers	10d.-16d.	,, ,,	Corvers	9s.-10s.	,, ,,

The coal leaders who filled the waggons and then led them to the harbour were paid according to the distance travelled from each pit. Thus at Howgill, using horses provided by the employer, the leaders were paid 1¾d. per waggon from King Pit, 2½d. from Ravenhill, Thwaite, Howgill Head and Fish pits, 2¾d. from Kells and Hind pits, 3d. from Fox and Gameriggs pits and 4d. from Scalegill. At Whingill, where the horses belonged to the leaders, higher rates were paid, 10d. per waggon from Carr Pit and 9d. from Jackson Pit.[82]

Work in the pits was extremely dangerous and fatal accidents occurred frequently. Men were killed by explosions, falls of rock or coal and by gas. Others died by drowning in pit sumps or by falling out of baskets as they rode up or down the pit shaft. Many of the miners were heavy drinkers, a habit which endangered their own lives and those of their colleagues. The Corporal Pit disaster in 1737 was, according to Spedding, caused "chiefly by the workmen's getting too great a Qu^{ty} of Drink at the Fair from which wretched Practice most of the great misfortunes we have ever had has arisen". Many of the twenty-two victims were killed, not by the explosion of fire-damp, but suffocated by the afterdamp which followed the explosion. An accident in 1753 was attributed to "Jos. Iredale's being in liquor which cost him and 3 more men and a boy their lives, from his pulling out a Stop in the head of a Board Race between two Pitts and keeping there candles in the course of the foul air that was lodg'd below the Stop contrary to my orders and his own judgement". In the following year Iredale's nephew and two more men were killed in exactly the same way.[83] The total number of Cumberland miners killed at work can never be known, but some of these deaths have been recorded. Of the miners who died because of their work there are no records at all. There are no statistics of the many miners who died, after years of suffering from respiratory diseases like pneumoconiosis, caused by working in an atmosphere polluted by dust and gas. Others suffered from nystagmus or were crippled by rheumatism, "beat knee" and "inflammation of the joints". Their working conditions were such that the miners aged prematurely and for most of them their life expectancy was short.

Miners, particularly those who depended entirely upon their wages for a living, were often improvident. It was said of them that "if they're but a week out of work they are ready to starve" and it was their lack of means which made it impossible for them to sustain a prolonged strike. Many of them found it necessary to work, if it was physically possible, even in old age and at lower wages. At Whitehaven in 1756 fifteen old men were paid 4s. or 5s. per week to clean and keep open the levels to ensure a clear passage for the water. Another twenty old men were employed to assist the masons in quarrying stone and in building walls. Other "old men worn out in the collierys" were retained to help to deal with emergencies requiring additional manpower or to take the place of absentees. In providing employment for aged colliers there was an element of charity combined, it was alleged, with empire-building by the overseers who. "like Ministers of State, are Seldome for lessening the Number of their Dependents, & hate to turn any of them off, being mighty Charitable & Compassionate to such folks, when themselves are at no expence in doing it".[84] In 1732 Sir James Lowther ordered the construction of three dwellings, each containing four rooms, to house twelve retired pitmen or colliery widows. In addition to free accommodation each of them received a pension of 5s. per month and a coal allowance of 6s. 8d. per annum; altogether this cost him £40 a year.[85]

COSTS AND PROFITS

During the first half of the eighteenth century the cost of mining coal increased notably; deeper pits, longer distances from the face to the shaft, improved ventilation, rising wages and the provision of steam engines, staithes and waggon-ways accounted for the increase. Small landowners who had obtained coal from outcrops or from shallow pits could not meet these growing costs and those who had borrowed money to finance their collieries, often on the security of their estates, were forced to sell out. Sir James Lowther, however, could afford not only to work the collieries he inherited but also to add to their number by investing much of his income in the acquisition of more collieries and coal royalties. Having no need to borrow money he was spared the burden of interest payment imposed upon many of his competitors. His ability to build up a consolidated colliery empire depended upon long term planning and this in turn depended upon his great wealth. Unlike his less fortunate competitors, he was rich enough to wait, if necessary, for a return upon his mining investments.

Mining costs, and therefore profits, varied with the thickness of the seams being worked, coal from the thick Main Band seams at Whitehaven being removed at lower costs than coal from the thinner seams of the inland collieries. As pits went deeper the costs of winding, drainage and ventilation increased. Similarly, as mining operations extended laterally, it cost more to move the coal from the face to the shaft. Since most of the coal mined in Cumberland was sold in Ireland a major item in production costs was the cost of leading to the nearest harbour. This varied with the distance of the collieries from the ports and fortunately for the Lowthers many of their pits were situated near Whitehaven harbour. Nevertheless, until the adoption of waggon-ways, leading charges accounted for a large proportion of total expenditure. An examination of the weekly wage bills at Whitehaven in 1709 shows that it cost 1s. to raise a ton of coal to the surface and 9d. per ton to cart it to the ships. By 1756 the cost of leading coal from the Howgill pits had fallen to between 1¾d. and 3d. per waggon. In 1749 coal from Henry Curwen's colliery at Workington was led to the harbour at a cost of 10d. per ton; from his Harrington Colliery the cost of leading was 1s. 6d. per ton. At that time coal from both of these collieries was sold to the ships at Workington for 3s. 4d. per ton.[86]

With the sinking of many new pits and the development of existing ones the total annual cost of mining coal at Whitehaven Colliery rose from £3,413 in 1701 to £14,321 in 1750. The opening out of the Whingill division of the colliery after 1712 added appreciably to production costs. The cost per ton ranged from 1s. 8d. in 1719 to 2s. 11.3d. in 1732 and after 1723 it always exceeded 2s. per ton.

Whitehaven Colliery Costs

Year	*Total Cost*			*Cost per Ton*	
	£.	*s.*	*d.*	*s.*	*d.*
1701	3,413	3	2	2	1.6
1710	3,847	12	6	1	10.9
1720	4,561	18	0	1	9.2
1730	10,868	6	0⅝	2	7.8
1740	11,171	17	11	2	3.5
1750	14,321	18	1¾	2	3.2

The Colliery Abstract Books show that expenditure changed from year to year but they do not distinguish between normal working costs and capital investment so that for particular years the expenditure is inflated by development costs. Between 1729 and 1732 costs rose annually from £9,611 to £12,400 and much of the increase must be attributed to the cost of sinking Saltom Pit. Pit explosions could cause enormous damage and consequently fluctuations in production costs, but these "Abstracts" do not separate the cost of repairing such damage from everyday expenditure, nor do they take depreciation into account.[87]

R. W. Moore in his manuscript "History of Whitehaven Collieries"[88] analysed the cost per ton of working coal at the Howgill pits between 1734 and 1737 and provided these figures:

Year	*Getting*		*Other Expenses*		*Leading*		*Total*	
	s.	*d.*	*s.*	*d.*	*s.*	*d.*	*s.*	*d.*
1734		11.9	1	1.7		3.4	2	5
1735		10.8	1	1.3		3.1	2	3.2
1736		11.5		11.4		3.1	2	2
1737	1	0.2	1	1.6		3.3	2	5.1

Since the coal was sold to the ships at 3s. 4d. per ton there was a large margin of profit for Sir James Lowther.

When Sir James Lowther, later Earl of Lonsdale, inherited the Whitehaven estates detailed "Explanations of the Coal Accounts" were drawn up for his benefit in 1756 and these not only explain the accounts but also provide information about the materials needed in the collieries. In particular, vast quantities of timber were required. The Whitehaven shafts, oval in shape, had

> "a Timber Partition fixed in the middle of them on the Shorter Diameter, lin'd on each side of it with close joynted Deals, so that the Air may pass down one side & up the other; These Pitts are also Cribb'd with Oak Timber let into the Sides of the Pit, at every 3 or 4 foot distance from one another, & Deals also close joynted are nail'd to those Cribbs quite round the Pitt, from Top to Bottom."

In addition to lining pit shafts, timber was needed for underground roads, partitions in the air courses, waggons, waggon-ways and boarded corves. The Saltom Pit air course, constructed in 1732-33 at a total cost of £950, required 5,000 deals costing £300 for its two partitions. Pit explosions could wreck air courses. An explosion at Thwaite Pit in 1743 "threw down all ye partitions &c in its way" and John Spedding reckoned that it would take nearly 2,000 deals to make partitions "to carry off so much of the foul air as to enable the men to breath in the rest". Large sums were spent on timber and deals for the the Lowther mines but Sir James, in the early 1750's, was able to prune the cost by obtaining supplies directly from Norway, thereby eliminating the profits of middlemen and saving "several hundred pounds a year".

There was also considerable expenditure on iron and rope. Iron was needed for tools, waggons, chains, pistons, cranks, horseshoes and nails. The bar iron used was bought from Peter How & Co. at 17s.-18s. per cwt. and was either imported Swedish or Russian iron or forged at the company's Low Mill ironworks. Iron waggon wheels and other cast iron work were obtained from the Clifton furnace at a cost of 14s. per cwt. Boiler plates for the pumping engines had to be imported from Sweden and these cost £24-£25 per ton. More and deeper pits, generally working both day and night, required large quantities of ropes and they had to be of the finest quality hemp and very carefully made. The quality of the winding ropes was of prime importance to the miners who rode up and down the pit shafts and they insisted "upon having the Ropes fresh & Good, their Lives depending upon it". In addition, rope traces were needed for the many tram horses underground and waggon horses above ground. The estimated charge in the colliery accounts of each of Lowther's thirty-two waggon horses was 10s. per week, although this was less than the full cost of their maintenance.[89]

In some years costs were swollen by extraordinary items of expenditure. For example, between 1732 and 1733 the new coal staith built on the west wharf at Whitehaven cost £611. 1s. 3¾d. and the waggon-way from Swinburn and Corporal pits to this staith cost £588. 19s. 2d., a combined total of just over £1,200.[90]

Because of the lack of records it is not possible to state how much was spent by Sir James Lowther on all his collieries between 1706 and 1755. There is no information about the cost of his collieries at Dearham and Scalegill, or of Seaton Colliery after 1732, or of Whitehaven Colliery after 1750. Nevertheless, the greater part of his massive expenditure on his mining enterprises can be defined.

Sir James Lowther's Colliery Expenditure 1706-50

Years	*Colliery*	£	*s.*	*d.*
1706-50	Howgill	326,219	9	2½
1714-50	Whingill	37,069	4	4⅛
Howgill and Whingill Total		363,288	13	6⅝
1706-08	Lattera	1,318	9	6
1714-37	Parton	15,997	1	10⅛
Total Expenditure		380,604	4	10¾

Whitehaven Colliery was a source of increasing profit to the Lowther family. Between 1701 and 1750 the proft increased sevenfold, from £986 to £6,983, the highest annual profit being £8,512 in 1749. Profits per ton ranged from 3.24d. in 1733 to 1s. 8.27d. in 1744. The Howgill pits consistently showed a profit, but at the Whingill division of the colliery annual costs always exceeded annual receipts. The quarterly accounts show that only on nine occasions during the years 1714-50 did Whingill produce a profit balance and each time it was very small.

Whitehaven Colliery Profits

Year	*Total Profit*			*Profit Per Ton*	
	£	*s.*	*d.*	*s.*	*d.*
1701	986	16	8		7.40
1710	1,047	1	7		6.24
1720	3,570	13	5½	1	4.61
1730	2,202	0	7⅞		6.45
1740	6,921	11	11	1	5.08
1750	6,983	6	11¼	1	1.28

During these fifty years the colliery's financial record was, for the most part, one of growing profits, although in some years development costs ate deeply into profits. By 1726 the colliery's profit had risen to £4,588, but it then fell year by year until 1733 when it amounted to only £1,154, the lowest annual profit since 1711.[91]

As the volume of coal exports from Whitehaven increased so did Lowther profits. Although the shipmasters opposed Sir James Lowther when he raised the price of his coal it was the ability of the latter to control export prices which made possible his substantial colliery income. In view of his many and prolonged absences from Cumberland the success and profitability of Whitehaven Colliery owed much to the continuing presence of the Spedding brothers, to John's outstanding business ability and to Carlisle's uncommon technical skill.

Lattera Colliery from 1706 to its closure in 1708 made an annual loss, but Lowther's other collieries, although much less profitable than Whitehaven Colliery, added to his revenue. Both the Parton and Dearham collieries yielded profits in most of the years

Sir James Lowther's Colliery Profits[92]

Years	*Costs*			*Receipts*			*Balance*			
	£	*s.*	*d.*	*£*	*s.*	*d.*	*£*	*s.*	*d.*	
1706-50 Howgill	326,219	9	2½	528,528	12	9⅜	202,309	3	6⅞	
1714-50 Whingill	37,069	4	4⅛	13,051	5	9½	24,017	18	6⅝	Loss
Howgill and Whingill	363,288	13	6⅝	541,579	18	6⅞	178,291	5	0¼	
1714-37 Parton	15,997	1	10⅛	18,016	0	6½	2,018	18	8⅜	
1723-55 Dearham							2,028	11	9¼	
1737-41 Scalegill							1,876	17	2¾	
							184,215	12	8⅝	
1706-08 Lattera	1,318	9	6	85	4	7	1,233	4	11	Loss
						Total	182,982	7	9⅝	

they were working. Unfortunately, information about the Scalegill and Seaton collieries is incomplete, but for the few years for which records are available after 1735 they showed annual profits.

Sir James also had a two-thirds share of Seaton Colliery and was therefore entitled to two-thirds of its profits, but very few records have survived to show its profitability. In 1736 the colliery made a profit of £719. 7s. 11½d. and £1,236. 1s. 6½d. in 1743.[93]

Lowther's colliery income far exceeded that of his competitors. Henry Curwen (*d.*1725) preferred to lease rather than work his collieries which, in his last years, brought in a rent of £315 per annum.[94] His successors managed the collieries themselves and their efforts brought increased rewards. According to Dr J. V. Beckett the Curwen collieries made a profit of £793 in 1736, £2,142 in 1742 and £2,218 in the following year.[95]

Other colliery owners in Cumberland during this period were much less fortunate. Broughton Colliery, handicapped by inefficient management and high leading costs, could yield but small profits. In 1749, when worked by Ewan Christian, it earned a profit of only £41. 1s. 5d. on an output of 7,680 tons.[96] Humphrey Senhouse's colliery at Ellenborough, despite years of effort, proved so unsuccessful that he was obliged to cease mining in 1736.

THE COAL TRADE

In the eighteenth century the coal trade expanded greatly to supply the needs of Ireland's growing population. Between 1701 and 1750 coal exports from Whitehaven Colliery more than trebled.

Whitehaven Colliery Coal Shipments

Year	*Tons*	*Loads*
1701	30,466	1
1710	35,129	4
1720	46,559	5
1730	81,228	2
1740	96,543	3
1750	117,618	3

After 1713 the colliery's shipments were increased by the output from the new Whingill division of the colliery, reaching 61,097 tons in 1717. This was followed by a trade recession and not until 1725 did coal exports again exceed 60,000 tons. Thereafter coal shipments rose to 100,314 tons in 1746 and to a maximum of 117,618 tons in 1750.[97] In this period between seventy and eighty per cent of the coal exported from England to Ireland came from Whitehaven.[98]

As the traffic in coal increased so did the delays in loading and improvements in harbour facilities at Whitehaven were badly needed. The situation was exacerbated by the inadequacy of Parton harbour and its virtual closure between 1718 and 1726 which compelled more vessels to take on cargoes at Whitehaven. Between 1710 and 1754 much was done to provide better accommodation for the collier fleets – the harbour was deepened, the old pier lengthened, four breakwaters were constructed and a new pier built. As a result of this work Whitehaven had a new outer harbour and congestion in the port was eased by the availability of many more berths.

Trade at Parton was hindered by the poor condition of the harbour. After the passage of the 1706 Act the trustees made the mistake of leaving the repair and management of the harbour to Thomas Lamplugh who seemed more concerned with collecting harbour duties than with repairs. A pier was built but it was so badly constructed that for years it threatened to fall down and block the harbour, a matter of great concern to James Lowther who resumed mining at Lattera in 1714, the coal being shipped from Parton. When the temporary duties lapsed in 1716 the Parton trustees agreed to continue them for twenty-one years, and to allocate the first £500 of the proceeds to repairing the harbour. Little was done and in June 1718 it was feared that the harbour would "fall to rubbish". Six months later this fear became a reality when a storm destroyed the pier which "tumbled into the Harbour".[99] Much to Lowther's chargrin Lamplugh made no effort to rebuild the pier. In the struggle to have the harbour repaired Lowther was joined by John Brougham, who, after buying the Moresby and Distington estates in 1722, was anxious to have the harbour re-opened to enable coal from his Moresby pits to be exported from Parton. They co-operated to procure an Act of Parliament in 1725, Lowther himself being largely responsible for its speedy passage. By this Act the duties on coal to pay for harbour improvement were renewed. Both Lowther and Brougham lent equal amounts of money to help to pay the cost of repairs to the harbour. Work on re-building the pier began in 1726 and was completed before the end of the year, a task which cost £1,900 in addition to the duties collected since 1 May.[100]

As the coal trade developed it became necessary to regulate the order in which vessels were loaded. In the Cumberland ports priority was given to those regularly engaged in the trade, vessels trading to foreign parts had to take their place at the end of the queue if they wished to enter the coal trade. At Whitehaven coal leaders breaking turn at the harbour caused friction and disturbances were so frequent that instruction had to be issued, 21 September 1731, to control the loading of vessels:

> "Whereas great disorders are occasioned by the Coal Leaders putting up their Coals Irregularly at the Ships, whereby great damages frequently happens to their Horses & Carts, It is ordered by James Lowther, Esq.,
>
> That all Carts as well as from Howgill as from Whingill shall sling their Coals in Turn as they come to the head of the Ships that are Loading; and that any Leader who discharges their Carts otherwise than in Turn as aforesaid shall forfeit and pay for Every offence one shilling.
>
> And it is further Ordered that all Masters of Ships and others concern'd in Takeing of Coals who refuse to discharge the Coal Leaders in Turn as aforesaid, or preferr one Leader to another, whereby the good order of unloading in Turn as aforesaid is in any ways Obstructed by their default; such Masters or persons concern'd in Takeing Coals shall for Every offence Forfeit the Value of the Coals in Each Cart so refused to be taken in Turn as aforesaid, The said Penaltys to be applied to the relief of Poor Coal Leaders."[101]

When loading coal at Whitehaven the shipmasters, with the connivance of minor customs officials, evaded payment of full customs duties on it. The custom house and a watchtower belonging to the Lowthers were leased to the Commissioners of Customs and when the lease was due for renewal in 1745 Spedding urged that the watchtower should on no account be included in a new lease because from it customs officers could observe "every waggon that goes into a Ship, and Charge the Ships wth Duty to the full."

The landwaiters, who were aware of the total quantity of coal being loaded, used this knowledge to "Extort 1^s a Voyage . . . wch the Masters are obliged to Comply wth when they know they have them in their power to make some of them pay twenty or 30 times more in the Duty".[102]

Although coal exports from Whitehaven increased notably the story of the coal trade was not one of smooth, uninterrupted progress; it was beset with difficulties and conducted with a lack of business efficiency. Bad weather or contrary winds could hold up vessels for days and crowded harbours could delay loading at Whitehaven or unloading at Dublin. Strikes of miners or coal leaders or a shortage of carts interrupted coal supplies to the ships and every increase in the price of coal led to a refusal by the shipmasters to buy coal for some time. When it was announced, 23 September 1714, that the price of coal would be raised the following week from 2s. 8d. to 3s. per ton the angry shipmasters at Whitehaven immediately joined together against the increase and for over five weeks they took on no coal before giving way and accepting the new price. The same thing happened in 1721 when it was decided to fix the price of Whitehaven coal at 3s. 4d. per ton between October and March and at 3s. per ton for the rest of the year. The new price of 3s. 4d. came into force 4 January 1721 and again the shipmasters refused to load coal, their strike lasting until 12 February when they had to submit to Lowther's terms. The pricing system had to be modified in 1723. The winter increase to 3s. 4d. per ton was delayed until 1 November of that year but 29 November the price was lowered to 3s. 2d. for the winter months and to 2s. 10d. for the summer months. These differential prices were later abandoned; from October 1724 until October 1728 the price of coal remained unchanged at 3s. 2d. throughout the year. At the end of October 1728 the price was raised to 3s. 4d. per ton and another buyers' strike began, the Whitehaven shipmasters combining to stop the coal trade. Potential strike-breakers were warned off by threats of violence and not until 3 January 1729 did the shipmasters end the strike. Colliery accounts show that from this time until 1751 the price of coal to the Whitehaven ships remained constant at 3s. 4d. per ton.[103]

Coal owners in the Workington area had also to face opposition from local shipmasters when they followed Lowther's example in raising the price of coal to 3s. 4d. per ton in October 1721. For over a month the Workington masters refused to buy coal at this price. To overcome this opposition James Gorton freighted some Whitehaven vessels to carry his coal at 9s. 6d. per ton, but failed to hire Irish ships for the same purpose. The lessees of the Harrington pits went further and bought several ships of their own to transport their coal to Ireland.[104]

The Cumberland colliery owners were not bound by price-fixing regulations when selling their coal. Thus Ewanrigg coal was sold at Maryport at 3s. per ton in 1750 and at other ports coal was then being sold at 3s. 4d. per ton. There were no marked differences in price at the various ports; they were too close together to permit wide variations in price and the self-interest of the coal owners made a common policy desirable. None thought of a price-cutting war. Such a struggle, had it taken place, would have been won by the Lowthers. What small differences in price there were resulted from the difference in the quality of the coal. The colliery owners did act together on occasion when price changes took place and, as might be expected, the lead usually came from Lowther or his agents. Spedding reported in 1738,

"I determin'd to Raise the price of Seaton Coales lett the Consequence be w^t it would . . . I went then to the Pitts & gave positive directions to Raise the price imediately to 3s. 4d. . . . The Masters imediately applied to Mr. Curwen, who promised to Load them all at the old price & measure, but when I came back on Wedny I happen'd to meet with him, and having Consider'd better of it he came roundly into raising ye price 4d. per Ton from that moment."

Shipmasters often delayed paying for their cargoes of coal and there was some co-operation among the coal owners in tackling this problem. In 1726 James Lowther made an agreement with Henry Curwen.

"that if any Ship is in arrear for Coales & avoids paymt on pretence of getting a new Master or being sold & goes from one Harbor to another to load they shal have no Coales til they have pd the other Coal Owner for the former Voyage wch I hope all the Coal Owners wil joyn in."[105]

Although the Lowthers controlled the supply of coal at Whitehaven they did not dominate the coal trade and their relations with the traders were often difficult. Apart from opposing price increases the shipmasters were sometimes ready to abandon the Irish trade and make voyages to Norway, Virginia and other places. This usually happened in the summer months when low prices in Dublin made trading unprofitable. As a consequence there was a shortage of vessels for the trade with Ireland which in turn could lead to restrictions upon the output of coal. A letter from John Spedding shows this,

"By the Inclosed List the Ships now in the Harbour you will see we are quite at a Stand as to the Coal Trade. I have gon over this List of our Ships and dont find above 37 Sail of Ships belonging to this place and Parton, but are already engaged for the Virga, Norway, and other Trades, and a good many of that Numbr are of the Smallest Sort, so that we have a bad prospect of the sumr Trade . . . The Steath is pretty full and Incapable of holdg such quantities as to bear any proportion to our present Gettings, so that unless Irish Ships will come and help the Exportation twill be impracticable to Gett so many Coals as we do."[106]

An attempt was made to persuade the owners of Irish vessels to carry Whitehaven coal to Ireland but it had little success.

Whitehaven shipmasters who proved troublesome to Lowther or his agent could be made to suffer by being deliberately supplied with the worst coals available. Those who were more amenable were given preferential treatment. Spedding informed his employer,

"I have incourag'd a great number of small ships & others that are friendly to Load out of turn, wch is a great plague to the big ships; I also take all the care I can to lett Friends have the best Coals, so that the others may see plain enough they are to have no favour till they mend their ways."

A stop of trade in April 1736 made Lowther decide different measures were necessary to break the opposition. He already owned shares in half a dozen small vessels; if he could control a number of large vessels he would then be able to ensure the continuance of the coal trade should the shipmasters refuse to load coal or should there be a lack of shipping because of the growth of the Virginia trade. Further, vessels owned by him could direct the coal trade to ports other than Dublin. He wrote to Spedding that the best way to circumvent the masters would be to buy "Commanding Shares of large floaty Ships to follow the Coale Trade all the year round" and that "when any Ships are to be sold that wil do for the Coale Trade I should buy 9/16 of some of them to keep them constantly in the Coal Trade". With Spedding's help Lowther bought shares in more ships, many of which he controlled by owning a majority of the shares. Lowther was joined by Walter Lutwidge, his partner in Seaton Colliery, in the buying of vessels. One of these was the *Cumberland* launched at Ravenglass in April 1737, in which Lowther had a half-share at a cost of £1,230. This ship was built in accordance with Lutwidge's plan to assist the export of Seaton coal, but proving too large for Workington harbour she traded with coal from Whitehaven. With the outbreak of war seamen became scarce and because it was increasingly difficult to man so large a vessel "she was Lett out for the Virg^a trade for 3 voyages, in the last of wch she was taken by the French on her return from Virg^a to London 1745". Between 1722 and 1749 Lowther bought shares in twenty-four vessels at a cost of nearly £11,500; but these were not all exclusively engaged in the Irish coal trade, some traded to Norway and Virginia. However, the investment of the Lowther family in ships compared with collieries was always small; their capital was much more profitably employed in coal mining than in the coal trade.[107]

Other coal owners in Cumberland saw the advantages of owning both collieries and ships. Eldred Curwen invested £2,317 in ships in the hope of earning profit in the coal trade as well as providing transport for his coal. He was the sole owner of three vessels, the *Charming Jenny*, the *Workington* and the *Lark* and also had shares in six more. Coal shipments from Maryport increased rapidly after an Act was passed in 1749 to improve the harbour there. Humphrey Senhouse II soon saw the possibilities of profit and began to buy shares in ships. Between 1751 and 1754 he bought shares in eight vessels and in later years added to his investment in shipping.[108]

The direction of the coal trade lay with the masters of vessels sailing out of the Cumberland ports and not with the colliery owners. It was the master who bought the coal, shipped it, sold it, and, after paying all expenses, shared out trading profits among the owners of the vessel. The masters were inefficient in their conduct of the coal trade and seemed incapable of organizing the sale of their coal on a sound and more profitable basis. Although they were selling an essential commodity to a large market in Dublin, which they dominated for much of the eighteenth century, the trade produced meagre profits. They insisted on trading with Dublin and were unwilling to develop trade with other Irish ports or with English west coast ports; had they sought alternative markets they would have benefited both there and in Dublin. Thus high prices in Dublin were soon followed by a glut of Cumberland coal and rapidly falling prices. In 1715 Spedding reported, "The Price of Coals at Dublin is fallen to 13s. 6d. a ton wch is so very low that the Masters can but just make the ends meet, & yet there is very few of them that show any disposition to goe to other places". In the following year the supply of coal to Dublin exceeded the demand to such an extent that "the Masters cannot get them off at any price. The Gabards are all full & many of them have laid under their Coals this Six weeks wch makes the Gab^d owners insist on Demurrage & several ships have staid at Dublin for 10 days before they could get Gabards to take in their Coals".[109]

From the letters of John Spedding the rapid fluctuations in the price per ton of Whitehaven coal in Dublin can easily be seen:

	s.	*d.*		*s.*	*d.*
21 Sept. 1712	17	0	12 Jan. 1718	24	0
24 Dec. 1712	22	0	14 Feb. 1718	21	0
8 Feb. 1713	28	0	5 Mar. 1718	28	0

	s.	d.		s.	d.
4 July 1714	13	6	14 Apr. 1718	16	0
7 Aug. 1714	16	0	20 Apr. 1718	22	0
3 July 1715	13	6	14 May 1718	17	0
23 Nov. 1715	17	0	23 May 1718	15	0
13 Apr. 1716	13	6	9 Nov. 1718	22	0
15 Mar. 1717	18	0	21 Dec. 1718	24	0
24 Mar. 1717	20	0	17 Jan. 1720	17	6
24 May 1717	16	0	5 Feb. 1720	16	6
10 June 1717	15	0	10 Apr. 1720	16	0
16 June 1717	14	6	26 Apr. 1721	15	0
22 Sept. 1717	18	0	21 Oct. 1722	15	6

On resuming the coal trade after the end of their strike in January 1729 the Whitehaven shipmasters, much to their delight, were able to sell their coal in Dublin at 30s. per ton. Their joyous mood did not last long for the lord mayor used his powers to slash the price to 16s. per ton.[110]

At this time the "ton" was a measure and not a weight, the weight of the ton varying from one port to another, but it may be noted that the Dublin ton weighed 21-22 cwts. and the Whitehaven ton about 14 cwts. The larger measure of the Dublin ton accounts for part of the increase in the price per ton of Whitehaven coal in Dublin. William Gilpin in 1695 examined the measures in use at Whitehaven and found that a ton at the pit measured 264 gallons and a ton sold to the ships 192 gallons, so that eight pit tons equalled eleven shipping tons; in his own words, "every 8 Tun at the Pitt makes out 3 more vizt ll at the Ships." The Dublin ton measured 320 gallons and therefore twelve Dublin tons equalled twenty Whitehaven tons, i.e. a Whitehaven ton amounted to three-fifths of a Dublin ton. But Gilpin added the rider that "most Commonly by Reason of bad Measure at Whitehaven they (i.e. the coals) do not hold out in proportion as 20 to 12 but as 20 to 11." After 1735 the Whitehaven shipping ton weighed 17 cwts;* the Whitehaven ton at the pit, as distinct from the shipping ton, weighed about 21 cwts.[111]

Repeatedly Spedding tried to persuade the masters that it would be to their advantage to ship coal to ports other than Dublin and to promote the diversification of the trade he instituted a bond system in 1719 whereby masters were bound in the sum of £20 or £30 not to trade with Dublin and in return were allowed to load their vessels out of turn. Any master who broke his bond by sailing to Dublin had to pay the appropriate penalty. Spedding recorded one such instance when he refused to allow a ship to load coal until "they pay the penalty for Carrying Coals to Dublin contrary to Bond". At the same time he was well aware that the glutting of the Dublin market could benefit his master by driving his competitors out of business and when the time seemed ripe he worked towards that end. Thus in 1726 he wrote,

> "As it seems now to be to yor Interest to run down other Collieries by shipping great Qu^{tys} from your own, we shall pick up all workmen we can. And I hope with the Qu^{tys} you'l be able to Lead in a few months G^{rt} Distress will fall upon the other Coal Owners; I see nothing so likely

* "The NewCastle Chaldron being 53 Cwt by Act of Parl^{t} & the London Chaldron being always reckon'd ½ of the NCastle Chaldron but now falling short, I reckon'd our Tun at 17 Cwt & so a Tun and a half w^{ld} be about a London Chald." (Lowther, 23 May 1738).

to pinch them as glutting the Trade now when yor Coals are so much superior to theirs in Goodness."

Vessels in which Sir James Lowther held majority shares were used to divert the coal trade from Dublin to other Irish markets such as Cork, Limerick and Youghal. In an attempt to diversify the trade further Lowther, in 1738, sent printed advertisements proclaiming the merits of Whitehaven coal to the leading Channel ports. His sales campaign had little success. Ten years later he hoped to develop a foreign trade by having his coal delivered to ports in southern Ireland for transhipment to France, Portugal and Spain. His plan does not seem to have achieved very much, nor was it soundly based for it had little to commend it to the Whitehaven shipmasters who preferred to continue thier customary voyages to Dublin.[112]

In selling their coal the masters were frequently cheated by the pursers or factors who acted for them in Dublin. Irish pursers were generally believed to be dishonest rogues but when they were replaced by Englishmen there was little improvement in the standard of business morality. Spedding in 1735 wrote,[113]

"The Whitehaven masters were mighty fond last Summer of having Whitehaven pursers and to employ none of the Irish . . . but the Irish pursers have . . . found means to Corrupt the Morralls of those that were Employ'd from Whitehaven, by Leading them into Gameing, Whoreing & other Vices, and two or three of our Whaven folks have been stript of all they had & are made off, w^{rth} the loss of a great many Hundred Pounds to the owners . . . to the amount of 5 or 600£ in the whole. So they find it no longer practicable to employ any but the Dublin people to Sell their Coals, who have got good Substance from Sharping the Masters formerly, & they must be Contented to be cheated a little in every voyage rather than to have their own people run away with whole Cargoes together."

In 1737 the coal trade was said to be in a "ruinous state" and in August of that year, in an attempt to restore profitability, the masters and owners of ships at Whitehaven set up a committee of eleven members to regulate the trade. A resident agent at Dublin was appointed to control the price of coal, examine pursers' accounts and generally safeguard the trade under the direction of the committee. This was Thomas Bacon who received £150 per annum, office expenses and house rent. Under him were twelve pursers, each paid £40 per annum, who reported to him daily with their accounts and the money they had collected in payment for coal. To meet these costs 4d. per ton was to be paid into Bacon's hands by masters of Whitehaven ships who discharged their coal into gabards or made use of a purser; masters who discharged coal at the quays without the assistance of a purser had to pay 2d. per ton. A month later a similar committee was established at Workington to co-operate with the Whitehaven committee in the regulation of the trade. The hope of bringing order and profit out of confusion and loss proved misplaced for it was impossible to enforce total compliance with the rules laid down. Some masters sold coal below the price fixed by the committee and in doing so undermined the basic aim of the new marketing system, the maintenance of profit. Others sold their coal direct to customers without making use of the pursers. Coal owners such as Lowther and Curwen were anxious, primarily as colliery owners, but also as shareholders in trading vessels, that the coal trade should proceed smoothly and profitably and used their influence to try to ensure that the masters obeyed the trading regulations. Spedding, a member of the Whitehaven committee, believed that the sale of coal should be kept entirely out of the hands of the Irish pursers who, he said, "cheat us to our Faces as much as they please" and whose "pretended Skill cannot possibly make the Consumers burn one Ton a y^{r} more than they have occasion for". By April 1738 the cost of running the agency already exceeded income by about £250 and to meet this deficiency the committee increased the tonnage dues payable to Bacon from 4d. and 2d. to 6d. and 4d. per ton. To pay off existing debts quickly it was decided to impose a forced loan. All masters, for one voyage only, were to pay an additional 12d. per ton on all coal shipped to Dublin and this money was to be repaid with interest out of the committee's fund. Some of the masters continued to disregard the rules about fixed prices and the use of pursers when selling their coal and in less than a year the selling agency came to an end.[114]

Given Dublin's dependence upon Cumberland for its coal supplies it should have been possible to organize a marketing scheme that would have ensured higher profits to the trade. Little was done to control the volume and rate of flow of coal exports to Dublin so that the pattern was of scarcity in Dublin leading to high prices, followed by excessive imports and rapidly falling prices. Had a few wealthy merchants controlled the trade the picture would have been different. As it was the coal sold in Ireland was sold on behalf of innumerable small shareholders in Cumberland colliers who were incapable of organizing themselves or the coal trade. In an age of bad communications it was difficult to weld together into a coherent body the many masters who were responsible for coal shipments. When prices and profits fell all they could think of was to make fewer voyages to restrict supplies and thereby force up prices.

War added greatly to the difficulties which afflicted the coal trade. Attacks by enemy vessels, especially the privateers which infested the Irish Sea, deterred the masters from sailing. Many Cumberland ships were captured, but the usages of war then permitted a captain, by the payment of a ransom, to regain his ship. So many vessels were captured that Lowther decided that the matter of ransom money required some regulation, "If any of our Ships with Coals should be taken I think the Master should not engage for above 1/6 or at most $\frac{1}{4}$ of the value for a Ransome but shal advise & consider further". Sometimes collier fleets were escorted by warships but if such protection was not available the masters often refused to sail and the trade could be held up for long periods. In July 1706 nearly forty loaded ships in Whitehaven habour remained there for some time because their masters refused to leave port without a naval escort. Trade was hampered not only by enemy action but also by the government when vessels were commandeered to serve as transports thereby causing a shortage of ships for carrying coal. In times of peace a lack of seamen could restrict coal exports; when the demand for coal was high there were not enough sailors to man the colliers and ships had to be laid up. The press-gangs which operated in both peace and war were a serious threat to the trade when they seized seamen from the colliers and fear of the press was enough in itself to cause a similar shortage of seamen. If seamen were being pressed in Dublin Whitehaven sailors refused to leave harbour and if the press appeared likely in Whitehaven they deserted and hid in the country. Spedding in 1718 described the actions of a press-gang in Dublin which, "began to press Seamen & swept all before them, from some they took 6, others 8 and some 14 men wch has quite disabled the Ships from comeing home" and when this was known in Whitehaven the local sailors, fearing a press there, fled in large numbers "so that 'tis certain we shall have an intire stop of Trade till the Men can be protected from the press". Any threat to the coal trade was a threat to the Cumberland coalfield since the bulk of its output was exported and both Spedding and Lowther did their utmost to remove any hindrance to coal shipments. Another press in Dublin in 1726 again threatened to stop the trade "for not a man will stirr out of ye harbour till they have protections or till the press warrants are call'd in" and also seemed likely to put an end to coal leading to Whitehaven harbour where twenty loaded ships were already help up. A "protection" was a certificate from the Admiralty exempting the crew of a particular Vessel from the press. To safeguard the trade Spedding had to make repeated requests to Lowther to obtain protections for Whitehaven vessels and the latter was most assiduous and successful in obtaining them. His political influence together with a little bribery of minor officials accelerated the grant of protections.[115]

With the growth of the coal trade later wars brought the same problems in an intensified form. When war broke out with France in 1744 Spedding was alarmed at the prospect of French privateers "swarming amongst us imediately" and at the certainty of inadequate convoys to protect the coal trade. It was feared that the French would attack Whitehaven itself and again Lowther showed his ability to defend his interests by arranging the transfer of a company of soldiers to guard the town and harbour. The lack of naval escorts led Spedding to consider arming a Whitehaven ship with sixteen or twenty guns which would be constantly employed in protecting the colliers, its costs being met by a levy of 12d. per ton on all coal shipments. Nothing came of this, but the *Howard*, a ship of ten guns belonging to Peter How of Whitehaven, convoyed a collier fleet to Dublin in 1744, the masters paying, with extreme reluctance, the 6d. per ton charged for this service. It was estimated that it would cost £549. 6s. to fit out the *Lowther*, a brigantine of 70 tons, "for a Privateer or Convoy for the Coal Trade with 12 Car^g & 12 Swivel Guns, Small Arms &c for 50 Men" and the cost of working the ship would amount to £170 per month. With few warships available to protect the trade the Whitehaven shipowners were obliged to arm their vessels themselves. In 1744 the *Cumberland* and the *Augusta*, both partly owned by Lowther, were equipped with guns borrowed from Mr How and from the battery at Whitehaven.[116]

Although it was a growing trade the export of coal to Ireland was not very lucrative. According to Spedding in 1720,

> "As to the prices that the Masters w^d be willing to settle at & not to sell under I think they could not go under 16^sh a Ton in Summer & 18^sh in Winter & by all the Ships A^ccts that I have yet seen their profit at those prices wil but just keep the ships in repair if they come to no Accidents, and the lesser sort of Ships can but barely save themselves at those Rates and are many of 'em out of pocket."

A fundamental cause of the trouble was the practice of selling small shares in vessels, for the ownership of a vessel could be divided into sixty-four shares. This led to multiple ownership and this meant, of necessity, that the trading activities of a ship were usually directed by its master who often lacked the requisite skill to sell his coal to the best advantage. Too many vessels were engaged in the trade to make substantial profits possible and excessive concentration upon the Dublin market further reduced profits. Matters were not helped by the ships employed in the tobacco trade, which, on their way to Virginia, carried a cargo of coal to Dublin. Apart from picking up a cargo on their outward voyage their masters had no real interest in or knowledge of the coal trade and in order to unload quickly were often ready to lower the price of coal in Dublin to the detriment of those constantly engaged in the business. Operating costs were higher than they might have been because of the inability to obtain return cargoes; all too often the colliers returned in ballast to their home ports. A contemporary pamphleteer made the additional point that the shipowners

> "Being the only persons concerned in the Sale of Coals at the several places of consumption who run the Hazard of their fortunes . . . As they have no Freight but what arrises from the profits of the Coals; they are, in that Respect, different from most other proprietors of Shipping who are intituled to the Freight agreed for, exclusive of any Hazard in Regard to the Goods, or Price they may bear at the market."[117]

The Lonsdale estate accounts, which detail the work and financial record of Lowther vessels, show that the coal trade generally was not highly profitable although some colliers did earn sizeable profits. The *Liffey*, in which Lowther had a three-quarters share at a cost of £581. 12s. 11½d., made 166 coal voyages between 1738 and 1753 earning £1,113. 0s. 6¾d. in dividends. The *Hibernia* in the course of 97 coal voyages between 1738 and 1747 made a profit of £503. 7s. 3d. Lowther's two-thirds share of this vessel, built at Workington in 1735, cost him £685. 4s. 7d. The varying fortunes of some of the vessels in which Lowther owned shares are illustrated in a rental account of 1751 which makes it clear that his investment in ships did not always yield a high return:

Vessel	*Share*	*Received more than 1st Cost* £ s. d.	*Wants of 1st Cost* £ s. d.	*Date of purchase of shares*
Bonetta	5/8	17 17 5½		1723
Saltom	1/8	17 18 0		1733
Liffey	3/4	461 14 3¾		1737
Bee	all		198 10 11½	1736
Globe	3/4		243 0 0	1737
Howgill	3/4		514 6 3	1742
Devonshire	3/4	311 10 0		1737
Shannon	5/8		295 16 1¼	1737
Augusta	7/8		974 13 3	1738
Dublin	3/4	176 8 1		1741
Resolution	5/8		232 5 0	1747
Felicity	3/4		308 4 2½	1742
Friendship	1/4		32 0 0	1742
Experiment	1/8		35 0 0	1746
Boyn	1/2		352 0 0	1749
		985 7 10¼	3185 15 9¼	
Ballance wanting of 1st Cost		25 March 1751	2200 7 11	

Lowther, like all part-owners of vessels, had to pay his proportion of the costs of repairing and rebuilding ships in which he had shares. This could be expensive. In 1739 the rebuiling of the *Union*, later named *Augusta*, cost him £1,050 and in 1743 he had to pay £756 towards the cost of rebuilding the *Globe*. Investments of this kind ended abruptly when ships were lost at sea. The *Lucia*, for example three-quarters owned by Lowther, was sunk in a storm in 1742 and three years later the *Cumberland*, in which he had a halfshare at a cost of £1,230, was captured by the French. Considering his wealth Lowther suffered comparatively little from the vicissitudes of the coal trade; whatever happened to the traders when they sold their coal in Dublin he controlled the selling price in Whitehaven and increasing exports provided him with growing profits.[118]

James Lowther hoped to monopolize the coal industry in west Cumberland and the coal trade. To do this he would have to take over the coastal collieries and control the ports. He asserted his intention "to bid for anything on the Coast from Wh^{n} to Elnfoot"; by acquiring rival collieries he would be able to dominate the coal trade. As he informed Spedding, "I wil buy more Collierys to have it in my power to increase or lessen the Trade at different harbours as I see cause, so you wil do well to make the enquiry you propose about Sanderson's Colliery at Crosby or any other that is like to be sold". His land purchases gave him control of mining in the Whitehaven area and of the port of Whitehaven. Using the same methods he gained control of Parton harbour and its adjacent collieries. After buying Brougham's estates in Moresby and Distington in 1737 he paid Brougham £2,040. 4s. for his "19 Harbour Bonds & Interest from 1st May 1736 to this day the 22^{d} Dec", a transaction which, in effect, made the harbour his. A few months later he was able to write, with some satisfaction, "The more I think & hear of the Coal Trade the greater reason I have to be thankful there is none to interfere with me now either at W*n* or Parton".[119]

Further north he failed to establish a monopoly although he did manage to take over collieries at Seaton and Dearham. Lowther had a chance to lease the Curwen collieries at a rent of £340 per annum but he foolishly threw away a valuable opportunity by refusing to pay more than £300; he miscalculated if he believed that by holding out he could compel Curwen to accept the lower rent. He made several unsuccessful attempts to obtain Broughton Colliery which, despite inefficient management and insufficient development, was then said to be "one of the best places for great plenty of Coal" in Cumberland. It may be that Lowther's success in creating a monopoly in the Whitehaven area made it impossible for him to achieve a monopoly of coal in Cumberland. His territorial expansion led people in Whitehaven to make loud outcries against a monopoly and inspired jealousy in the minds of other colliery owners and it is possible that these factors helped to account for his inability to gain possession of the Curwen and Broughton collieries. Humphrey Senhouse was certainly jealous of Lowther's success and showed it by refusing to allow the latter's Dearham coal to be stored near the harbour at Ellenfoot. In addition, he obstinately refused to do anything towards improving the harbour because he feared Lowther would benefit more from such improvement than he would himself.[120]

During the first half of the eighteenth century most of the coal shipped from Cumberland was exported from Whitehaven but coal shipments, although modest by comparison, were increasing at the other Cumberland ports. As the Curwens opened out their pits at Workington, Harrington, Lowca and Clifton they added to the volume of coal moving to Ireland. In the year ending 13 February 1755 Henry Curwen exported 21,494 tons of coal and 22,354 tons in the following year. Although few statistics of Workington coal shipments are available it is likely that a large proportion of them came from Seaton Colliery. Between 1728 and 1732 annual shipments from the colliery ranged from 4,005 to 9,866 tons. In the next twenty years output and exports from the colliery greatly increased, 77,760 waggons of coal being shipped between March 1741 and March 1748. With the growth of trade harbour improvements became necessary and preparations were made in 1719 and again in 1731 to obtain an Act of Parliament authorizing the imposition of duties to finance such improvements. These plans were abandoned and improvements to the harbour were made following agreements in 1721, 1723 and 1734 between the coal owners and shipmasters of Workington about the collection of duties and their allocation to the maintenance of the harbour. The harbour was deepened, damaged portions of the retaining walls were rebuilt and in 1751, because of the money laid out on it, the harbour was said to be in so good a condition that there was no thought of applying for an Act of Parliament.[121]

Little evidence has survived to illustrate the early history of the coal trade at Ellenfoot (Maryport) but coal was being shipped from there in 1699 and Sir John Lowther, because of his mistaken belief that the Whitehaven pits were exhausted, was then of the opinion that the coal trade must move there from Whitehaven. Coal from neighbouring collieries was carried to Ireland and the Isle of Man in small, Irish vessels of about fifteen or twenty tons according to a report in 1719, the "inhabitants near Ellenfoot not concerning themselves in Shipping only in digging & Leading the Coal in Sacks on horseback to the Harbour". Coal exports in quantity were not possible without greatly improved harbour facilities and these were long in coming. As early as 1719 a broadsheet entitled "*Reasons for Building a Pier and making a Harbour at Elnefoot in Cumberland*" pointed out:

1. Whitehaven, Workington and Parton are unable to meet the Irish demand for coal and this allows Scottish and Welsh coal to compete in Ireland to the detriment of Cumberland.
2. There is abundant coal near Elnefoot and good roads to lead it to the ships if there were harbours available.
3. Coal deposits near Whitehaven are almost exhausted, (This was a completely erroneous supposition.) but the Dublin demand for coal will increase.
4. Elnefoot, near the centre of the county, is ideally situated for a port.
5. If local coal owners refused to help with the scheme it will be easy "to procure an Act of Parliament to oblige all to pay, which will meet with no opposition, when espous'd by the two Noble Dukes, whose great Collieries in that part of the County will by this means be prodigiously rais'd in value".*

Despite the obvious advantages to be derived by developing Ellenfoot it remained a small creek for the next thirty years, capable of accommodating small vessels only and there were "seldom above one or two in the Harbour at a time" which had to load their coal from the open shore. Early in 1749 Humphrey Senhouse II (1706-1770) and other local landowners were able to obtain an Act (22 Geo. II, cap. 6) to erect a new pier, enlarge and deepen the harbour at Ellenfoot and to levy duties for its maintenance. Until

* This broadsheet is undated but a letter from John Spedding, 22 Nov. 1719, shows that it was then in circulation. The two "Noble Dukes" were the Dukes of Wharton and Somerset.

1833 the trustees of the harbour were nominated by the lord of the manor and in this way the Senhouses controlled both the town and harbour of Maryport. Improvements to the harbour quickly followed the Act and these were accompanied by increasing coal shipments. In 1749 3,560 tons of coal were shipped there from Ewan Christian's colliery at Broughton. In 1750 the collector at Whitehaven reported to the Commissioners of Customs that there was "a Considerable Trade in Coals" at Ellenfoot with vessels arriving almost daily and that thirteen or fourteen ships at a time were often to be seen in the harbour. This, in his view, made necessary the appointment of two or three additional customs officials to deal with the growing traffic. Nine years later he noted the great increase in the coal trade at Maryport, but suspecting that vessels were taking more coal than they paid duty for, recommended the removal of the tidesman and two boatmen there for failing in their duties. The number of vessels arriving at Maryport went up from 147 in 1750 to 201 and 216 in the next two years. Records of coal shipments from Maryport at this time do not exist, but that they increased is shown by the growth of the harbour duties from £22. 2s. 10d. in 1750 to £189. 2s. 5d. in 1755.[122]

Compared with exports to Ireland coastwise shipments of coal never absorbed more than a tiny fraction of Cumberland's coal output. A little coal, brought by ship, was used at places on or near the coast which, because of prohibitive transport costs could not have been supplied by land. From Whitehaven 793 chaldrons of coal were shipped coastwise in 1688 and 207 chaldrons in 1708; in the years 1724-1727 these shipments were successively 340, 557, 384 and 328 chaldrons. Coal from the collieries at Broughton and Dearham was loaded at Ellenfoot and carried up the River Eden for consumption in Carlisle. Coal had not yet become the only fuel for industrial and domestic use in the county; timber and peat were still used in areas remote from collieries, but as these materials became scarcer and dearer consumers turned to coal if it could be shipped to some creek nearby. A report from the Commissioners of Customs, 9 May 1729, shows that the woollen and leather manufacturers of Kendal were suffering because turf, hitherto their usual fuel, was exhausted and the expense of fireing had "caused almost the entire loss of their trade." According to this report 308 chaldrons of coal had been shipped at Whitehaven and discharged at Milnthorpe in the past five years and upwards of 1,000 chaldrons at Grange, Penny Bridge and Rampside, all four of these places being within the jurisdiction of the port of Lancaster. Fifty years later peat was continuing to compete with coal as a domestic fuel in Carlisle; the northern part of the Cumberland coalfield was still largely undeveloped and much cheaper means of transport were essential before coal could become the only fuel used in the county.[123]

The expansion of the coal trade changed Whitehaven, once an obscure fishing village, into a busy and flourishing port. By 1725 Defoe could describe Whitehaven as

> "grown up from a small place to be very considerable by the coal trade, which is increased so considerably of late, that it is now the most eminent port in England for shipping off coals, except Newcastle and Sunderland, and even beyond the last, for they wholly supply the city of Dublin, and all the towns of Ireland on the coast; and 'tis frequent in time of war, or upon the ordinary occasion of cross winds, to have two hundred sail of ships at a time go from this place to Dublin, loaden with coals."

In 1748 it was said of Whitehaven that "The vast supply of coals in this neighbourhood has rais'd this port from a fishing town to be the next after Bristol and Liverpool on the west sea".[124] For its growth Whitehaven is indebted to Sir John Lowther and his son James, both men of ability, industry and perception, who made Whitehaven Colliery one of the most important collieries in the country. Since James Lowther was absent from Whitehaven for long periods the success of the colliery after 1706 depended not only upon its owner but also, and to a large extent, upon the men on the spot, particularly John and Carlisle Spedding whose long, continuous service was of vital importance. There was a time, however, when it seemed likely that Carlisle Spedding would leave the colliery. Early in 1721, dissatisfied with his salary of £30 per annum which would "but just get his Family bread & that wth the utmost Frugality", he was seriously considering accepting better paid employment elsewhere. The persuasion of his brother and an increase of salary to £35 per annum induced him to remain at Whitehaven. Lowther realized how much he owed to the Speddings and on his death in 1755 they were handsomely rewarded. To John he left £1,000 and to Carlisle £500. His successor, Sir William Lowther, immediately increased Carlisle's salary to £150 and that of his son James to £100 per annum. Carlisle Spedding did not enjoy his new salary for long because he was killed by an explosion of fire-damp 8 August 1755.[125] Throughout his career he had shown courage, skill, inventiveness and, as his correspondence shows, a willingness to learn as well as to teach. Mining methods at Whitehaven in his time were the most advanced in the kingdom and his pioneering efforts earned for him the admiration and gratitude of all who worked in coal mines.*

Notes and References

1 For detailed accounts of the geology of the coalfied see T. Eastwood, *The Geology of the Maryport District* (1930) and *The Geology of the Whitehaven and Workington District* (1931).

2 W. A. J. Prevost, "'A Trip to Whitehaven to visit the coal works there in 1739' by Sir John Clerk", CW2, lxv, 312.

3 Local historians have accepted Fletcher's statement, *op. cit.*, 278, "1685 Edward Spedding settled at Whitehaven as principal steward of the Lowther estates". He was never principal steward of these estates. Between 1666 and 1692 Lowther's principal steward was Thomas Tickell and he was succeeded by William Gilpin. Spedding was a tenant farmer at Akebank from 1687 until his death in 1706. The Lowther estate accounts also show that he supplemented his income by working as a carrier. An entry, dated July 1698, records, "On Monday the 4th Instant Edward Spedding with a Servt, a Cart of one pr. of large wheeles and 4 Horses departed from Akebank

to Newton 15 Miles		Returned	8th to Hexham 16	
5th to Miltown 24	Miles	Laden	9 to Haltwessel 12	Miles
6 to Hexham 19		C Q lb	10 Rested there	
7 to New Castle 16		12 2 20	11 to Carlisle 16	
			12 to Newtown 15	
			13 to Akebank 15	

Expended in the Journey (as he alledges) upon Men and Horses £2. 5s. od. The Worst way between Harrington and Whitehaven. Paid to him for the Carriage £3. 4s. od."

4 D/Lons/W, Copy of the will.

5 Spedding, 13 Feb. 1708.

6 Spedding, 19 April, 27 Oct. 1706, 30 March 1707. Lowther, 24 April 1707.

7 Spedding, 21 April 1710.

* "There is Scarce a Viewer or Enginear of Notice in either Tine or Wear but has been Solissting me to take a trip with them to see you and your works." (William Brown to Carlisle Spedding, 10 July 1751 – Letter Book of William Brown).

[8] Hutchinson, ii, 69-70.

[9] R. L. Galloway, *Annals of Coal Mining and the Coal Trade* (1898), i, 346. *VCH*, 361.

[10] D/Lons/W, Misc. Colliery Papers, "Extract of the Charge of Saltom Pitt", "Extract of the Charge of Driving the Air Race at Saltom".

[11] Prevost, *op. cit.*, 308.

[12] Spedding, 2 June, 10 Nov., 3 Dec. 1732, 15 Nov., 1 Dec. 1734, 21 Feb. 1735. D/Lons/W, Colliery Abstracts. Fletcher, *op. cit.*, 288. *VCH*, 362.

[13] D/Lons/W, Collieries List 2, 31-46. Fletcher *op. cit.*, 288. Ferguson *op. cit.*, 397.

[14] Brown, Carlisle Spedding to William Brown, 25 Oct. 1750.

[15] D/Lons/W, Colliery Abstracts.

[16] D/Lons/W, Deeds registers. D/Lec 79.

[17] Lowther, 14 July 1737.

[18] Spedding, 20 Oct. 1706, 22 Nov. 1732, 17 Jan. 1739. Lowther, 3 Sept. 1723, 28 Jan. 1735. D/Lons/W, Deeds registers; W/Collieries, 11; Bateman's report on collieries 1781; Misc. Colliery Papers, "Mr Lysons' Queries".

[19] Spedding, 20 Nov. 1726, 4 Nov. 1741, 29 Oct. 1742, 28 Dec. 1744. Lowther, 24 Nov. 1733. Forster, View Book of Matthias Dunn. The school governors at this time were: Joseph Smith, Sir James Lowther, Joseph Ritson, John Ponsonby, Thomas Patrickson, John Spedding and Thomas Lutwidge. The names of these governors were kindly provided by Mr W. Fox, Bursar of St. Bees School.

[20] D/Lons/W, Estate Memo Books, James Lowther's "Advice about Parson".

[21] D/Lons/W, Deeds register. Spedding, 11 Nov. 1722, 24 May 1723, 2, 18 Dec. 1726. Lowther, 16 Mar 1723.

[22] D/Lons/W, Colliery Abstracts; Deeds registers.

[23] *VCH*, 372

[24] Spedding, 17 Sept. 1710, 2 March, 12 Oct. 1712, 17 May 1714, 24 July 1715, 19 May 1728. Lowther, 1 Nov. 1722, 11 April 1730.

[25] *VCH*, 372-3. D/Lec, Coroners' reports.

[26] Spedding, 15 Oct. 1721, 14 July, 29 Sept. 1725. Whitehaven Custom House letter books, 9 Sept. 1720.

[27] Spedding, 16 July, 29 Aug., 22, 29 Sept. 1725. Lowther, 29 July 1725.

[28] Spedding, 8 Aug., 5 Sept., 17 Nov. 1725. Lowther, 19 May 1726.

[29] Spedding, 14 Feb., 31 March, 14 April, 5 May 1728, 29 Jan., 28 March 1729.

[30] Spedding, 10 Nov. 1734, 11 Feb., 9 May 1736. D/Cu Misc. Colliery Papers, 'Account of Coals sold'.

[31] D/Cu, Coal Accounts 1750-57. C. R. Hudleston, "An 18th Century Squire's possessions", CW2, lvii, 130.

[32] Spedding, 28 Oct. 1722, 2 Jan. 1723, 24 Dec. 1727. D/Lons/W, Collieries, List 2, 13-30; Deeds register 1709-59.

[33] D/Lons/W, Collieries, List 2, 13-30; Collieries 25. Spedding, 11 March 1737, 27 Jan. 1744. D/Lons/W, Seaton Colliery Book. *VCH*, 373.

[34] Spedding, 10 May 1730, 13 Dec. 1732. *VCH*, 373. D/Lons/W, Collieries List 2, 13-30; Letter Book of Walter Lutwidge, Lutwidge to James Gorton, 1 Jan. 1741, to John Spedding, 23, 29 Dec. 1746, to Sir James Lowther, 17 Jan., 10 April, 3 May 1747. E. Hughes, *North Country Life in the Eighteenth Century: Cumberland and Westmorland 1700-1830* (Oxford, 1965), 52.

[35] Lutwidge Letter Book, Lutwidge to Sir James Lowther, 23 Dec. 1746, 11 Aug. 1747. D/Lons/W, Collieries List 2, 13-30. Spedding, 5 Jan. 1728.

[36] D/Lons/W, Deeds register 1709-59; Deeds 32/1/13.

[37] H. A. Fletcher, "The Archaeology of the West Cumberland Iron Trade", CW1, v, 10. Spedding, 8 Sept, 1 Oct. 1721, 30 Aug. 1723.

[38] Spedding, 22 Dec. 1725, 9 Feb. 1726, 26 March 1735, 9 May 1736. D/Lons/W, Memo Books, Copy of the draft of an agreement between Edlred Curwen and Lutwidge and Spedding; Collieries List 2, 13-30.

[39] D/Lons/W, Colliery Abstracts; Deeds register 1709-59. Spedding, 29 Sept 1723.

[40] T. Pennant, *A Tour in Scotland and Voyage to the Hebrides* (Chester, 1744), ii, 52.

[41] Hughes, *op. cit.*, 134-7. Spedding, 15 April 1730. D/Lons/W, Estate papers, List 2, 22.

[42] D/Ben, Christian colliery records. Spedding, 21 Nov. 1722, 22 Dec. 1727. Lowther, 10 Sept. 1723.

[43] Spedding, 23 Aug. 1723, 2 May 1742. D/Lec, 65/2-5.

[44] D/Lons/W, Deeds registers; Deeds 32/1/13; Dearham Colliery Accounts. Spedding, 30 Jan. 1723. VCH, 376.

[45] Spedding, 2 Dec. 1722, 27 Nov. 1726. Lowther, 8 Dec. 1722. D/Lons/W, Deeds register.

[46] Spedding, 28 Oct, 11 Nov. 1722, 2 Jan. 1723. Lowther, 1, 3 Nov. 1722. D/Lons/W, Account Books.

[47] Spedding, 20 Nov. 1726, 30 Jan. 1740. Lowther, 12 Dec. 1734. D/Lons/W, Estate Papers, List 2, 22. D/Lec, 41. Forster, View Book of John Barnes, Barnes's report on Lord Egremont's collieries, 2 Oct. 1773. "Bishop Nicolson's Diaries", CW2, iv, 6, 36.

[48] Spedding, 19 May 1723. Hughes, *op. cit.*, 134-7. D/Lec, 65/5.

[49] Spedding, 30 Aug. 1723, 4 Jan. 1727, 18 April 1736. *VCH*, 372. D/Cu, Lease of Broughton Colliery 1755.

[50] Lowther, 23 Feb. 1738. D/Lons/W, Account Books; Colliery Abstracts.

[51] D/Lons/W, Colliery Abstracts.

[52] *General Magazine*, May 1755.

[53] *VCH*, 352. Brown, Carlisle Spedding to William Brown, 4 May 1750, 20 March 1752; Brown to Spedding, 30 Oct. 1750. *General Magazine*, May 1755.

[54] Spedding, 13 April 1712.

[55] *VCH*, 352. D/Lons/W, John Spedding's notebook.

[56] D/Lons/W, Three Quarters Band Colliery account book 1675; "Explanations of the Coal Accounts Nov. 1756". Sir John Clerk, "A Trip to Whitehaven . . . in 1739", CW2, lxv, 309. *VCH*, 353.

[57] Spedding, 9 June 1732. D/Lons/W, "Explanations of the Coal Accounts Nov. 1756". Brown, Carlisle Spedding to Brown, 11 Oct. 1752. *VCH*, 353.

[58] D/Lons/W, Sir John Lowther to Thomas Tickell, 7 March, 9 Dec. 1676; Tickell to Lowther, 14 Dec. 1676.

[59] Spedding, 31 March 1723. D/Lons/W, 'Explanations of the Coal Accounts Nov. 1756'.

[60] Galloway, *op. cit.*, i, 350.

[61] D/Lons/W, "Explanations of the Coal Accounts Nov. 1756".

[62] I am indebted to Mr Alan Smith of Neuilly-sur-Seine for providing a copy of this broadsheet. D/Lons/W, Spedding-Lowther correspondence, Nov. 1734-May 1735.

[63] *VCH*, 353, D/Lons/W, T. Tickell to Sir John Lowther, 16 Oct., 6 Nov. 1676; Lowther to Tickell, 21 Nov. 1676. Spedding, 5 Oct. 1712.

[64] *VCH*, 354. Spedding, April-June 1717, 15 Feb. 1727. D/Lons/W, Colliery Account Book 1716. Lowther, 17 Jan., 18 Feb. 1727.

[65] D/Lons/W, Misc. Correspondence Bundle 38. Lowther, 6 June 1732. Spedding, 5 Dec. 1736, 20 Feb., 8 July 1737. D/Lons/W, "Explanations of the Coal Accounts Nov. 1756". Sir John Clerk, 'A Trip to Whitehaven . . . in 1739', CW2, lxv, 308.

[66] D/Lons/W, Thomas Tickell to Sir John Lowther, 19 Nov. 1682; Gale to Lowther, 15 Aug. 1697. Lowther, 20 Nov. 1733. Spedding, 22 Dec. 1727, 22, 24 April 1730.

[67] Lowther, 30 Dec. 1732. D/Lons/W, Plans, Folder (5), "An Acco[t] of the Damp Air at Saltom Pitt". Spedding, 31 Jan., 14 Feb. 1733, 18 April 1736.

[68] *VCH*, 357. Evidence of John Buddle in the *Report of the Select Committee of the House of Lords on the State of the Coal Trade* (1829), 48.

[69] *Gentleman's Magazine*, xvii (1747), 327. *General Magazine*, May 1755. Spedding, 27 March, 18 May 1743. I. Fletcher, *op. cit.*, 307.

[70] D/Lons/W, Misc. Colliery Papers, "Extract of the Charge of Driving the Air Race at Saltom"; "An Acco[t] of the Damp Air at Saltom Pitt"; "Explanations of the Coal Accounts Nov. 1756". *Wesley's Journal* (ed. N. Curnock) iv, 314 quoted by T. S. Ashton and J. Sykes in *The Coal Industry in the 18th Century* (Manchester, 1929), 49. Lowther, 27 Feb. 1731. Brown, William Brown to Carlisle Spedding, 30 July 1754.

[71] Spedding, 5, 10 Aug. 1737, 27 Feb. 1740. Lowther, 18, 19 Aug. 1737. Brown, Carlisle Spedding to Brown, 28 Oct. 1753, 19 May 1754.

[72] Spedding, 7 Jan. 1743, 28 Dec. 1744.

[73] Spedding, 8 Feb., 30 March 1716, 13 Dec. 1732, 15 Jan., 7 Feb., 2 March, 2 May, 16 Nov. 1735. Lowther, 3 Dec. 1734, 9 Jan. 1735. D/Lons/W, Misc. Colliery papers, "Extract of ye Charge of the Wagon Way from Swinburn & Corp[l] Pitt to ye new Coal Steath". J. Dixon, *An Account of the Coal Mines near Whitehaven* (Whitehaven, 1801), 108-110. *VCH*, 362. Hutchinson, ii, 53.

[74] Brown, Carlisle Spedding to Brown, 19 May 1754, 15 June 1755. *VCH*, 362.

[75] Spedding, 20 Jan., 23 July 1710.

[76] Spedding, 8 March 1723, 10 April, 30 Nov. 1726, 7 Feb. 1729.

[77] Brown, Brown to Carlisle Spedding, 30 April 1750. Spedding, 20 Nov. 1726.
[78] D/Lons/W, Sir John Lowther to Thomas Tickell, 10 July 1666, 5, 27 Sept. 1670; Colliery Account Books.
[79] Spedding, 5, 17 March 1710.
[80] D/Lons/W, Colliery Account Books. Spedding, 29 June 1707.
[81] Spedding, 14, 28 May 1710, 1 Oct. 1712, 11 Jan. 1727, 7 Feb. 1729.
[82] D/Lons/W, "Explanations of the Coal Accounts Nov. 1756". Brown, Carlisle Spedding to Brown, 11 Oct. 1752.
[83] Spedding, 5, 10 Aug. 1737. Brown, Carlisle Spedding to Brown, 28 Oct. 1753, 19 May 1754.
[84] Spedding, 5 March 1710. D/Lons/W, "Explanations of the Coal Accounts Nov. 1756".
[85] Lowther, 3 June 1732.
[86] I. Fletcher, *op. cit.*, 276. D/Lons/W, "Explanations of the Coal Accounts Nov. 1756". D/Cu, Misc. Colliery Papers.
[87] D/Lons/W, Colliery Abstracts.
[88] Formerly at the Carlisle Record Office but withdrawn by Lord Lonsdale's estate.
[89] D/Lons/W, "Explanations of the Coal Accounts Nov. 1756"; Misc. Colliery Papers. Spedding, 27 March 1743.
[90] D/Lons/W, Misc. Colliery Papers.
[91] D/Lons/W, Colliery Abstracts.
[92] D/Lons/W, Colliery Abstracts; Dearham Colliery Accounts.
[93] Spedding, 11 March 1737, 27 Jan. 1744.
[94] Spedding, 14 July 1725. D/Lons/W Collieries, List 2, 13-30.
[95] J. V. Beckett, *op. cit.*, 81.
[96] D/Lec, 65/5.
[97] D/Lons/W, Colliery Abstracts.
[98] L. M. Cullen, *Anglo-Irish Trade 1660-1800* (Manchester, 1968), 82.
[99] D/Lons/W, Estate Memo Books, James Lowther's "Advice about Parton", June 1718. James Lowther to Gilpin, 14 June 1715. Spedding 11 Jan. 1716, 3 Dec. 1718.
[100] Lowther, 11-27 March 1725. Spedding, 12 May, 2, 11 Dec. 1726.
[101] D/Lons/W, Misc. Colliery Papers.
[102] Spedding, 29 March 1745.
[103] D/Lons/W, Colliery Abstracts. Whitehaven Library MS 30 John Spedding's Token Book. Spedding, 24 Sept, 30 Oct. 1714, 8 Jan-12 Feb. 1721, 1 Nov. 1728-3 Jan. 1729.
[104] Spedding, 15 Oct.-24 Nov. 1721.
[105] D/Ben, Christian colliery records. Spedding, 28 May 1738. Lowther, 19 May 1726.
[106] Spedding, 24 March 1736.
[107] Spedding, 21 Dec. 1720, 4 March 1724. Lowther, 20, 22 April 1736. D/Lons/W, Estate Accounts.
[108] D/Cu, Misc. Papers. D/Sen Shipping 7, Account Book of H. Senhouse 1751-65.
[109] Spedding, 3 July 1715, 11 April 1716.
[110] Spedding, 9 Feb., 4 May 1729.
[111] D/Lons/W, Misc. Colliery Papers, "Observations about the measures of Coales made by Mr Gilpin". Cullen, *op. cit.*, 80. W. H. Makey, "The Place of Whitehaven in the Irish Coal Trade 1600-1750" (London University M.A. thesis, 1952), 300.
[112] Spedding, 10 Dec. 1721, 30 Dec. 1726. Lowther, 18, 20 May 1738.
[113] Spedding, 28 Feb. 1735.
[114] D/Lons/W, "Committee to Regulate the Coal Trade". Spedding, 21 April 1738. Lowther, 20 April, 27 May, 17 Oct. 1738.
[115] Spedding, 14 July 1706, 11 Jan. 1708, 16 April 1718, 11 Feb. 1726. Lowther, 17 Nov. 1739.
[116] Spedding, 28 March, 4, 18 April, 9 May, 27 July 1744. D/Lons/W, Misc. Estate Papers.
[117] Spedding, 13 April 1720. Lowther, 27 June 1721. D/Lons/W, Thomas Bacon, "Some Reflections on the Coal Trade" (Dublin, 1738).
[118] D/Lons/W, Estate Accounts; Rental 1771.
[119] Lowther, 8, 13 Dec. 1722, 22 Dec. 1737, 27 May 1738.
[120] Spedding, 23 Aug. 1723, 22 Oct. 1725, 15 April 1730.

[121] D/Cu, Henry Curwen's Account Book 1749-78. D/Lons/W, Collieries 25; W/Collieries, List 2, 13-30; James Lowther to William Gilpin, 17 Nov. 1719. Lowther, 3 Aug. 1723, Jan.-Feb. 1731, 21 March 1734, 3 Jan. 1751. Spedding, 10 April 1721, 27 Feb. 1723.

[122] D/Lons/W, Sir John Lowther to Lord Lonsdale, June, 3 July 1699. Whitehaven Custom House Letter Books, Collector to Board, 8 July 1719, 20 July 1750, 30 July 1759. D/Lec, 65/5. D/Cu, Misc. Papers. Information provided by the late Dr J. W. Crerar of Maryport.

[123] Nef, *op. cit.*, App. D iii. Whitehaven Custom House Letter Books, 7 Feb. 1728. Spedding, 21 Nov. 1722, 30 Jan. 1723. *CTB&P 1729-30*, 66. Hughes, *op. cit.*, 133.

[124] D. Defoe, *Tour Through England and Wales* (Everyman ed.) ii, 273. *Gentleman's Magazine*, xviii (1748) 5.

[125] Spedding, 5 March 1721. Lowther, 25 March 1721. R. W. Moore, "Whitehaven Collieries", *Trans. of the Federated Institute of Mining Engineers*, vii (1893-4), 627. Brown, Carlisle Spedding to Brown, 15 June 1755.

CHAPTER III

1756-1815

WHITEHAVEN DISTRICT

On the death of Sir William Lowther in 1756 the Whitehaven estates passed to Sir James Lowther (1736-1802), later Earl of Lonsdale, known locally as "the bad earl". An egoist of enormous wealth and almost insane arrogance he overrode opposition without regard for his equals or for the miners compelled to depend upon him for their livelihood. To defend himself and his interests he resorted to the dangerous pursuits of duelling and litigation. It was said of him that he showed "a determination to oppress, by means of wealth, and under colour of law, all who were obnoxious to him", a criticism justified by the series of lawsuits he initiated.* He continued his family's policy of buying up land to extend the Lowther control of mining in Cumberland. In addition to many small parcels of land he bought Seaton manor in 1758 for £55,000 from Charles Pelham and also acquired the manors of Millom, Whicham, Greysouthen and Clifton.[1]

For many years the Lowthers had coveted the Bransty estate, about 150 acres in extent, which adjoined their Whingill Colliery. The surface area belonged to the Ellison family and the underlying coal to Lord Egremont. It was not the intention of the Lowthers to work the coal at Bransty but to deny its use to others who might compete with their collieries at Whitehaven. A lease of this coal for thirty years was granted to William Thynne in 1719. Thynne, however, did not work any coal at Bransty. Nor did Peter How who leased the coal in 1753 for twenty-six years at an annual rent of £10. After How's bankruptcy in 1763 "Lord Egremont Bought in the Colliery Lease to prevent its Falling into improper hands". John Barnes in 1773 surveyed the Bransty estate and examined the costs Lord Egremont would have to meet if he decided to sink and work a colliery there. In the following year Egremont bought the Bransty estate from Henry Ellison for £16,000 and soon afterwards had the estate surveyed again, this time by John Smeaton. The main difficulty appeared to be that of determining exactly the place of export for Bransty coal. If the coal went to Whitehaven harbour it would go to a harbour wholly controlled by the Lowther family. Alternatively, Lord Egremont could construct a new harbour on his Bransty estate from which the coal could be shipped. This would be expensive and probably explains Egremont's decision not to proceed any further in the matter although Barnes had calculated that on an estimated

* A Whitehaven shoemaker recorded one example of Lowther's methods, "23 Sept. 1780 S[r] James Lowther called here today desiring me to revoke a promise I with many others had made to Jn[o] Pennington & wished to have us to vote for Sir James and no other man, which was refused by most in Town. Whereupon he ordered his Steward not to furnish me (with a long list of others) with any more coals". (Whitehaven Library – Notebook of John Bragg).

output of 40,000 tons per annum a colliery at Bransty would show a profit of £1,541. 13s. 4d. In 1781 the estate and its coal were sold to Sir James Lowther for £26,000.[2]

James Spedding (1720-1788), succeeded his father, Carlisle Spedding, as manager of the Whitehaven Collieries in 1755 and three years later, following the death of his uncle John, combined this post with that of agent to the Lowther estates. In his time more new pits were sunk – Wolfe, Davy, Lady, George, North, Bateman, Howe and Scott pits on the Whingill side and Croft and Wilson pits on the Howgill side. Of these the most important was Croft Pit sunk in 1774 which ultimately reached a depth of 910 feet and was worked continuously to 1903. At Saltom Pit the mining of submarine coal went on and Gabriel Jars who inspected the pit in 1765 was much impressed to find that the workings then extended one and a half miles from the day-hole, some of them being three-quarters of a mile under the sea.[3]

In 1781 Spedding resigned and was succeeded as general manager of the Whitehaven collieries by John Bateman who had been in charge of the Howgill division. At that time there were eight working pits at Whingill — Lady, George, Davy, North, Pearson, Bateman, Howe and Jackson pits; on the Howgill side there were five – Duke, King, Kells, Croft and Saltom pits. On assuming office Bateman drew up a detailed report on the collieries and apart from describing the existing situation he had much criticism to offer about his predecessor. He alleged that:–

1. Spedding had thwarted an agreement made with the men which would have reduced hewers' wages at Croft and King pits.
2. The number of workers at Howgill was excessive – "there is now above four times the Number of Labourers employed that is necessary". Further, the overseer in charge was frequently drunk and, on Spedding's instructions, paid no attention to Bateman's orders.
3. In the granary yard the men wasted their time drinking and when at work were indifferent workers and their work unnecessarily expensive. In addition they were frequently engaged in Spedding's private business rather than that of Sir James Lowther.
4. Tenant farmers supplied with household coal at a fixed rate per annum were using three or four times the quantity they paid for.[4]

For ten years Bateman did what he could to increase the efficiency of the colliery and the growth of coal exports shows that his efforts were successful. But such improvmeent did not save him from abrupt dismissal following an accident 31 January 1791. As coal was being worked under the town of Whitehaven two men, a woman and five horses were drowned by a rush of water liberated from some old workings. This resulted in damage, by subsidence, to Somerset House, then occupied by H. Littledale, and to other property; in all eighteen houses were destroyed. Littledale brought a successful action for damages against Lord Lonsdale who, in a fit of temper, thereupon closed his mines at Whitehaven, regardless of the pitmen and others who, directly or indirectly, depended upon the continuance of coal mining. Not until he had received a petition signed by 2,560 people, begging him to re-open the mines and promising to indemnify him against all such actions in future, was worked resumed.[5]

Thomas Wyley then took over as manager but he proved unequal to the task, a fondness for liquor interfering with his work. Coal Shipments declined sharply, from

64,032 waggons in 1790 to 45,314 waggons in 1801. Under him colliery discipline was slack and various malpractices flourished which reduced both output and profits. His regime, however, was not entirely without merit. He began the sinking of James Pit in 1797 and reached the coal in 1800. It was he who was largely responsible for the introduction of steam winding engines at Whitehaven Colliery.[6]

Complete statistics of the output of coal at Whitehaven in the latter part of the eighteenth century are not available but it has been estimated that the average annual output between 1755 and 1780 was about 150,000 tons and about 160,000 tons between 1780 and 1800. The Howgill division of the colliery produced the following quantity of coal:

Year	*Tons*	*Year*	*Tons*
1759	80,245	1779	94,153
1760	80,177	1780	100,991
		1781	89,700
1775	115,337	1782	83,950
1776	104,766	1783	98,060
1777	97,074		
1778	88,887		

By 1814 output at Whitehaven Colliery had increased to 116,619 waggons, each waggon containing 42 cwts.[7]

John Bateman returned to his former post in 1802 at a salary of £500 per annum determined to prove his worth as colliery manager and to demonstrate that his earlier dismissal had been a grievous mistake. He found that a great deal of laxity and dishonesty had grown up in the colliery during his predeccesor's tenure of office which he found extremely difficult to eradicate. Nevertheless he made his presence felt within a year. In view of the shortage of labour he had to walk warily, but by the dismissal of the worst offenders and by a mixture of firmness and persuasion he was able to secure the diminution, if not the complete cessation, of dishonest practices. He discovered that, "The greatest abuse is Masters of Ships giving money to the waggoners & also the men at the Banks to procure large Measure". When exact measure was insisted upon the shipmasters protested volubly, but their clamours were silenced when a test of several waggons showed conclusively that they had no real cause for complaint.[8] However, the bribery of waggoners to give excess measure continued until Lord Lowther, who considered the masters who paid bribes to be more culpable than the waggoners who received them, issued orders that masters guilty of this offence should be denied cargoes of coal.[9] Other forms of fraud were the insufficient filling of baskets and pay claims for more coal that had been cut; at Croft Pit in one week the banker, hooker, fillers and hewers tried to obtain payment for 62 baskets of coal more than had been sent to bank.[10]

Bateman's main concern was to increase output as quickly as possible. This, together with an increase in profits, would be the most direct way of impressing his employer. Realizing that mining practice at Whitehaven might benefit by the adoption of methods employed in pits outside Cumberland he sent three of his senior staff to Newcastle to learn what they could of the work done in that area. In this he had the full support of Lord Lowther who was well aware that he could profit from the knowledge thus acquired. Bateman's energy and determination soon led to greater output and reversed the downward trend in coal shipments. In less than a decade coal exports from the colliery were more than doubled, from 47,906 waggons in 1802 to 100,121 waggons in 1810.[11] He was responsible for the sinking of William Pit on the shore near Bransty which was begun in 1804 and completed in 1812. Sinking was hampered by a large feeder of water bursting into the Whingill division of the colliery and John Buddle and Thomas Ramsay were called in to investigate. Believing the influx of water came from some old workings they recommended Bateman's plan of fixing a frame dam in the stone drift which connected the two divisions of the colliery. This was soon followed by the erection of a 100 h.p. atmospheric pumping engine. Coal was first shipped from this pit 10 March 1806 and the pit worked continuously for over a century.[12]

Despite the improvement in output and profits and the restoration of pit discipline Bateman was dismissed a second time. There was an element of farce in his departure and the cause was his favourite nephew John B. Longmire, viewer at Howgill Colliery. Longmire, a young man of talent but of greater ambition, assumed that he would succeed his uncle as manager of the Whitehaven collieries and suffered from the delusion that he was indispensable. Without giving notice and without his employer's permission he left Whitehaven 6 March 1811 and did not return until 19 August. Bateman tried to conceal his nephew's absence for some time and when pressed by Lord Lonsdale for an explanation indulged in evasions or proffered very unlikely excuses. What really angered Lord Lonsdale, who tried to help Longmire's career, was information from Whitehaven that Longmire's return was made an occasion of public rejoicing. Longmire, although clearly at fault, was given a hero's welcome. On reaching Whitehaven he was greeted by cheering crowds, flags were flown and guns fired in his honour, the festivities culminating in a lavish dinner attended by many people. Lord Lonsdale, who had paid Longmire's salary throughout his period of absence, ordered his immediate dismissal and Bateman had then to write what was probably the shortest letter in his life, "Agreeable to your Lordship's orders on the 22nd Instant John B. Longmire is suspended from the Service". Believing that Bateman had supported the demonstration in favour of his nephew Lord Lonsdale then dismissed Bateman himself.[13]

Coal mining at Cleator Moor began when William Walker leased an Egremont royalty in 1788.* This lease included all coal from the Ehen to Keekle and was later extended for a period of 31 years from Michaelmas 1792. Output at this colliery, which ceased production in 1805, amounted to:–[14]

Year ending		*Tons*
29 Sept.	1791	3,790
29 Sept.	1793	3,933
29 Sept.	1801	4,887
2 Oct.	1802	4,745
1 Oct.	1803	3,955
29 Sept.	1804	4,131
29 Sept.	1805	266

Bowthorn Colliery, Cleator Moor (a Lonsdale royalty) was worked by John Litt & Co. from 1802 until 1815 when it was laid in.[15] Nothing more was done in this area until the iron-masters took an interest in it some twenty years later.

* Moore (*VCH* 366) is mistaken in stating that the first collieries at Cleator Moor were John Litt's at Bowthorn and Mr Dean's at Keekle Grove in 1802.

WORKINGTON DISTRICT

Henry Curwen did much to develop the Moorbanks Colliery at Workington. Between 1771 and 1774 the colliery produced in successive years, 1,701, 10,742, 12,949 and 23,600 tons of coal. From "An A/ct of all the Coals raised from Moorbanks Colliery, Workington, from 5 October 1771, in the customary tons, taken from the Fawcett papers and the subsequent A/ct" we learn that the colliery between 1771 and 1801 raised a total of 859,260 tons of coal from seventeen working pits.[16]

To move coal from Moorbanks Colliery to Workington harbour it was necessary to construct a waggon-way through the Banklands estate where another colliery was already working. Accordingly Curwen agreed, 27 July 1771, to pay the Banklands Colliery Co. 40s. for every acre of Banklands that his waggon-way would traverse. It was also agreed that the company would be allowed to make a branch waggon-way from their colliery to link up with the projected waggon-way from Moorbanks to Workington harbour. They would be allowed to use the waggon-way from Banklands to the harbour and the three additional hurries that would be erected. For this they were to pay £300 towards the cost of constructing the waggon-way and hurries, an annual rent of £100 and half the cost of repairing the hurries and the waggon-way from Banklands.[17]

The Banklands Colliery Co. fell into difficulties and the estate was purchased by Anthony Bacon for £4,400.* He continued to work coal there and by an agreement made with the executors of Henry Curwen's estate he was allowed to use the waggon-way to convey Banklands coal to the harbour up to a limit of 30 waggons per day. Bacon had to pay one shilling for each waggon of coal raised and disposed of from the colliery. He came to believe that this arrangement was unfair since, he complained, "I am limitted to ship no more than Thirty Waggons a day, and am to pay one shilling a Waggon for the use of your Fire Engine and way leave, by which I shall pay you about five hundred pounds per ann. when I am able to raise sufficient to ship my Morsel". His complaint had some effect for the terms of the agreement were modified. By a new agreement (25 February 1783) he continued to use the waggon-way from the colliery to the harbour and to have the benefit of the pumping engine to drain Banklands. He had to pay an annual rent of £100 and 6d. per waggon of coal raised, but he was still restricted to 30 waggons of coal per day.[18] After his death in 1786 Banklands Colliery was managed by William Walker under the direction of Thomas Harrison of Whitehaven, a trustee of the estate. In 1786 there were four working pits at the colliery – Great Band, Stephen, High and Thompson Pits – and five in 1794.[19]

It has been said that all the coal from Workington Colliery until 1794 came from the Banklands or eastern division, but in that year J. C. Curwen, advised by John Grieve, an Edinburgh engineer, began the Chapel Bank or western division of the colliery when he completed the sinking of Lady Pit on the shore to the Main Band at 84 Fathoms.* But the Curwen records show that mining was going on at Chapel Bank before this date.

* This famous iron-master was engaged in the tobacco trade at Whitehaven before he amassed a fortune at his ironworks in South Wales. In 1764 he became M.P. for Aylesbury in place of the expelled John Wilkes and was re-elected in 1768, 1774 and 1780. On his death Banklands Colliery was worked for the benefit of his five illegitimate children (Anthony, Thomas, Robert, William and Elizabeth) by Mary Bushby. In his will he acknowledged these children as his own.

* *VCH* 366. Presumably based upon Parson & White, *History and Directory of Cumberland and Westmorland* (1829), 284n.

The Chapel Bank estate belonged to Robert Gibson but the underlying coal to the Curwens and in 1770 Henry Curwen was working coal there. A letter shows that operations had hardly begun when the colliery was drowned; it refers to "the injudicious working of Mr Curwen's new colliery at Chapel bank on the South side of Workington Harbour. The Sea burst into it on the 22nd Inst. & has drown'd the whole, so that he cannot now keep 1/20 of his friends ships employ'd, without either contracting for Coals from some other Colliery or opening a Colliery of his own upon some new ground". A week later Spedding wrote to Sir James Lowther, "Mr Curwen got to Workington last Friday evening. His learned Stewards are advising him to Erect another Fire Engine upon his Colliery at Chaplebank which they assure him will effectually drain it. They will be mistaken (I shall not hinder them from making the attempt) the Sea ebbs & flows in the pitts, the water is very brackish & will make salt in their Boilers. I apprehend the Sea Water communicates with the Colliery through a Sand Dyke which they have wrought too near".[20]

Understandably Curwen in the following year (1 August 1771) was willing to lease the colliery to John Hodgson, Edward Stanley, John Falcon and John Stead for eleven years. After the first eighteen months the lessees were to pay Curwen £300 for the use of the waggon-way and hurries and 6d. for every ton of coal raised throughout the whole period of the lease. This venture was short-lived. The agreement bears this endorsement, "Be it remembered (January 9th 1773) the above unfortunate lessees gave up their Rights and Lease of the said colliery to Henry Curwen, Esquire. He at the same time generously forgave them all Rents and payments due to himself and took the colliery into his own possession again together with all the materials &c". Curwen bought the Chapel Bank estate from Mr Gibson in 1775 for £2,000 and worked the colliery there between 1773 and 1778 in partnership with William Hodgson and others. He was more fortunate than his predecessors for his half-share of the profits amounted to:[21]

Year	*£*	*s.*	*d.*	*Year*	*£*	*s.*	*d.*
1773	180	10	6	1776	120	1	0
1774	185	16	6	1777	290	4	6
1775	253	0	3½				

He continued coal mining at Harrington and between May 1757 and July 1760 shipped 8,253 tons of coal from Harrington harbour. To facilitate exports he had a new quay built at Harrington in 1760 and thereafter coal shipments increased:[22]

	Tons	*Lo.*
28 July 1760-24 January 1761	2,483	4
10 February 1761–9 August 1761	3,962	7
18 August 1761–27 May 1762	5,105	1
1 June 1762–17 September 1762	2,705	6
21 September 1762–26 December 1762	1,598	2
11 January 1763–26 March 1763	1,066	5
8 April 1763–10 June 1763	2,270	5
14 June 1763–1 September 1763	3,886	2
6 September 1763–7 December 1763	3,256	5
12 December 1763–9 June 1764	6,679	
12 June 1764–15 September 1764	4,746	
18 September 1764–14 January 1765	5,541	

Between 1770 and 1780 annual coal shipments from the Curwen collieries ranged from 35,319 to 45,936 tons.[23] Henry Curwen died in 1778 leaving personal estate valued at £15,000, real estate valued at £2,000 a year and his collieries at Moorbanks, New Lands, Harrington and Chapel Bank which were reckoned to produce "one year with another" £5,500 a year.[24] The heiress to the Curwen estates was his daughter Isabella who, in 1782, married her cousin John Christian of Ewanrigg. His marriage was undoubtedly Christian's most successful venture; controlling the combined resources of both families he was able to play a large part in the mining development of his native county. In 1790 he adopted the Curwen name and it is as John Christian Curwen that he is known to history.

He carried on exploiting the coal seams at Banklands. Engine and Crosthwaite pits were working there in 1787 and in the same year Hope Pit began production. These were followed by Henry Pit (1789), John and Elizabeth Pits (1790) and Bowness Pit (1799). In 1800 these seven pits produced 31,689 tons of coal. As the Banklands pits declined Curwen devoted much energy and capital to winning coal at Chapel Bank. After Lady Pit went into production the sinking of Union Pit was begun in 1795 and completed in 1798. These two pits in their early years yielded the following quantities of coal:[25]

Year	*Lady Pit* *Tons*	*Union Pit* *Tons*
1795	6,197	
1796	19,725	
1797	22,253	
1798	19,120	3,107
1799	17,379	6,970
1800	21,062	11,091
1801	17,863	10,027
1802	18,756	14,670

Further exploratory work was undertaken when the sinking of Isabella Pit was begun in 1807 and by 1814 it had reached the Moorbanks Seam at 97 fathoms. Sinking was resumed in 1816 in an effort to reach the Main Band, but at a depth of 128 fathoms a fault was encountered and not until 1822 was the Main Band found after driving a stone drift through the "nip". This pit proved most expensive; by the end of 1817 £62,100 had been spent on it.[26]

Workington Colliery produced 53,543 tons of coal in 1780[27] but after J. C. Curwen took over there was a distinct improvement which continued for many years. Increased output, however, could not continue indefinitely and it appears from the decline in coal shipments that production began to diminish about the year 1807. Curwen continued the development of Harrington Colliery which in 1783 consisted of seven pits – McCall, Tarn, Dyke, Lowes, Lonsdale, New and Lonsdale 2nd Pit. Between 1783 and 1792 Curwen was responsible for the sinking of eleven new pits there.* Subsequently the

* These were: Drape, McMillan, Hodgson, Old Baskett, Natty, Bella, Udale, Fox, Laybourn, Henry and John Pits. Moore (*VCH* 368) is mistaken in stating that Hodgson Pit was sunk in 1804; pay books show that it was working in 1787.

number of working pits was reduced, to five in 1804 and to three in 1812. These three were the Hodgson, John and Henry Pits. By 1802 output at Workington Colliery had risen to 65,309 tons and shipments to 104,940 tons. Harrington Colliery shipments rose from 12,516 waggons in 1783 to 29,387 in 1795 and fell to 15,391 waggons in 1814.[28]

Sir James Lowther in 1734 had agreed not to work any of his coal in Clifton or Greysouthen for the next twenty-one years. Production was resumed after 1755. In 1761 Lowther was mining coal at Reelfitz Pit, Greysouthen at a depth of thirty-four fathoms. Twenty years later he had two pits working at Clifton, two at Flimby and four at Seaton.[29] John Cookson was also working coal at Clifton and Greysouthen at this time to supply fuel for the Clifton furnace. He leased Greysouthen Colliery in 1761 from Lord Egremont for twenty-one years at a rent of £10 per annum and 4d. per ton on all coal raised above 300 tons each year. The daily output in 1773 was 30 tons. Working costs amounted to 1s. 8d. per ton and the coal was sold at 2s. 3d., leaving a profit of 7d. per ton.[30] In 1781 Sir James Lowther discovered that by a clause in the lease he had granted to the owners of the Seaton Ironworks he was obliged, as long as he worked coal within four miles, to supply the ironworks with coal at shipping prices. At once he closed all his pits at Clifton, Flimby and Seaton, an action which compelled the closure of Cookson's collieries at Clifton and Greysouthen since they were flooded by the water hitherto lifted by Lowther's engines. As a result there was no mining of any consequence at Clifton for the next twenty-two years. In has been estimated that prior to their abandonment in 1781 the Clifton pits yielded over two million tons of coal.[31]

A large portion of the coal under Greysouthen was leased in 1787 by William Walker & Co. who carried on a profitable business for eighty years. Another colliery was opened there in 1800 by John Wilson & Co. As frequently happened in Cumberland when adjoining royalties were being worked, the question of trespass arose. Disagreement culminated in litigation and in 1807 Wilson & Co. at the Carlisle Assize Court obtained over £16,000 damages from Walker & Co. who had worked a large quantity of coal belonging to the former company.[32]

The Lowther pits at Seaton were again at work in 1800 to supply coal to the Seaton Ironworks. Lord Lonsdale resumed mining at Clifton in 1803 when he opened out a colliery on the west side of Lostrigg Beck and worked it until 1815.[33]

Little is known of coal mining in the manor of Dean before the latter half of the eighteenth century. The Leconfield MSS. show that in 1711 William and Henry Stockdale were given leave to search for coal there. Some colliery development must have taken place for fifty years later Gabriel Griffith leased Dean Colliery for twenty-one years from Lady Day 1761 at an annual rent of £10 and a tonnage rent of 2d. on all coal raised above 1,200 tons per annum. This was an unprofitable undertaking and after Griffith's bankruptcy John Fearon agreed to pay off the arrears of rent and to work the colliery from Michaelmas 1766. The colliery produced very little coal. Between 1767 and 1772 the annual output ranged from 210 to 745 tons and Fearon, having incurred heavy losses, had to be released from further payment of rent by Lord Egremont. John Barnes, who in 1773 wrote a detailed report on the Egremont estates, was of the opinion that the colliery could not be profitable since it was too far from the sea for export purposes (although only about five miles). Leading costs to Workington harbour were very high and access was possible for only a few months in summer. For purposes of land sale demand was small, the price was only 2s. per ton and there was competition from Cookson's colliery at Greysouthen and Lowther's colliery at Bridgefoot.[34]

Others worked the colliery with no more success. Richard Fawcett leased the colliery for seven years from Michaelmas 1774 at a yearly rent of £10 10s. He was succeeded by William Waite who in the course of nine years lost all his money and ended up in gaol. His release was obtained by his brother-in-law, John Fisher, who took over the colliery himself and became bound for the payment of the colliery rent. He leased the colliery for twelve years from Michaelmas 1789 at an annual rent of £10. He too fell into difficulties and before his lease expired owed three years' rent.[35]

Mining continued at Distington throughout the latter half of the eighteenth century. A colliery was working at Boonwood in 1768 and in 1781 Mr Crosthwaite was working Moss and High Pits in his own land, Mr Walker had another pit to the north of Moss Pit and Sir James Lowther's Gunnerdine Colliery was producing coal. Gunnerdine Colliery yielded 6,581 tons of coal in 1806 and between 1811 and 1814 its output in successive years was 5,815, 3,258, 4,746 and 4,478 tons. It continued production until 1815 when it was superseded by Bottom Bank Colliery which was carried on for the next thirty years. A colliery at Winscales belonging to J. C. Curwen was worked from 1783 to 1806.[36]

MARYPORT DISTRICT

According to R. W. Moore, "In the manor of Ellenborough Mr Senhouse of Netherhall began to work coal in a small way about 1740".[37] This is incorrect. Senhouse, by an agreement made with Sir James Lowther in 1736, had covenanted not to work any of his coal in Netherall or Ellenborough for the next twenty-one years. Following the 1749 Act to improve Ellenfoot harbour Senhouse was anxious to be freed from this compact in order to resume coal mining in his estate. A letter from John Spedding to Senhouse, 27 April 1751, shows that Lowther was prepared to allow Senhouse to work coal in Netherhall or Ellenborough provided that Lowther could be assured of wayleave for all his coal and sufficient steath room at Ellenfoot for a long term of years.[38] There is no evidence to show that coal was worked in Senhouse land between 1736 and 1769. In 1769* Senhouse leased his coal in Ellenborough to George Monkhouse and Richard Laws for twenty-one years, the lessees to pay 4d. per ton on all coal raised and to provide free fuel for the Senhouse family at Netherhall throughout the term of the lease. A clause in the lease stipulated that the lessees should "pay attention and regard to the recommendation of the said Humphrey Senhouse . . . by preferring, loading first, and dispatching the Ships of such of his Friends as he shall request, when and as often as such his Inclination and desire shall be signified to the said George Monkhouse and Richard Laws".[39] This venture cannot have been very successful for the colliery was taken over in 1772 by Senhouse himself who in the next twenty years sank many small and short-lived pits in Ellenborough.

Colliery accounts show that in 1772 there were four working pits – Sally, Betty, Mally and Good Intent Pits – employing a total of fifteen haggers who were paid 6d. per ton.

* Not 1767 as stated in *VCH* 375.

Other pits were sunk: New Pit in 1773 and the Far, Hopewell and Schoolhouse Pits in 1774. By 1782 all these pits had gone out of production, the pits then at work being the Great, Peat and Fletcher Pits. At least a dozen other pits were sunk at Ellenborough before the end of the century.[40]

By an agreement, 15 June 1780, Joseph Jackson of Gilcrux took an eighth share in Ellenborough Colliery which he was to manage. He was to pay one-eighth of the working costs and to receive one-eighth of the colliery's profits. The partnership was far from amicable. Jackson's long absences from the colliery and disputes over money received from the sale of coal led to the dissolution of the partnership, which should have lasted twenty-one years, in 1783. Jackson surrendered all claims in the partnership in return for a cash payment of £200 and an annuity of £60.[41]

The Senhouses continued working Ellenborough Colliery until 1808. Some statistics of the output and shipments of this colliery have survived:[42]

Year ending March	*Output Tons*	*Shipments Tons*
1787	24,158	
1788	25,163	
1789	13,382	
1790	13,268	
1791	12,662	
1792	11,185	15,099
1793	3,438	5,016
1794	14,693	21,888

In the adjacent manor of Ewanrigg John Christian worked a few small pits* but his most successful operations were in the Broughton area. The Broughton pits, situated between Maryport and the Derwent and some two miles from the sea, were denied access to Maryport harbour by Senhouse and Christian lands. It was impossible to export Broughton coal without the consent of these landowners, for it would be too costly to cart coal by a route which would avoid their territory.[43] In 1755 an understanding was reached by the landowners when the collieries in Great and Little Broughton were leased by Lord Egremont, lord of the manor, to John Christian, Humphrey Senhouse, John Gale, Thomas Hartley, Edmund Gibson and James Postlethwaite for twenty-one years at a yearly rent of £200 and 10s. royalty per Ten† of coals raised above 400 Tens per annum. John Christian was to have a quarter share in the colliery company and his five

* Gin, Harry, Water, Thompson, Union No. 3 and Union No. 4 Pits were working 1762-3.

† Not per "tun" as stated by E. Hughes, *North Country Life in the Eighteenth Century*, ii, 138, 146. John Barnes in his "View Book 1717-78" when dealing with this lease defines a Ten in these terms: "A Ten to contain 22 waggons, each Waggon 19 Bowles, each Bowle 4½ Bushels Winchester measure, and each Bushel 8 gallons Winchester measure". From the account he gave of the tentale rent owing to Lord Egremont, it is clear that the lease called for 10s. per Ten, not per tun. W. Tiffen in a letter to C. Udale, 8 February 1794, said, "In a note of Mr Addison a Ten is said to contain 22 waggons, each waggon containing 19 bowls of 36 Gallons each (Dry Measure). By the lease of 1781 from Lord Egremont to Mr Curwen I perceive a Waggon is considered as equal to three Tons – if so & if Mr Addison's note is correct the Ten is 66 Tons". (D/Cu 3/68). Nef's statement (J. U. Nef, *The Rise of the British Coal Industry*, ii, 377) that a Ten "May, I believe, invariably be taken to mean 10 Newcastle chaldrons' would therefore appear to be incorrect.

partners were each to have an equal share in the remainder.[44] Before this lease was signed Christian and Senhouse had agreed to provide the necessary wayleaves to the harbour at Maryport. Christian had also promised to allow the partners to use the levels or drains in his existing collieries and to extend them to Broughton to drain the mines there. For this he was to receive £150. In addition Christian undertook to pay half the cost of a waggon-way through his grounds from the Broughton pits and was to receive a yearly wayleave rent of £78. 15s. Similarly Senhouse would pay one quarter of the cost of a waggon-way through his lands and would receive an annual wayleave rent of £10. William Williamson contracted to build the waggon-way from Broughton to the harbour for £900, and Richard Laws was appointed overseer of Broughton Colliery at a salary of £50 per annum.[45]

In less than a year the colliery was producing coal. By 1781 it had yielded 765,631 tons of coal, annual output ranging from 7,175 tons in 1756 to 45,665 tons in 1769.[46]

By an agreement, 19 September 1763, John Christian took over the lease of the colliery from the other partners, agreeing to pay them 9d. for every ton for coal shipped and 1½d. for every ton sold locally. As Professor Hughes has pointed out,[47] it is difficult to understand why Christian's partners should withdraw from an apparently successful concern. He suggests that lack of ready cash was the probable cause of Senhouse's withdrawal. Four of the partners (Gale, Hartley, Gibson and Postlethwaite) were also partners in the Maryport Ironworks. It seems likely that they were having difficulties with their furnace at this time and therefore felt it necessary to reduce their financial commitments. Further, the colliery would be better managed by one person than a number of partners living at a distance from each other.

The new arrangement benefited Christian's partners considerably, for in the first four years, Michaelmas 1763–Michaelmas 1767 they received £4,863. 7s. 10½d. and £4,611. 3s. 8½d. for the three years 1767-1770.[48] But Christian did not live long to enjoy the growing success of his colliery. He died in 1767, and since his son and heir John was a boy of twelve, the management of the estate, including the coal mines, fell to Charles Udale who had served the family faithfully since 1755.

It was during the minority of the heir that a dispute arose over the amount of tentale rent payable to Lord Egremont. As stated above, he was entitled to 10s. per Ten for every Ten of coals above 400 Tens per annum, according to the Newcastle measure. If he was to obtain all he was entitled to an accurate account of the output of Broughton Colliery, measured in the Newcastle manner, would be necessary. By 1773 he had received nothing since the beginning of the lease because "Lord Egremont's agent (Robert Baynes) has hitherto declined to receive any Tentale rent, upon a doubt he has whether the Accounts so delivered in are just and true. He has great reason to believe that the accounts delivered are short; but he cannot ascertain how much they are so, without being informed what is the customary way of measuring a Ten of coals at Newcastle; vizt. Whether the same ought to be heap measure or strike measure, meaning corn upon a fair even streak". Baynes had been remarkably dilatory and had clearly failed in his duty to ensure full and prompt payment of the rent. In 1773 Messrs. Walton and Barnes were called in to inspect the records of the colliery and to measure the corves in use there. This they did and their investigations showed that by 1773 Lord Egremont was owed £1,825. 2s. 2½d. as tentale rent for coal. In the next four years Egremont was entitled to £788. 16s. 10½d., making a total of £2,613. 19s. 1d. for the years 1756-77.[49]

Coal had been raised from seven pits at Broughton between 1755 and 1763 and from eighteen more pits sunk by John Christian after he had taken over the lease from his partners. After his death in 1767 Charles Udale was responsible for the management of all the Christian collieries and his success can be measured by the net income derived from them:[50]

Year ending Sept.	*Net Income* £	*s.*	*d.*	*Year ending Sept.*	*Net Income* £	*s.*	*d.*
1768	1,738	14	5¼	1774	1,117	16	0¼
1769	2,131	17	8	1775	1,375	13	2½
1770	1,842	17	1½	1776	2,614	8	11¼
1771	1,522	9	6	1777	1,775	1	7
1772	1,452	1	5½	1778	2,110	2	6¼
1773	1,856	12	5¾	1779	828	2	6½
				1780	695	4	9

On attaining his majority young John Christian found that he had inherited problems as well as property, and although from now on the decisions were his he continued to rely heavily upon the advice and experience of Udale. His first major move was to break away completely from his father's partners in the Broughton Colliery and to negotiate a new lease with Lord Egremont, for by the agreement of 1763 these partners had become merely sleeping patners and had been more than amply rewarded. Udale had never liked this agreement by which Christian was obliged to pay his partners 9d. per ton on a hundred tons per day whether he shipped so much or not and in 1773 he wrote, "The 100 Tons per day from Mich. 1763 to the end of the present lease @ 9d. per Ton amounts to £20,834. By sinking and working 18 pits only £9,474 has been paid of that sum. There yet remains to be paid £11,360. Then according to that proportion there will be nearly 22 Pits yet to be sunk & wrought to afford the Company that sum.

Now instead of 22 pits required, Certain it is that at this Time, it is out of the power of any Agent or workman employed in this Colliery to shew where even 4 pits can be sunk and wrought at a Common expence".[51]

The 1755 lease* was due for renewal in 1781 and Lord Egremont was prepared to renew provided his higher terms were accepted. By a lease of 1 June 1781 John Christian became the sole lessee of the Broughton collieries for twenty-one years on payment of a yearly rent of £400 and 3d. for every ton of coal for land sale above 12,800 tons and 7½d. per ton for coal shipped above the said 12,800 tons. Egremont's decision to grant the lease to Christian alone appears to have followed the discovery that the original partners in the Broughton Colliery had made a secret agreement between themselves in 1758 to deny wayleave to coal from his father's colliery at Birkby and in 1781 their successors were still refusing to provide this wayleave. Bitterly disappointed that the agreement by which they had profited so much between 1763 and 1781, without risk or expense to themselves, had come to an end Humphrey Senhouse and his associates tried to thwart Christian by indulging in protracted litigation. Attempts were made to prevent him using a road from Ellenborough to Maryport for moving his coal and to have the

* The 1755 lease for 21 years had been extended for a further five years in 1760.

1781 lease declared invalid. Ultimately, on the major question of the validity of the new Broughton lease Christian obtained a verdict in Chancery with costs.[52]

Christian also leased, 1 June 1781, from Lord Egremont the coal in the manor of Birkby for twenty-one years at a total annual rent of £100. Some twenty years earlier Udale had successfully and cogently argued against a suggestion that the Broughton Company should lease this colliery. However, mining operations began in 1785. Pits were sunk on the north bank of the River Ellen to a 30 in. seam at a depth of ten fathoms and were drained by a level from the Ellen. This colliery achieved very little and when the lease expired ceased production. Like his father, Curwen throughout this period worked Broughton Colliery in conjunction with his colliery at Ewanrigg. In 1791 his collieries at Broughton, Ewanrigg and Birkby brought him a profit of £1,396 10s. Broughton Colliery proved much the most productive and under the new lease lucrative for both lessor and lessee. J. C. Curwen renewed the lease of the colliery, 13 May 1802, for another twenty-one years and again the lease terms were raised. He had now to pay a dead rent of £1,000 per annum and 2s. 6d. for every waggon of coal beyond 8,000 waggons, each waggon containing three Cumberland tons.[53]

Little is known of the Lowther colliery at Dearham in the latter half of the eighteenth century but the Lonsdale MSS. show that in the year ending Lady Day 1772 it made a profit of £75. 5s. 10¾d. Mining on a small scale was carried on at Dearham in the early years of the nineteenth century. James Penn was working the Rattler Band from a pit thirteen fathoms deep in 1803 and five years later the Dearham Row Colliery was at work.[54]

After buying William Blennerhasset's Flimby estate in 1772 Sir James Lowther began his Flimby Park or Woodside Colliery. From old mining plans it seems that he was working coal in the same area as James Gorton in the 1720's whose Flimby Colliery had been leased from the Blennerhasset family. Lowther's colliery in 1781 consisted of two pits and continued until 1791. Moore states that the colliery "was stopped in 1781", but this must be a misprint for a document dated 30 September 1791 says, "His Lordship's Colliery at Flimby is nearly worked out". (Sir James Lowther became Earl of Lonsdale in 1784). In 1781 William Allanby leased his coal mines at Flimby to the owners of the Maryport Furnace for twenty-one years at a rent of £52 per half-year or 4d. per ton on all coal raised, at the option of the lessor, and 1s. per ton on all ironstone. This firm achieved very little either at Maryport or Flimby and in 1784, because of financial difficulties, sold the furnace and its materials to Humphrey Senhouse. Lord Lonsdale's Flimby Park Colliery was re-opened in 1802 by John Walker who worked it until 1825. In 1802 it had a daily ouput of seventy tons and employed twenty-three men. A report on the colliery in the following year shows that it then consisted of five pits, one of which was an old pit opened by Walker to provide an air course to Smithy Pit. At the Lowther Pit, where the Three Quarters seam was being worked, coal was raised to the surface from a depth of eight fathoms by men using a jack-roll. At the other pits this task was performed by gin horses. For the five years 1809-13 annual average coal exports from the colliery amounted to 4,000 waggons. By 1816 the colliery consisted of sixteen pits.[55]

In the parish of Gilcrux coal was first worked by the Dykes family of Dovenby Hall in 1740. The colliery, which adjoined the River Ellen, was taken over by Messrs Jackson & Co. about 1770 who worked the Yard, Ten Quarters and Three Quarters seams at depths of 8, 16 and 21 fathoms respectively. Coal was wrought and laid above ground at a cost of about 12d. per ton. Four pits were in operation at this colliery in 1784, two owned by Miss Dykes, lady of the manor, and two by Mr Sealby, but the output was small.[56] Like other inland collieries coal at Gilcrux was raised to satisfy the needs of people living within a short distance of the colliery and with few inhabitants there was little incentive, even had it been possible, to embark upon large scale mining development.

Aspatria Westmoor or Oughterside Colliery was leased by Lord Egremont in 1753 to the Lawson family for seventeen years at an annual rent of £15. The Lawsons never worked this colliery but they made use of the lease to cut a dam on Egremont's common to convey water to the engine of the colliery which they worked on their own land at Oughterside. To obtain access to the sea and thereby to a much larger market it was proposed that a canal should be cut from the River Ellen to the sea at Blue Dial, near Allonby. This scheme, designed to facilitate the export of coal from this district, had to be abandoned since it proved impossible to obtain the consent of all the many landowners through whose lands the proposed canal would go. Nothing was done at Westmoor Colliery until 1788 when it was leased by the Bolton Co. (Messrs Crosthwaite & Co.) for twenty-two years at a rent of £20 per annum and 3d. per ton of coal over 1,600 tons per annum. The company worked the colliery until the lease expired in 1810 but there are no records in existence to give details of output, profits or losses, or of mining methods employed.[57]

Few records have survived to illustrate the history of Bolton Colliery from the Elizabethan age until the late eighteenth century, Bishop Nicolson noted in his diary his two visits in 1707 to Mr Robinson's colliery at Bolton. The collieries in the manors of Bolton and Westward were leased for twenty-six years from Lady Day 1753 at a yearly rent of £60. Two of the lessees, Samuel Chappelhow and John Marshall, managed Bolton Colliery and worked coal to a depth of 21 fathoms with the help of a "fire engine" with a 36 inch cylinder which had cost about £1,400. In 1773 the colliery had a weekly ouput of about 60 tons. Working costs amounted to 3s. 10½d. per ton and since the coal was sold at 4s. the profit was only 1½d. per ton. On an estimated annual output of 3,000 tons the colliery would therefore show a profit of only £18 15s. – a meagre return for a year's work. Another pit at Weary Hall was more lucrative; coal worked at 2s. 4d. per ton was sold for 3s. 4d. yielding a profit of 1s. per ton which, on an estimated ouput of 1,200 tons would amount to £60 per annum. In addition the lessees sub-leased the colliery in Westward to Robert Monkhouse at £40 per annum. It was computed that this colliery earned a profit of 10¼d. per ton or £76 17s. 6d. per annum on an output of 1,800 tons. Messrs. Fawcett, Crosthwaite & Co. leased Lord Egremont's collieries in Bolton, Westward and Wigton for thirty-one years from 25 March 1779 at an annual rent of £300. This must have been a successful enterprise for they renewed the lease in 1810 for a further twenty-one years at a rent of £1,000 per annum.[58]

Until the nineteenth century collieries in Cumberland were carried on largely by landowners who exploited their own coal deposits and leased others when they could not buy out the owners. The Egremont estates (later the Leconfield estates) were a notable exception. Yet even here the important collieries were developed by landowners such as the Christians and Curwens who had sufficient capital for the purpose. Iron-masters were responsible for a little progress in coal mining e.g. Messrs. Cookson at

Greysouthen and Clifton and four of the partners in the Maryport Ironworks also had shares in Broughton Colliery. But compared with the landed gentry their contribution to the growth of coal mining in the county was small; it was the wealthy landowners who were best able to finance and develop the collieries.

MINING METHODS

Towards the close of the eighteenth century mining practice in the Cumberland coalfield was much the same as it had been fifty years earlier. At a number of pits miners continued to reach the coal face via bearmouths; in 1815 there were three bearmouths at Howgill and four at Whingill. At Whitehaven in 1801, after completing the sinking of a pit, a "level is first driven for 400 or 500 yards, in a north and south direction; and then a passage is cut through the several strata, to the surface of the earth, as a road for the men and horses; and also to admit a fresh current of air into the mines". The bord and pillar system of working coal was still being employed but as greater depths were reached the pillars increased in size. By 1800 pillars in the deep mines at Whitehaven were 18-20 yards square, only about one-third of the coal being extracted, the other two-thirds being left to support the roof. At Broughton Moor in 1812 the pillars were seven yards wide by twenty-two yards long, the bords or working-places being five yards wide. Some of the Whitehaven mines were still the deepest in the kingdom; in 1793 King Pit, Howgill had been sunk 160 fathoms. Apart from having the deepest pit in the country Howgill Colliery with an area of 2,400 acres, was the most extensive in Great Britain.[59]

At the close of the century the wooden rails in collieries were being replaced by cast iron rails. Professor Hughes has shown that there is direct evidence that iron rails were being used at Workington by 1800. After the adoption of tram plates the corves were placed on low trams fitted with small plain wheels which were "trailed" from the workings to sidings in the main roads. At the sidings hand cranes were used to lift the full corves from the small trams on to larger trams, mounted on flanged wheels, which were then drawn by horses to the shaft bottom. Between 1802 and 1811 John Bateman was busily engaged in replacing the wooden railways underground at Whitehaven by cast iron railways and one such metal tramway at Wilson Pit, completed in 1806 would, he believed, save £1,560 a year in horse labour compared with the old method of conveying coal underground by means of sledge trams. By 1815 Lord Lonsdale's large iron foundry was constantly at work to provide rails for his pits and over twenty miles of them had been laid in the underground ways at Whitehaven. The use of horses underground was possible only in the thicker seams and as late as 1801 in some of the Whitehaven pits coal was moved from the face to the shaft, not by horses, but by men.[60]

After the use of cranes was abandoned in the Main Band collieries at Whitehaven and Workington light wooden bogies fitted with small wheels, drawn by men, were used to bring the coal from the workings to "stears" or turning places. There the bogies were emptied into baskets which were placed on larger bogies capable of carrying two baskets each, and the bogies, after being made up into trains or "rollies" were drawn by horses to the pit bottom. It is probable that mechanical underground haulage was first adopted in Cumberland in 1818 when a high pressure engine was erected in William Pit; this became necessary as the main underground roads were extended to the dip and horsework became correspondingly heavier. Thus the steam enginge was applied to drawing trains of tubs along these roads by means of hemp, and later by wire ropes.[61]

Deeper pits required more powerful pumping engines. The two pumping engines at Saltom were replaced in 1782 by an engine then regarded as a mechanical wonder. It was an atmospheric engine with a cylinder of seventy inches diameter and a stroke of six feet and it continued to work until 1866. According to R. L. Galloway a steam engine was introduced underground at Whitehaven to raise water from the dip workings in 1776 and this he describes as the first recorded instance of an engine being erected in the workings of a mine. This statement, apparently based on "A Treatise on the Steam Engine" by John Farey (London, 1827) p. 238, seems doubtful. I have found no evidence among the Lonsdale MSS. or elsewhere to confirm this statement. Such an important innovation, if this account is correct, would almost certainly have been reported in the local newspaper, the *Cumberland Pacquet*.[62]

Changes took place in the means employed to wind coal out of the pits. In 1787 the first recorded departure from winding coal by horse gins in Cumberland was made at George Pit, Whitehaven, where, on the same shaft as the rope rolls, an overshot water-wheel driven by water pumped by the "fire engine" was erected.* This awkward pumping arrangement could not have been a great success, but this water gin for raising coal continued working until 1803.[63] The first direct application of the steam engine to winding coal in Cumberland was made at the Hope Pit, Workington. J. C. Curwen in the 1780's became increasingly aware that in winning coal at greater depths the old methods of winding were inadequate and that mechanization was essential. Accordingly he turned for help to the famous firm of Boulton & Watt in Birmingham. By an agreement of 1 July 1788, between John Christian (as he was then) and Messrs Boulton & Watt, the latter undertook to provide at Workington Colliery

> "a Steam Engine of the said James Watt's invention as aforesaid and of dimensions following, that is to say the Cylinder of the said Engine to be fourteen inches in diameter and four foot long in the stroke and the piston thereof to be acted upon by the force of Steam both in its ascent and descent with a rotative motion of his, the said James Watt's invention applied thereto for the purpose of drawing coals and water from the said Colliery."[64]

This was a 6 h.p. engine which cost £607. 5s. – £251. 13s. 7d. for materials supplied by Boulton & Watt and £355. 11s. 5d. for workmen's wages August 1788-July 1789. In addition, Christian had to pay an annual premium of £30 for the use of the engine since the firm charged £5 per horse-power.† This engine was "calculated to wind Coals at the rate of 4½ feet per second" and its success led Christian to order more engines from Birmingham. Thus we find Charles Udale writing to Messors. Boulton & Watt (6 June 1789), "Mr Christian is desirous to be furnished with directions for Erecting another of your patent Engines as soon as can be convenient . . . . The Engine should be adapted so as to draw water by pumps as also to wind Coals, it should therefore have power to take out the water in a few hours of the 24, that the rest of the time they may be apply[d] to the winding up of Coals and other matter that may be necessary to take up".[65]

* This engine was set to work 12 May 1787 and is described in the *Cumberland Pacquet*, 16 May 1787.

† A paper entitled "Memorandums of Mr Curwen's Engines for Mr Lawson", dated 13 Oct. 1798, says this engine was "put together" at Hope Pit by James Murdock in 1788, (Boulton & Watt MSS.) But from the accounts of money paid out in wages when erecting this engine it would seem that it did not begin working until 1789.

Messrs Boulton & Watt recommended a 16 h.p. engine for which they said "our annual premium (£5 a horse) would be £80 a year". This suggestion was adopted and a 16 h.p. engine was erected at Elizabeth Pit in 1789 at a total cost of £1,300. 8s. 3½d. the materials supplied by Boulton & Watt costing £462. 9s. 1d. This engine and its predecessor gave satisfaction as Udale reported

> "I now enclose you Mr Curwen's draft on Mr Joseph Christian London, for Four Hundred & Sixty two pounds nine shillings, which is the amount of your charge received here, some time ago, for the materials of the latter Engine . . . . This Engine performs the work we have at present for it very well, and goes with the pumps now 70 fathom, at the rate of 24 or 25 Strokes per minute when necessary – As to its Work in drawing Coals, it has very ample power – The little Engine also continues in very good order, and does much business.
>
> We are much amused here with an acc^{t} of A. Heslop's improvements made in those kind of Engines, whose powers are said to be equal to more than 10 pounds per inch effective in a single Engine, and that he has not only obtained a patent for his invention, but has received large orders for the Erection of such Engines."[66]

Additional Boulton & Watt engines were bought later. A winding engine was installed at Lady Pitt in 1793, presumably of 17 h.p. since the annual premium amounted to £85. This was followed by a pumping engine, erected in 1795, which worked a 9¾ inch pump; the annual premium for this engine cost £200. Materials for a 160 h.p. Boulton & Watt pumping engine with a cylinder of 66 inches diameter were moved to Workington by Messrs Fenton & Murray of Leeds in 1811. This engine, at that time the most powerful pumping engine in Cumberland, was set to work at Isabella Pit in August 1813 and operated six sets of pumps down to a depth of 130 fathoms. By 1814 Boulton & Watt winding engines were at work at the Lady, Union and Isabella Pits. Lowther's Whitehaven pits in 1815 employed eleven pumping or winding engines, five at Howgill, five at Whingill and one at Scalegill. In addition, a steam engine was in use at his iron foundry.[67]

Udale had referrd in pejorative terms to Heslop's engine, a view shared by the firm of Boulton & Watt.* Nevertheless the Heslop winding engine, patented by Adam Heslop in 1790 and made at the Seaton Ironworks, was adopted at a number of local collieries. At Whitehaven a Heslop engine was erected at Kells Pit in 1793; similar engines followed at Davy Pit (1794), Lady Pit (1795), James Pit (1802) and William Pit (1806). At least three Heslop engines were bought by Curwen, one at John Pit and one at Harrington for which he paid annual premiums of £27 and £45 respectively and a third of 14 h.p. erected at Isabella Pit to draw coal from the Moorbanks seam. Heslop's engine was provided with two cylinders, one at each end of the beam, named respectively the hot and cold cylinders, the former resembling the cylinder of an ordinary Newcomen engine, while the latter was used as a separate condenser. Matthew Boulton was sufficiently concerned by this engine to have a copy sent to him of the drawings and specifications of Heslop's patent. But if it was, as has been alleged, an infringement of the patent rights of his partner James Watt, nothing was done about it.[68]

* James Watt, junior, to J. C. Curwen, 2 Nov. 1798, 'We may have acted wrong in not noticing Heslop, but if so it has proceeded from an error of Judgement & in suposing his Engine too contemptible. Had we thought it necessary to notice his Engine, we certainly should not have commenced with those he has erected for you". (Boulton & Watt MSS.)

Horse gins, however, continued to be used in some pits. At Whitehaven in 1802 horse gins in twenty-four hours could raise 440 baskets of coal from a depth of 120 fathoms at Wilson Pit and 100 baskets at Croft Pit from a depth of 118 fathoms. Winding engines, by displacing horses, reduced costs considerably. Until the adoption of engines horses were both numerous and expensive; at Howgill and Scalegill Collieries in 1787 there were 72 underground tram horses, 87 gin horses and 24 other horses. Of these 183 horses 131 belonged to Lord Lonsdale. In addition there were 32 waggon horses. Hutchinson estimated that an engine capable of drawing coal from 90 fathoms could be erected for £600, and that in lieu of an engine, the purchase price of horses necessary to draw the coal would amount to more than half of this figure. Moreover, the horses would have to be fed and wages paid to their drivers and grooms, a heavy item in production costs; in 1803 each horse at Whitehaven cost between 27s. and 29s. per week and at Workington the feed of each colliery horse was estimated to cost 30s. per week. Winding engines made gin horses unnecessary but many horses continued to be used both underground and on the surface for haulage purposes. A list of horses, dated 30 December 1799, gives a total of 153 horses then employed at Howgill Colliery and sixty-nine more were needed. After the adoption of the winding engine it was calculated (in 1802) that a yearly output of 80,000 waggons of coal at Whitehaven would require 306 horses and 182 hewers. A shortage of horses could still cause serious difficulties and this explains the Lonsdale practice of letting farms on condition that the tenants provided horses for the collieries; farmers who failed to provide their quota of horses were fined.*[69]

Winding engines were not only quicker and cheaper than horse gins but they could also raise coal and men from greater depths; at the same time they created another problem, that of providing ropes safe enough and strong enough to lift from these depths. Bateman at Whitehaven was experimenting in 1803 with the flat ropes invented by John Curr of Sheffield. At first he did not find them satisfactory, nor did he like the high price charged for them, a flat rope of 130 fathoms costing £67 4s. compared with £20 for an ordinary rope. However, flat ropes were adopted and Whitehaven's example was followed at other Cumberland collieries. A letter from William Swinburn, Curwen's agent, to John Curr, dated 7 October 1812, shows this:

> "We are induced to make one fur. trial of your ropes – you may therefore if you please send us 4 Ropes of 234 yards long each & 2 of 200 yards long each, by the usual conveyance *as soon as you can.*
>
> What we have to complain of in the former Ropes are first the *Sticking together* which very frequently opens in many parts – and next the quallity. Which in the late ropes we have had are very little if any superior to the common round; as you will see by the quantity drawn at a pit 180 yards deep, the Basket & Coals weighing 8 cwt. 1 qr. I think if I do not mistake our manager.
>
> The two ropes I am now about to state the quantitites upon went about 6 months each drawing on 2625 Score of such Baskets and the other 2601. If we are so unfortunate as to get any more

* In 1794 a letter was sent to all Lonsdale tenant farmers who had provided insufficient horses, "According to your agreement on the 8th day of March 1794 with Mr Geo. Wheatley agent to the Earl of Lonsdale for supplying your Number of Horses and is not come. If you do not send your full Number you will be fined according to Mr Brownrigg's account of Lost Work without further Notice". (D/Lons/W Collieries List 2, 33-39).

from you of this inferior quallity we must change our Pit ropes for the round as any thing you State to return is nothing near adequate to the loss."

Apparently the new ropes were acceptable since four more flat ropes were ordered (14 January 1814) – two ropes, each about 230 yards long and two about 200 yards long each. In the following year four more ropes of the same dimensions were ordered from the same source.[70]

At Whitehaven changes took place in the movement of coal to the harbour. According to a contemporary writer, "In 1813, the waggon-ways, which were before of wood, were laid with cast iron, and on the Howgill side a self-acting inclined plane was constructed, 290 yards long with a perpendicular altitude of 115 feet. Since this alteration, three waggons, coupled together, are conducted by one man and horse, whereas, formerly it was necessary to have man and horse to each waggon".[71] Another writer elucidates further–

"The descent of the waggons down the western hill has been simplified by a contrivance which saves the labour of forty men and as many horses. From the point where the declivity of the hill commences a regularly inclined plane has been cut, down which three waggons in a string roll at a time. The loaded waggons pull up three empty ones in their descent, but this counterpoise does not sufficiently reduce their speed, and they are therefore made to give motion to some machinery which requires considerable power, working a huge pair of bellows that force air through a valve into a receiver. As the receiver fills, more power is necessary for the depression of the bellows, and the waggons are thus duly checked as they acquire force in their descent."[72]

Fifteen years later a new inclined plane, about 1,400 yards long, was constructed from Croft Pit, down which twelve loaded waggons at a time descended by their own weight and when emptied were drawn back to the pit by a small steam engine erected at the top of the plane.[73]

THE MINERS

An adequate supply of labour to meet the growing demand for coal was not readily available in sparsely populated Cumberland and colliery managers had to make strenuous efforts to recruit workers from other areas. A substantial proportion of the manpower in Cumberland mines came from immigration into the County. Ireland, Scotland and the Newcastle district provided many of these immigrant miners. Bateman reported in 1804, "A great number of Irishmen from the County of Down have very lately come into Cumberland – we have got several to work. Some offer well and others were very indolent and went to country work, several Scotchmen have come over since Whitsuntide as wages there are very much come down the last term. We had rather have one Scotchman than five Irishmen, if we can keep the Scotchmen a few weeks they are sure to turn out well".[74] At Workington Curwen employed "a great number of Irish, much to the dissatisfaction of his English workmen, who protest against them generally as foreigners and papists, and are particularly irritated against them for sneakingly submitting to receive such wages as they can procure".[75] These feelings precipitated a serious riot by the Workington miners in 1814 which had to be put down by troops. Of the 453 employees at Howgill Colliery in 1802 149 were immigrants. Their numbers and places of origin were:[76]

Ireland	44	Lancashire	14
Scotland	41	Durham	13
Wales	2	Northumberland	8
Isle of Man	5	Yorkshire	4
Westmorland	17	Leicestershire	1

Ireland, whose rapidly growing population outstripped the home demand for labour, had to export its surplus manpower and Whitehaven, a short voyage away, was certainly well known to Irishmen. The Scottish contingent, many of whom came from Galloway, did not have to travel far in search of employment. The other immigrants also, apart from the two Welshmen and one man from Leicestershire, were short-distance migrants from neighbouring northern counties. Like many miners during the Napoleonic War the non-Cumbrians hired at Whingill Colliery were extremely mobile in their search for higher wages. Between October 1810 and May 1811 forty-three employees left the colliery and not one of them was a native Cumbrian; of these absentees nineteen were Irish, nineteen from Newcastle, two were Scots, and Wales, Alston Moor and Lancashire had provided one each.*[77]

Until the late eighteenth century Scottish miners and their families were in a state of bondage, bound to work in their collieries for life. An Act to liberate them, passed in 1775, was only partly successful and not until the passage of another Act in 1799 was their legal emancipation completed. Many Scotsmen escaped from the servitude of the Scottish collieries to find employment in the Whitehaven pits. Scottish mineworkers, who described themselves as "all workers in your Honour's Coalworks",[78] petitioned Sir James Lowther in 1763 to remit the annual ground rent on their Presbyterian meeting house. An examination of the ages and lengths of service of the Scots on the Howgill payroll in 1802 makes it reasonably certain that at least some of them were refugees from Scottish collieries.

In Northumberland and Durham it was customary to hire miners for a year at a time and this "yearly bond" was a severe restriction upon the liberty of the collier. Miners from these counties undoubtedly made their way to Whitehaven and other Cumberland collieries in search of employment and the greater personal freedom which they obtained there. A letter, 31 July 1789, from Thomas Allen of Newcastle to John Losh deals with the case of nine Newcastle pitmen alleged to have broken their bonds and to have been employed at Cumberland pits.

"Dear Sir,

The undermentioned Pitmen are bound Servants to Mr Liddell & Partners to work in Greenwich Moor Colliery – they have absented themselves and we are inform'd they are employ'd by Mr Christian. I will be much oblilged to you if that Gentleman is in Carlisle tomorrow if you will apply to him requesting their discharge & sent immediately again to Greenwich Moor Colliery.

* Cf. A. Redford, *Labour Migration in England 1800-1850* (Manchester, 1926), 49, "In general, the evidence for any strong influx of labour into coal mining is not plentiful. A large part of the supply of new labour required by the expansion of the industry probably came from the natural increase of a notoriously prolific section of the population."

We are very much in want of these Men & sho'd Mr Christian not be in Carlisle I beg you will be so good as forward this Letter to him – when I make not the least doubt but he will give immediate orders for their discharge."[79]

During the Napoleonic War the shortage of labour in the mines became an acute problem for colliery owners in Cumberland. Sickness, enlistment in the armed forces,† competition among employers to obtain mines and the rapid turnover of workers aggravated the situation. The mobility of labour is shown by the number of absentees from Workington Colliery noted by the viewer, Edmond Bowness:

23 July 1806	–	30 May 1807	60
30 May 1807	–	30 March 1808	49
30 March 1808	–	6 July 1808	32
6 July 1808	–	31 August 1808	25

Between 1809 and 1812 the colliery lost in successive years 147, 108, 81 and 50 employees, but gained in the same period 154, 102, 119 and 111 employees.[80] To attract mineworkers advertisements were inserted in Newcastle, Scottish and Manchester newspapers as well as local publications and newcomers to the collieries were offered cash bonuses.

At a time when the industry was expanding and more miners were badly needed the Cumberland colliery owners could not be too scrupulous about the antecedents of additions to their labour force. One of the most intractable difficulties was the widespread practice of colliery owners poaching miners from each other. Bateman informed his master, Viscount Lowther,

"I afterwards found Mr Curwen's people had been trying to get hold of some of your good men. I think it would be an agreeable and beneficial thing both for your Lordship & Mr Curwen if you would draw a line not to employ each other's men but under certain restrictions – their running from Colliery to Colliery makes them very unmanageable."

Nevertheless, Bateman himself was very willing to hire men belonging to the Curwen collieries – the prospect of increasing output at a time of rising prices and profits removed any scruples he might have had. A letter of his refers to a visit from Curwen's agent,

"about some of their men we got last week, & wanted me to promise not to take any more, & they would engage not to take any of your Lordship's men, but I found they were full of men & Shipping a Thousand Waggons a week which is more than we are Shipping here & we used to Ship more than all the Northern ports together. I told him if he would allow us to fill up our Number wanted, then we would comply with his request, this he objected to; then I asked him if they would give up all the Colliers that formerly belonged to these collieries; this he also objected to; then I told him we must remain at liberty to take all that offered untill we have enough & then we would treat with them . . . .

I think it would be very wrong to stop taking their men untill we have enough, in fact we might be some years in getting to the extent we aim at, as they are better situated for getting the Newcastle Colliers than we are, they call there first, as their Colliery is in their way to this Place."

† "The late Militia Ballot and an Epidemic fever carried off nearly one fifth of our Haggers." (Bateman, 20 June 1808).

Curwen's men continued to try to tempt miners away from Whitehaven and in 1803 they were offering,

> "the Haggers 4/6 per day certain, if they could not make that sum at their price of Hagging Mr Curwen would make it up to them, the women fillers of Coals he offered 12s. per score (*above double our price*) . . . they would give in hand on their arrival at Workington from one to five guineas for each Hagger. What Mr Curwen can mean by such conduct I cannot guess. He knows I dare say that all your Lordship's People are hired Servants, & that he cannot legally keep one of them."[81]

Such a state of affairs could not continue indefinitely. Reason prevailed when an agreement was made, 11 June 1803, by the rival employers not to employ each other's men for the next seven years although each would retain workmen acquired prior to this date. Agreements of this kind did much to reduce the pressure on harassed colliery managers. After the agreement with Curwen Bateman could write, "Our Haggers' Conduct is very much altered for the better since they find they cannot when loose go to Workington, they follow their work better & are far more moderate in their expectations about the Prices of their Works". If negotiations did not stop the poaching of colliers it was sometimes necessary to resort to stronger measures. In 1804 when Wilson & Co. of Greysouthen Colliery were employing Whitehaven miners, Batemen threatened to open a colliery at Clifton which would compete with their trade. The threat sufficed, the firm agreeing to cease hiring Lowther's workmen.[82]

A letter from Lord Lowther shows that he had some sense of the proprieties in these matters, although at times, if he were the injured party, he did not hesitate to sanction drastic action against erring competitors,

> "I am induced to write to you at present in consequence of a letter from William Bryham relating to a charge against him for attempting to hire some of the men to follow him to Newcastle. This he denies & whether it is true or not I have no wish to enquire. But he goes on to say, 'I may endeavour to get some men from Mr Curwen's works, as in that case I am doing your Lordship no Injury but throwing the Balance of Power in your Lordship's Favour'. I wish you would inform him that I shall consider any attempt on his part to entice any of Mr Curwen's men from their Employment as no less improper than his practising the same Plan at Whitehaven, as his situation under me has furnished him with the means of doing it & as I think it highly dishonorable in any persons to encourage such attacks upon their neighbours, and not to take the necessary steps to defeat them – I should therefore think it proper that Mr Curwen's Agents should be apprized of his design, if he does not at this Representation entirely abandon it.[83]

In 1810 Bateman found that his labour supply was being reduced by Mr Christian, who, by offering higher wages than were paid at Whitehaven, was taking away many of his best hewers to Broughton Moor Colliery. Bateman wrote to Lord Lonsdale, "Our men say they earn a Shilling an Hour at Mr Christian's works. With us they could earn a 5d. to 6d. per hour. With such difference of Wages all our Men will go if he can employ them".[84] At that time Christian was selling coal to the country at 14s. per waggon and to the ships at Maryport at 16s. 8d. per waggon. To meet this competition a scheme was put forward designed to put Christian out of business. It was suggested that pits might be opened at Clifton and Dearham to provide coal for country sale to rival Christian's coal. In addition, Mr Walker, then working a colliery at Flimby, should be persuaded to reduce the price of his export coal to 12s. per waggon and country sale to 10s. per waggon, Lord Lonsdale paying Walker the difference in price. But to increase his sales at Christian's expense Walker would have to enlarge his output and would need more hewers.[85] Lord Lonsdale* immediately approved of the plan,

> "I would therefore have you send to Mr Walker without a moment's delay & let him reduce his prices so as to get as much of the sale from Christian as he can both by the ships and the country. I think it would be adviseable to lend him a few haggers & also to let him tempt Christian's men by bounties & indeed it would be better to sink entirely all the profits of the Flimby Colliery than suffer such an upstart to carry his point. This proceeding will soon bring him to his senses or ruin him."[86]

However, it proved unnecessary to implement this plan since in November 1810 Bateman and Christian agreed not to employ each other's workmen and to discharge men they had already acquired from each other.†

Newcastle colliery owners also shared in this poaching of miners and attempted to thwart their employees' demands by hiring Cumbrians. In February 1804 a Newcastle basket-maker was trying to obtain carpenters and basket-makers from Whitehaven by offering them 21s. to 22s. per week, free house and coal and 40s. for removal expenses. In three weeks fourteen haggers left Whitehaven Colliery, most of them going to Newcastle, and it was discovered that a Peter Curry had taken a job in a Whitehaven pit in order to recruit men for work in Newcastle mines. But many Newcastle men were employed at Whitehaven – in March 1804 nineteen haggers and five sinkers or drifters were working at Whitehaven Colliery. These men, however, did not remain long for they found the coal hard to work and having broken their bonds at Newcastle they also feared arrest.[87]

Competition to obtain men undermined the authority of the employers; demands for higher pay, if rejected, were followed by strikes or threats of strikes, or by migration to other collieries. Such competition meant that "the men all become Masters, and more Money is given for work than it deserves, & they get Drunk with the Overplus".[88] Bateman found himself compelled to meet recurrent demands for higher wages, demands that were difficult to resist when the men could obtain better pay at nearby collieries. In 1805, within an period of six weeks, thirty haggers, about one-fifth of the total employed, left Whitehaven for more remunerative employment elsewhere.[89] The loss of haggers was particularly serious for each earned for his employer a profit of four guineas per week.‡ Men were also lost to the armed forces; in 1803 twenty-three colliers were in the militia, sixteen of these being hewers[90] and Bateman had frequently to urge Lord Lonsdale to use his influence to prevent coal miners enlisting. The only effective answer to demands for wage increases was an ample labour supply and Batemen took on all who offered since, in his own words, "having men to spare will be the first thing to quiet those we have".[91] Newspaper advertisements and premiums of up to four guneas failed to bring in the additional hewers required when the demand for miners reached

* Viscount Lowther was created Earl of Lonsdale in 1807.

† Bateman, 19 Nov. 1810, "If Mr C. had been able to have kept all our Men, our own Men were all ready I understand to have demanded an advance of 1/10 more which would have amounted to about 3000£ a year."

‡ "We want Haggers much, we could take on 20 or 30 more. The profit upon each Hagger until we have enough at 2 Waggons per day is 14/- or £4. 4s. per week." (Bateman, 12 Dec. 1802).

its peak during the Napoleonic War.* In his anxiety to obtain more men Bateman tried to secure the services of disbanded soldiers. A letter of his stated,

> "We wrote to our Agent, Mr Dixon, in Liverpool to be very active in procuring us all the Miners in the 9th Brigade which are expected to be disembodied very soon at Liverpool & we authorised him to give the Sergt. Major 5/- for every Miner he could influence to come hither, and 10/6 to every Miner as soon as the Liverpool traders were ready to sail to this Port, to pay their passage and to promise them 3 Guineas each if they remained 26 weeks in your employ with the further advantage of a free house."[92]

Unfortunately for Bateman this plan came to nothing for the troops were not demobilized.

A stable labour force was essential to maintain output and profits and to counter the excessive mobility of his workmen Bateman tried to establish the binding system. This is the only instance in the history of mining in Cumberland of an attempt to impose the type of servitude then common in the collieries of the north-eastern coalfield. He wrote to his employer (15 November 1802), "We have got all the Colliers bound except James's Pit men & we will prepare an article for them this week". The men at James Pit were bound and thus they were "safe for 11 months to come".[93] Two months later he persuaded the waggoners, carpenters, smiths, bankers, engine men, hookers and fillers to sign an agreement to remain until Martinmas on the old terms.[94] Some men, however, broke their bonds and found employment at the Curwen collieries. On the orders of Lord Lowther a warrant for their arrest was obtained and the men were compelled to return to their work at Whitehaven.[95] But Bateman was unable to make the binding system a permanent feature; it ceased with the expiry of these contracts. Presumably the miners, well aware of their scarcity value, were unwilling to enter into long agreements.

Labour relations at this time often degenerated into a trial of strength between two opposing sides. The Whitehaven miners, despite the Combination Acts, formed a trade union in 1811. Bateman reported,

> "I am sorry to say that nearly all our Haggers & Fillers are forming a very powerfull combination to advance their wages, report says that the advance is to be very considerable, they meet publickly now twice a Week to collect subscribers & to swear them to be true to each other. We are told that a part of their Oath is, not to tell what they want untill they have got the Number they want. At present they say Their Number is near 200 who have subscribed to their Rules . . . today we find that one of their Rules is, that if any Class of Workmen want an advance of Wages & it is not given them, all the Brotherhood are to lay Idle until their demand be complied with. By this plan they can stop the works whenever they please."[96]

In the following week the hewers and fillers at Croft Pit refused to work but Bateman soon crushed this strike. He wrote,

> "Last Wednesday Croft Pit Haggers & fillers made another stand off. On Thursday evening I sent for them to know their reasons for again laying the Pit idle. I told them they were not satisfactory & that unless they would promise to go to work on Fr. we were determined to lay the pit Idle, and they must all quit their Houses next day or we should be under the disagreeable necessity of throwing all their goods to the Door to make room for other men coming to seek work. This threat had the desired effect. Next day Fr. they *all* went to work and on Sat. likewise."[97]

* In 1804 Newcastle miners were being offered between ten and fourteen guineas binding money on signing articles to do eleven months' work. (Bateman, 24 Sept. 1804).

Occupation	*No.*	*Wages*	*Occupation*	*No.*	*Wages*
Agent	1	12s.	Fireman	8	10s. 6d.–11s. 8d.
Viewer	3	8s.–12s.	Smith	1	12s.
Farrier	1	Not given	Striker	1	8s.
Banksman	1	12s.	Picksharper	4	8s.
Hagger	81	Not given	Craner	5	11s. 6d.
Filler	19	5s. 2½d.–10s.	Waggoner	16	5s. 6d.–6s. 6d.
Tram Driver	70	5s.–5s. 6d.	Tracer	2	5s. 6d.
Hooker	8	5s. 7d.–7s.	Carpenter	2	£1 13s. 7d. between them
Gin driver	11	2s. 6d.			
Banker	8	10s.–12s.	Carpenter	2	£1 17s. 6d. between them
Attending way	28	4s.–8s.			
Bearmouth roader	7	3s. 6d.–4s.	Slatepicker	20	3s.–4s.
Engineman	5	12s.–14s.	Stage whipper	5	15s.
			Blasters at Kells P.B. drift	3	£1
Labourers undergound:-					
Mason	3	13s.	Lame persons doing nothing	2	5s.
Carpenter	1	12s.			
Labourer	22	7s.–15s.			
Hagger labouring	2	10s.			
Attending Fox Pit fire	2	8s.–10s.			
Drivers jobbing underground	3	5s.			
Labourers above ground:-			Groom	11	6s.–11s.
Basket maker*	1	10s.	Steathman	1	8s.
,, ,,	5	£3 12s. between them	Steathman's assistant	1	7s.
Mason	2	12s.	Waggoner	18	12s.
Carpenters upon the waggon way	4	£2 10s. between them	Gin driver	2	2s.–4s.
			Changer	2	1s. 9d.
			Gatewatcher	4	2s. 6d.
Smith	2	13s. 6d.–15s.	Fireman	1	11s.
Labourer	29	4s.–10s.	Filler	2	Not given
Granary cart	1	9s.	Trimmer	9	12s.
Cart	4	5s.			
Saddler	1	9s. 6d.	Waggoners that find their own horses	6	21s.–24s.

* By 1808 basket makers were being paid 21s. per week – (Bateman, 10 Oct. 1808).

During the first half of the eighteenth century wages at Whitehaven rose at a steady pace; at the end of the century wages increased rapidly in order to attract men to the mines. An account book shows that in 1757 hewers at Thwaite and Kells Pits received 1d. per corf, 1½d. per corf in the Bannock Band of King Pit, 2d. per corf at Fox Pit, 10d. per ton at Howgill Heads, 12d. per ton in the Yard Band of King Pit, 13d. per ton at Hind Pit and 14d. per ton at Saltom Pit. Fillers earned between 3s. 6d. and 4s. per week[98] and corvers 9s. to 10s. per week.[99] According to M. Jars hewers at Whitehaven in 1765 were earning 1s. 6d. per day for nine or ten hours' work.[100] The Howgill Colliery payroll for 1802 gives the occupations and weekly wages of its 453 employees and these are listed above. Since hewers' wages varied with output their precise earnings could not be given in this payroll; in 1803 the average hewer was earning 21s. to 24s. per week and the best hewers as much as 35s. per week.[101]

This payroll[102] tells us the agent and viewers were paid as follows:

John Bowman	agent	12s. per week
David Walker	viewer	12s. per week
Thos. Marley	viewer	12s. per week
Rich. Bryham	viewer	8s. per week

It seems inconceivable that men holding the senior positions in the colliery should be paid the same wages as smiths and carpenters and less than masons and stage whippers. A letter from Bateman, 22 December 1806, gives a very different picture of the situation:

	Salary	*Premiums yet due them on the profits being 30m£ & upwards in 1806*	*Yearly Income*
	£	£	£
Howgill			
Viewer J. B. Longmire	152	148	300
Overman Thos. Marley	52	88	140
Overman Geo. Shimwell	72	68	140
Whingill			
Viewer John Peile	152	148	300
Overman Rich. Bryham	82	58	140
Overman C. Hetherington	62	38	100

If the figures given in 1802 are accurate it is most unlikely that the wages of viewers or overmen would increase so much in four years, although it is true that wages were rising during this period. Either the agent and viewers at Howgill in 1802 were receiving bonuses not revealed in this payroll or, as seems more likely, it was deemed wiser not to show their precise earnings.

It will be noted that next to the hewers the most highly paid workers were the blasters with £1 per week and the stage whippers with 15s. per week; the lowest paid were the changers, gin drivers, slate pickers and gate watchers many of whom were females.

Of the 453 people employed at Howgill only a small proportion were actively winning coal and a much larger proportion were engaged in other tasks, particularly the movement of coal. There were only 81 hewers. The movement of coal, however, required 70 tram drivers, 40 waggoners, 35 persons to keep the roads clear and five stage whippers who were in charge of the pony drivers underground and who were responsible for the passage of the coal underground to the shaft bottom. In addition, four carpenters were employed upon the waggon way. The horses involved in moving the coal needed the services of a farrier, a saddler, and eleven grooms.*

* For details of the Howgill payroll see Oliver Wood, "A Colliery Payroll in 1802", CW2, lxxii, 303-319. According to Hughes, *op. cit.*, 178, "The total labour force in the Lowther pits at Whitehaven was under 200 in 1811". In 1811 Whitehaven Colliery exported more than twice the quantity of coal exported in 1802. This would have been impossible with a total labour force of less than 200. In 1811 the number of hewers and fillers alone amounted to about 200.

Wages at Whitehaven were usually lower than in other Cumberland collieries but this was offset by the fact that the Lowthers provided their workmen with rent-free cottages and with fuel for which they had to pay only the cartage cost to their homes. In the late eighteenth and early nineteenth centuries miners' wages increased throughout the coalfied. At Broughton in 1760 hagging cost 5d. per ton at Hopewell Pit; two years later it cost 6d. per ton at Hopewell Pit and 7d. per ton at the Well, Success and Delight Pits in the same colliery.[103] By 1783 hewers at Harrington were being paid 12d.–14d. per ton.[104] During the war years wages continued to rise sharply, the scarcity of manpower and the consequent loss of profit making it impossible for employers to resist repeated demands for higher pay. At Workington in 1796 the wages of hewers (who also trailed their coal) varied from 13d. to 22d. per ton. The daily wages of other employees were:[105]

Waggon drivers	12d.	Lampkeeper	18d.
Banksmen	22d.-24d.	Screenworker	21d.
Enginemen	24d.-30d.	Hooker	20d.
Overmen	24d.-32d.	Coal picker	5d.-8d.
Firemen	17d.-22d.	Doorkeeper	4d.-6d.

The miners worked six days per week and wages were paid fortnightly. By 1810 men at Broughton Moor were being paid 2s. 9d. per ton for hagging and trailing, bankers received 21s. per week and bank pickers 12s. per week.[106] At Harrington in 1812, haggers were paid 1s. 9d. to 2s. 2d. per ton, hookers 2s. 2d. to 3s. per day, engine men 2s. 6d. to 3s. per day, firemen 1s. 9d. per day and trappers 8d. per day.[107] In 1814 wages at Workington were:[108]

Hagging & putting the Main Coal	2s. 6d. per waggon
Driving rollies	2s. per day
Hookers and on-setters	3s. per day
Overmen	3s. 6d. per day
Shift work 2s. 6d. to 3s. to 3s. 6d. per shift of 8 hours	
Enginemen	3s. 2d. per day
Plugmen at Main Engine	3s. 6d.
Firemen, Waggon men and fillers	2s. 6d. per day
Bankmen	3s. per day

The Workington miners had to find their own candles, pay for their coal (which cost them about 1s. 6d. per week) and pay their own house rent which varied from £2 10s. to £5 per annum.

During the war Curwen established a warehouse where his workpeople could buy cheap meal and flour. This scheme was discussed by his senior officials at their regular meetings in 1801. To the question put forward for discussion, "Is Mr Curwen's carrying on a Warehouse for supplying the Colliers and others with Meal and flour at reduced prices, found to be generally beneficial to that Class of People?", the answer given was, "Under the present mode of Conducting such Warehouse it is not thought to be generally beneficial to the people". A week later (27 June 1801) the question was put, "Should Mr Curwen think proper shortly to discontinue his present warehouse (on giving Credit to such of the workmen as stood in need of it, for the amount of one 4th nights provisions, to be afterwards paid by small installments each pay day) would any inconvenience or hardship arrise by such discontinuance or would it be productive of any disturbance?" The reply was, "Under the mode proposed by the question it does not appear that any inconvenience or disturbance would arrise".[109] Nevertheless the warehouse continued and after the war it gave rise to much dispute because, it was alleged, it was part of a system of truck payments.

Truck payments were never widespread in Cumberland. There is no evidence to show that the Lowthers indulged in this practice. But a petition to J. C. Curwen, dated 15 May 1822, and signed by about a hundred tradesmen, attributed their diminished business to "the mode adopted of paying the workmen you employ in various articles of general consumption instead of money . . . we therefore trust, that upon this representation you will give instructions forthwith to your agents to pay the workmen in money only". It may be that the Workington miners had some reason for calling the pay-office the "Juggling Room", although one local newspaper claimed that the system enabled the employees to obtain meat, flour and other goods at low prices in bad times. It was asserted that a counter-petition, signed by 800 people, calling for the retention of the system was obtained by the threat of dismissing all those who would not sign.[110] Apparently truck payments were then abandoned.* The Report of the Committee on the Payment of Wages (1842) contains no evidence of the existence of truck in Cumberland.

By this time it was common practice for colliery owners to build houses for their workmen; where new collieries were being sunk they had to do so in order to obtain a labour supply. Thus in 1788 Lord Lonsdale built some two hundred cottages near Whitehaven for his miners which were let rent-free.[111] Originally the intention had been to build five hundred cottages but less than half this number were built. The provision of houses was not altruism on the part of the employers, it was necessary if they were to keep their collieries going. Further, miners living in cottages belonging to their masters were often, in effect, bound to their jobs, recalcitrant miners could be disciplined by the mere threat of eviction and strikes were broken by the use of this weapon.

In James Spedding's time something was done to help miners and their families when they were in difficulties. Sir James Lowther, in Pennant's words,[112]

> "always keeps filled a great granary of oats, which he buys from all parts; but never disposes of any while the markets are low; but the moment they rise above five shillings the Cumberland bushel, or three Winchester measures, he instantly opens his stores to the poor colliers and artificers, and sells it to them at five shillings, notwithstanding it might have cost him seven.

There was also a fund for the widows and orphans of miners, the widows receiving 8s. 4d. per month and, if sufficient accommodation was available, they were allowed to live in colliers' almshouses rent-free, priority being given to widows of men with long service or who had been killed in the mines.[113] By 1803 this fund no longer existed and Bateman favoured its revival. In addition, he favoured the payment of a certain sum to every miner's widow for each of her children employed in the pits, not only as an act of charity, but also because it would encourage the mothers to keep their children in the mines.[114] There is no

* Hughes, op. cit., 183-4, 370, does not appear to find any element of "truck" in Curwen's system. The 1822 episode he attributes to a malicious party political manoeuvre to discredit Curwen during a heated controversy with Lowther over improvements to Workington harbour.

evidence to show that this plan was adopted, but after the Kells Pit explosion in 1819 the families of the twenty victims were allowed to remain in their homes rent-free and received 1s. 6d. per week from the fund built up by weekly deductions from the miners' wages.[115] Injured men who were unable to resume their former work were given light jobs that might be done by women or boys in order that they might earn something.*

By means of friendly societies J. C. Curwen tried to help his miners to help themselves when afflicted by illness or injury. Under his patrongage a Colliers' Society was instituted at Workington 1 January 1792, its members contributing 1s. per week and Curwen £3 for every £10 collected by the Society. Disabled members received 5s. per week for the first thirteen weeks, 3s. per week for the next thirteen weeks, then 2s. 6d. weekly for a similar period and finally 2s. per week until they were fit to resume work. There was a steady drain of the Society's funds; by 1 January 1796 £376. 13s. 9d. had been collected and £375. 18s. 4d. paid out, leaving a balance of only 15s. 5d. A similar society was founded at Harrington in 1793, Curwen again advancing £3 for every £10 collected by the society. Sick members received 6s. weekly for the first twelve weeks of their illness and 2s. per week thereafter until they recovered. On the death of a member £5 was paid to his widow or nearest relative. If a member worked twelve days for any other employer he became ineligible for any benefit.†[116]

At Whitehaven in 1801 the miners normally worked nine hours daily six days per week, although when necessary work was limited to eight hours and three shifts worked to maintain continuous production. To raise twenty waggons of coal per shift from a pit 100 fathoms deep required eight hewers, two persons to fill the baskets, eight boys to drive the horses from the workings to the bottom of the pit, a hooker to remove the empty basket and to hook the full basket to the winding rope, a man at the surface to empty the baskets and two boys to drive the gin horses.[117]

By this time the employment of females in the mines was normal practice in Cumberland and at Howgill there were 124, over a quarter of the total number of workpeople, their ages ranging from 9 to 81. Their occupations were:[118]

Occupation	*No.*	*Occupation*	*No.*
Filler	19	Slatepicker	18
Hooker	6	Gin driver	7
Attending way	18	Drivers jobbing underground	2
Bearmouth roader	5	Cart	4
Tram driver	24	Groom	2
Waggoner	15	Gate watcher	3
Tracer	1		

As the demand for coal increased after 1750 a shortage of adult male labour in Cumberland mines became apparent and this in turn led to the growing employment of women

* The Howgill payroll, 1802, lists two men, both aged 66, as "lame persons" who were each paid 5s. per week and their occupations are given as "does nothing". Presumably they had been injured in a colliery accident and their wages were some form of compensation.

† From time to time the rules of these societies were amended and alterations made in the amount of weekly subscriptions and rates of relief. The rules of the Workington Colliers' Society (1805) are reprinted in Hughes, op. cit., Appendix B.

and children in a variety of colliery jobs which had formerly been done by the men. The above statistics of women employed in 1802 at one Cumberland colliery disprove the statement made by a recent writer,

> "The employment of females underground in England was, however, exceptional before the eighteenth century. Even then, the numbers employed in Durham, Northumberland and Cumberland, where the demand for labour was greatest, were small, and the practice had died out in these counties by about 1780. In 1842 females were employed below ground only in Scotland, parts of Lancashire and Cheshire, the West Riding of Yorkshire, and South Wales."*

The Howgill Colliery payroll shows that many children below the age of fifteen were then working there. The youngest were nine years old and there were only three of them – one was a tracer, another a gatewatcher and the third a slatepicker. There were six gin drivers aged 10 to 14 and 48 tram drivers aged 11 to 14. Others under 15 were variously employed as waggoners, labourers, grooms and changers. Hewing was usually the work of adults but there were five haggers under 21, one was 15 and another 16. This payroll makes it clear that it was possible to work in and about collieries for many years and at an advanced age; there were eighteen septuagenarians and six octogenarians, some of whom had worked at Howgill for over fifty years. It would also appear that the Lowthers were prepared to reward loyal service and to find jobs for aged employees whose best working days were long past.

Working conditions in the Cumberland coalfield in the early nineteenth century were harsh. To one horrified writer who visited the mines there the workpeople seemed little better than slaves, prematurely aged, diseased, wretched and almost sub-human. At Whitehaven, he says,

> "All the people whom we met with were distinguished by an extraordinary wretchedness; immoderate labour and a noxious atmosphere had marked their countenances with the signs of disease and decay; they were mostly half naked, blackened all over with dirt, and altogether so miserably disfigured and abused, that they looked like a race fallen from the common rank of men."

One old man working underground was

> "naked down to his waist, and exposed a body lean and emaciated: his hair was grey and his face deeply furrowed and seamed with lines made by streams of sweat that had trickled down his blackened skin . . . . This man was considered a very fortunate person, for he had worked forty-two years in the mines and never met with an accident. Few of the miners had served a third of this time who could not show some marks of the dangers of their employment."

In the writer's opinion,

> "The people in the mines are looked upon as mere machinery, of no worth or importance beyond their *horse power*. The strength of a man is required in excavating the workings, women can drive the horses, and children can open the doors; and a child or a women is sacrificed, where a man is not required, as a matter of economy, that makes not the smallest account of human life in its calculations."[119]

* A. R. Griffin, "The British Coalmining Industry", (Buxton, 1977), 82-83. At Whitehaven Colliery twelve women were employed underground in 1842.

What Ayton found particularly repulsive at Whitehaven was the employment underground of women and children. They worked as hookers, fillers, tended the roads and the horses drawing baskets to the shafts were "always driven by girls, all of the same description, ragged and beastly in their appearance, and with a shameless indecency in their behaviour". As to the women generally, "They lose every quality that is graceful in woman, and became a set of coarse, licentious wretches, scorning all kind of restraint, and yielding themselves up, with shameless audacity, to the most detestable sensuality". Immorality in the pits was widespread and incest not unknown. Child trappers were "inhumanly sacrificed" with "as little consideration for the injury that they suffer as is felt for the hinges and pullies of the doors at which they attend".[120]

In other Cumberland collieries children from the age of six were employed as trappers and to carry lights for the men, for as long as thirteen hours per day. An indenture[121] of 25 June 1804 shows James Wilkinson, a boy of seven, being bound apprentice at Workington Colliery for fourteen years as a coal hewer and engine-man,

> "And whereas the said Apprentice by reason of his Tender Age will not be fit to be employed in the business of Coal Hewer and Engine Hand for some years it is agreed, that in the meantime, and until the said Apprentice is of fit Age and Strength to be employed in the aforesaid business; He is to be employed in attending Trap doors, Cleaning of Roads, Driving of Horses under ground, or any other offices, in any of the Coal Works of his said master, and the said Apprentice shall and will find and provide for himself, Meat, Drink, Washing and Lodging and also all proper and necessary Wearing Apparel during the said Term of fourteen years."

The boy's weekly wage for the first four years was to be 5s., 6s. for the next four years, then 7s. for three years and 8s. for the last three years. Even if one discounts some of the assertions of a writer naive enough to be shocked by the fact that miners working underground were dirty, sweaty and half naked, Ayton's plea, as early as 1814, for government intervention to prevent parents from exposing their children to the hardships of the mines was sound. Long hours, loneliness, darkness and discomfort took their toll of these children. Inevitably this servitude impaired their physical and mental health; exhausted bodies and empty minds meant that for many of them life was "nasty, brutish and short".

Coal mining then, as now, was one of the most dangerous occupations in the country. Every year men, women and children were killed or maimed in Cumberland collieries, the number of victims increasing as the industry expanded and more pits were opened out. Yet mining accidents, although frequent, aroused little comment and the deaths of miners seemed hardly worth recording. It is impossible to obtain complete statistics of mining fatalities before 1851. The *Cumberland Pacquet* did not begin publication until 1774 so that printed reports of mining accidents before that date are rare. The *Gentleman's Magazine* reports an accident, 22 May 1769, whereby, "Four colliers at work at a pit near Whitehaven, were all suffocated by the foul air of an old adjoining work. They have left 26 children and four widows unprovided for". After 1774 accidents which involved few casualties were not even mentioned in local newspapers and contemporary letter writers who were concerned with mining, if they mentioned accidents at all, usually referred to them very briefly. An explosion at Wilson Pit, Whitehaven killed seven people[122] and all the *Cumberland Pacquet*, 23 February 1779, had to say about this accident was restricted to one sentence, "Yesterday morning the foul air in a pit near this town took fire by which some lives were unfortunately lost". The use of naked lights which ignited gas could cause disastrous explosions – at Whitehaven eleven people were killed in 1806[123] and two in 1810 by explosions of fire-damp.[124] Fatal accidents often resulted from the men's own carelessness – carrying naked lights, contrary to orders, into dangerous areas of the mine, dislodging stone, the insecure fastening or mishandling of baskets, taking insufficient care as they moved in baskets up and down the pit shaft. Whitehaven miners, despite frequent warnings, used bits of old gin rope to provide illumination, even in the most hazardous workings and Bateman had to obtain from Lord Lowther a positive order prohibiting, on pain of dismissal, the use of lighted ropes in any part of the colliery and compelling the use of lanterns where open lights could not be carried.[125]

It was a great shock to Bateman to find that when a woman hooker at Wilson Pit had been killed by a piece of stone falling down the shaft to the pit bottom, the coroner had deemed it necessary to summon a jury to enquire into the cause of her death. He described this as

> "*a thing never practised here in my memory*, such enquiry being supposed only calculated to frighten the ignorant & discourage them from going into the Pits; on this account the workmen were always forbid to even talk abt any accidents which happen[d] in the Pits; your Lordship can judge better than me how far it may be proper to *check this new Practice*."[126]

Coroners' inquests into mining fatalities were customary in the Workington district and records of these inquests show that twenty-nine persons were killed at Workington Colliery between 1757 and 1815. Of these ten were killed by falling down pit shafts (three of whom were boys aged 7, 8 and 12 respectively), five by gas explosions and five by falls of roof. In the same period three persons were killed at Harrington, two at Clifton and four at Greysouthen.[127]

COSTS AND PROFITS

An estimate[128] of the cost of sinking a colliery at Bransty in 1773 gives the following figures:-

	£	s.	d.
Erecting a Fire Engine finding all materials	1,600	0	0
Sinking a Pit 10 ft. Diameter at £4 4s. per fathom	84	0	0
Sinking with an Eng. 60 fathoms at £12 12s. per	756	0	0
Keeping the engine at the time of sinking 80 weeks with Leather Oil Candles Sapping and Wages at £5 per	400	0	0
Coals consumed by the Fire Engine Black Smith, Fire Coal etc. at 50 Cumb. tons per week	666	13	4
Timber for Cribs, Deals for Brattish etc.	600	0	0
Nails	50	0	0
To building Smiths Shop Store House & workmen's houses	200	0	0
Carpenter work at the time of Sinking	200	0	0
Coal Gin with 1 pair of Sinking and 1 pair of Coal Ropes	70	0	0
Carts employed leading Coals Timber etc. 100 weeks at £2 per	200	0	0
Sinking gear & other gear	50	0	0
Viewer and Stock Keepers Salliery 2 years at 70 & 30	200	0	0
	5,076	13	4

The cost of working and leading a score of baskets (equal to six tons) at this colliery was estimated in this way:-

	s.	d.
Hewing 20 Corves, each Corve to contain 64 Ale Galls	2	6
Putting	1	8
Filling the corves underground		4
Drawing	1	6
Corving 5d. Setting on Corves & Banksman 6d. Smith Wright 2½	1	1½
Brattish & Barroway Deals 3d. Sledds props & shovels 2½		5½
Sinking & Drifting	1	6
Laying Barroway carrying in Air & other Contingencies		8
Drawing Water by Fire Engine	1	6
Ropes and gins 3d. & Staith charges 3d.		6
Sallarys		6
Leading by Carts to the Ships	3	0
Reparing the Cart Ways		1½
	15	4½

Since a score of baskets would sell for £1 there would be a profit of 4s. 7½d. or 9¼d. per ton. At Whingill Colliery it was calculated in 1799 that to raise 900 waggons of coal per week from the Lady, Davy and George Pits and lead them to the harbour would cost £195. 15s. 4d. in wages and £60 for materials. The sale of the coal would yield £480 and on this output there would be a profit of £224. 4s. 8d.[129]

Working costs increased as the mining industry developed. A colliery account book shows that the total average cost of getting coal and raising it to the surface at Thwaite Pit in 1757 was 1s. 9·8d. per ton.[130] In 1775 and 1776 the total average cost of getting coal and leading it to the ships from Howgill Colliery amounted to 2s. 2d. per ton. Since the coal was sold at 3s. 4d. there was a profit of 1s. 2d. per ton.[131] A detailed analysis of working costs at the different pits at Howgill in 1781 gives the expenditure involved in raising and delivering a score of baskets to the steath at the harbour:[132]

	Duke P.B.		*King P.B. S. Side*		*King P.B. N. Side*		*Kells P.B.*		*Croft P.B.*		*King B.B.*		*Kells B.B.*		*Croft B.B.*	
	s.	*d.*	*s.*	*d.*	*s.*	*d.*	*s.*	*d.*	*s.*	*d.*	*s.*	*d.*	*s.*	*d.*	*s.*	*d.*
Hagging	2	1	2	3½	2	1	2	1	2	3½	3	4	3	9	3	9
Filling		4		4¾		4½		4½		4		4¾		4¾		4
Leading	4	0	4	0	3	6	2	7¼	1	8¼	3	6	2	2	2	2
Door opening		¼		½		¼		¼		¼		½		½		½
Hooking		1¾		2		2¼		2		1¼		2½		2¼		2
Winding	1	0	1	2¼	1	2¼	1	2¼	1	2¼	1	2¼	1	2¼	1	2¼
Banking		3½		4¼		4¼		4		2¾		4¼		4		4
Waymen		2½		3		2		2		¾		2½		3		2
Slate pickers		2		1		1		1½		½		1		1½		1½
Laying upon ye Bank	8	3	8	9¼	7	11½	7	0¾	5	11½	9	3¾	8	5¼	8	3¼
Lead^{g} p Waggon to ye Steath		7½		10		10	1	3	1	5½		10	1	3	1	5½
Costs p Score at ye Steath	8	10½	9	7¼	8	9½	8	3¾	7	5	10	1¾	9	8¼	9	8¾

These are not total costs e.g. they do not include the costs of pumping, materials and maintenance. By 1800 the total average cost of raising and delivering coal to the harbour from Howgill had risen to 2s. $5\frac{1}{2}$d. per ton,[133] the higher wages paid during the war years contributing largely to the increase in cost.

Costs varied from one colliery to another, the large and well-equipped collieries with their thick seams producing coal at a lower cost than their rivals. At Harrington in 1783 costs amounted to 2s. $1\frac{3}{4}$d. per ton, the shipping price was 3s. 6d. per ton, leaving a profit of 1s. $4\frac{1}{4}$d.[134] Coal was raised at the small inland colliery at Bolton in 1773 at a cost of 3s. $10\frac{1}{2}$d. per ton and sold for 4s. leaving a profit of only $1\frac{1}{2}$d. per ton.[135] In 1810 Broughton Moor coal was obtained at a cost of 10s. 9d. per waggon and sold to the country for 14s., earning a profit of 3s. 3d. per waggon. But the same coal, shipped at Maryport for 16s. 8d. per waggon, showed a profit of only 1s. 8d. because cartage costs were so high. To move a waggon of coal from Broughton to a staith near Maryport cost 1s. 3d., cartage from there to the ships cost a further 2s. and the upkeep of the waggon-way amounted to 1s. per waggon.[136] The high cost of moving coal by land is shown by the fact that in 1808 a waggon of coal shipped from Whitehaven to Millom cost £1. 18s. 6d., but if moved to Millom by land, a distance of 27 miles, it would have cost £3. 3s.[137]

Excessive wayleave charges often burdened the colliery owner if he had to move his coal across land belonging to someone else. Surface owners if they sold their land to provide a way for the coal charged between twenty and forty times its agricultural value. The Lowthers bought land wherever possible in order to obtain unhampered access to their mines; others, to avoid extortionate wayleave rents, built new roads, often very much longer than they need have been had the surface owners been reasonable in their demands.[138]

Eighteenth century writers who visited Whitehaven Colliery were much impressed by its size and therefore tended to exaggerate its profits. Jars,[139] in 1765, estimated that the colliery had an annual profit of over £15,000 and Pennant,[140] who visited Whitehaven in 1772, believed that Sir James Lowther's income from his colliery and property rents at Whitehaven amounted to £16,000 per annum. That large profits were made is true, but the Lowther records show that these estimates were inflated. A detailed account book for the year ending Lady Day 1772 gives the profit and loss of all the Lowther collieries and the total profit is less than the estimated profit for Whitehaven Colliery alone. The figures given are briefly summarized:-[141]

	Profit			*Loss*		
	£	*s.*	*d.*		*s.*	*d.*
Howgill and Whingill Collieries	9,164	14	7			
Scalegill Colliery	139	8	5			
Parton Colliery	815	19	$7\frac{1}{4}$			
Dearham Colliery	75	5	$10\frac{3}{4}$			
Seaton Colliery				653	9	$7\frac{3}{4}$
Clifton Colliery	264	9	$9\frac{3}{4}$			

Total Profit £9,806 8s. 8d.

Profits rose steadily thereafter. In 1775 Howgill Colliery made a profit of £9,218. 0s. $11\frac{1}{4}$d.

and £9,913. 5s. 10d. in 1776. During the Napoleonic War the profits of Whitehaven Colliery increased sharply:-[142]

Year	*Profit* £	s.	d.	*Year*	*Profit* £	s.	d.
1803	24,882	4	4¾	1812	40,455	12	1
1804	26,657	9	8	1813	45,979	2	7
1805	22,663	13	5¼	1814	58,687	5	11¾
1806	31,611	1	7	1815	54,704	16	8¾
1807	26,889	18	10½				
1808	48,495	16	2¼				

Further north the coastal collieries were less successful than Lowther's. In the year beginning March 1791 Humphrey Senhouse's Ellenborough Colliery made a profit of £143 7s. 9¼d. on an output of 11,185 tons, or 3·08d. per ton.[143] In the following year he estimated his colliery profit at "just over sixpence a ton".[144] His neighbour, John Christian, claimed in 1759 that the profit from his Ewanrigg Colliery was then "not less than 2/- a ton".[145] After his death in 1767 his collieries at Ewanrigg and Broughton, during the minority of his son and heir, were managed by Charles Udale. In the ten years 1771-80 total profits from the Christian collieries amounted to £15,347, or an annual average profit of £1,534. At this time the income from land rents came to £1,122 per annum.[146]

Christian's son John (later John Christian Curwen) added to his patrimony by acquiring control of the Curwen estates following his marriage in 1782 to Isabella Curwen, daughter of Henry Curwen and heiress to the Curwen estates. When Curwen died in 1778 his collieries at Workington and Harrington were making a profit of £5,500 a year.[147] Under his son-in-law these collieries were extended and for some years there was a substantial increase in profits. At Harrington Colliery, for example, annual profits rose from £2,644 in 1784 to £7,177 in 1796.[148] The Curwen collieries, however, were much less profitable than those of the Lowthers. J. C. Curwen's net income for tax purposes in 1798 was £10,141.2s. 1d., of which £6,875.15s. 5d. came from his coal mines.[149] In 1803 it was said that he did not clear £10,000 a year from all his collieries,* when he should have been making nearly twice that sum, largely because he paid such high wages.[150] His colliery profits fluctuated considerably as can be seen from the following table showing the profits and losses between 1806 and 1813 at Moorbanks, Chapel Bank, Harrington and Broughton Collieries, and from the sale of ironstone.† Losses are underlined.[151]

	Moorbanks Colliery £	s.	d.	*Chapel Bank Colliery* £	s.	d.	*Harrington Colliery* £	s.	d.	*Broughton Colliery* £	s.	d.	*Ironstone* £	s.	d.	*Total Profit* £	s.	d.
1806	3,952	14	8	9,627	13	2½	6,466	14	3	640	13	7¾	201	0	4	20,888	16	1¼
1807	649	3	11½	5,912	14	9½	6,078	18	1¾	314	2	8	72	17	11	13,027	17	5¾
1808	618	12	8½	1,716	2	7	6,782	17	3½	845	4	9¼	247	15	10	8,973	7	9¼
1809	347	14	2½	40	18	1½	3,452	17	4½	824	4	3	12	9	7	3,982	15	1½
1810	835	12	7½	2,217	10	11	4,531	7	5	907	8	0	410	15	9	2,796	7	7½
1811	2,794	1	4	3,218	14	6	5,517	17	0				102	8	0	11,633	0	10
1812	1,499	9	3	308	18	2	4,496	14	1	1,500			43	15	0	6,871	11	2
1813	2,791	18	8	4,254	17	10	4,719	0	11½	1,000			56	15	2½	4,312	17	0

* His financial position was made worse in 1803 by the settlement of a dispute which had begun in 1799. Curwen was compelled to pay £7,368.5s. damages for working coal under the Cloffocks, Workington which belonged to Lord Lonsdale.

† There was also a loss of £271. 19s. at Schoose Colliery which has been taken into account when calculating the total profit for the year.

But these figures do not include the costs of sinking the new Isabella Pit and the amount of profits should be amended as follows:[152]

Year	Cost of Isabella Pit £	s.	d.	Total Profit £	s.	d.	Total Loss £	s.	d.
1807	265	8	4	12,762	9	1¾			
1808	5,774	6	9	3,199	1	0¼			
1809	5,841	2	11				1,858	7	9½
1810	4,837	8	10½				2,041	1	3

Matters were made worse by an explosion at Isabella Pit in November 1810 which caused a loss estimated at £3,000.

Lord Egremont, unlike the Lowthers, Curwens and Senhouses, did not work his own coal, but was content to act as a rent-receiver. Between 1756 and 1777 Broughton Colliery earned for him £7,013. 19s. 1d. on an output of 612,082 tons, this sum including both the fixed annual rent of £200 and the tentale rent on coals. In later years Egremont's rent from the Broughton Colliery continued to grow; between 1793 and 1809 his annual rent exceeded £1,000, reaching a maximum of £1,754 in the latter year.[153]

THE COAL TRADE

Since only a small proportion of their output could be sold locally the Cumberland collieries continued to export the bulk of their coal. Between 1757 and 1761 the following quantities were shipped from Whitehaven:[154]

Year	Tons	Loads	Year	Tons	Loads
1757	115,303	7	1760	98,100	0
1758	117,075	5	1761	123,310	5
1759	101,513	5			

An attempt was made to reduce the dependence upon the Irish market by opening up trade with London. Coal shipments from Whitehaven to the capital were, however, very small:[155]

Year	Tons	Year	Tons
1746	9	1758	84
1755	49	1760	156
1756	395	1764	234
1757	72		

Other plans to diversify the trade proved unsuccessful. John Spedding tried, and failed, to obtain contracts for the supply of coal to the naval dockyards at Plymouth and Portsmouth.[156] Similarly Sir James Lowther's efforts to develop coal exports to France and Spain came to nothing.

A little coal was sent outside the United Kingdom; the *Cumberland Pacquet* of 26 January 1775 shows that the coal trade was flourishing, coal going not only to Ireland, but also to

Mediterranean countries, Virginia and other places in America. A report in the Whitehaven Custom House Letter Books, 21 November 1778, refers to coal being shipped from that port by Mr Bacon, "for the Use of His Majesty's Forces in America". Coal shipments to countries outside the British Isles were always minute compared with the Irish trade; from the Port of Whitehaven between 1790 and 1798 they amounted to:[157]

Year	*Chaldrons (Newcastle measure)*
1790	156
1791	209
1792	206
1793	34
1794	344
1795	43
1796	85
1797	37
1798	25

The importance of the coal trade is illustrated by the fact that in 1777 out of the 260 vessels belongs to the port of Whitehaven, all, except for 30, were engaged in it. Exports were increasing, but the statement that Whitehaven by 1777 was exporting over 200,000 tons of coal per annum must refer to the "Port" of Whitehaven and include shipments from Workington, Harrington and Maryport as well as from Whitehaven.[158]

Although coal shipments from the Cumberland ports increased in the latter half of the eighteenth century the trade did not seem to prosper, largely because of mismanagement by the masters from the Cumberland ports who tried to undersell one another. At a meeting of coal traders in Whitehaven in November 1755 it was decided to revive the committee to regulate the coal trade and it was resolved that the sale of all coal carried from Cumberland to Dublin "shall be taken intirely out of the hands of the Masters of Ships, and be sold and disposed of by Factors appointed for that purpose under the directions of one common Agent". The agent, who would reside in Dublin and be paid £100 per annum, would control all the factors engaged in the sale of coals – twelve factors for Whitehaven and Parton coals and eight factors for coal carried from Workington and Maryport. These factors would be paid £50 per annum. Coal could only be sold by one or other of the factors from the place from whence the coal was exported and it had to be sold for ready money. Out of the proceeds of these sales 4d. per ton was to be deducted by the factors and paid into the hands of the agent.

The selling price of coal in Dublin was fixed. The minimum price for Whitehaven coal all the year round was 16s. per ton, south side Workington and Parton coal 15s. 6d. per ton in summer and 15s. in winter and north side Workington and Maryport coal 14s. 6d. throughout the year. The maximum price for Whitehaven coal was to be 20s. per ton and for other Cumberland coal the maximum price would be proportionate to the Whitehaven price. Any variation in price between 16s. and 20s. per ton would be determined by the agent and the majority of the masters then at Dublin at a meeting to be appointed for that purpose by the agent.

Masters trading from Cumberland who broke the rules would be debarred from loading at any of the ports in Cumberland for twelve months after the offence. All ships belonging to "strangers" who did not conform to the regulations would be permanently excluded from the Cumbrian ports; this prohibition would continue to be applied to the masters of these vessels if they obtained command of other ships. To enforce the trading regulations a committee of eleven persons was set up to deal with Whitehaven and Parton and a similar committee of eleven persons to deal with Workington and Maryport. Breaches of the regulations were to be reported to the coal owners who might be asked by the committee to deny offenders the opportunity to trade.[159]

Thomas Manning was appointed agent at Dublin in February 1756. His task was not easy. Within two months one Parton shipmaster was found to be selling coal at Dublin without a factor. The committee requested the Cumberland coal owners not to supply him with coal for twelve months. Two Whitehaven captains who had broken a recent agreement regulating sailors' wages by offering higher wages were, if the coal owners agreed, to be disciplined by being refused coal for the next six months.[160] This attempt to control the coal trade was no more successful than the attempt twenty years earlier. Price fixing stirred up much resentment in Dublin and Manning and twenty factors were fined and imprisoned in December 1756. The committee retaliated by trying to stop all coal shipments to Dublin. This was impossible and by February normal trading was resumed and the price of coal reduced to 18s. per ton.[161] It was clear that the committee to regulate the coal trade had failed. Spedding reported to Sir James Lowther that the committee

> "are so Embroyl'd that they must be forced to put the whole direction of the Coal Trade into your Hands, & not only allow, but desire you to lay down your own Rules for its government . . . . The Irish are most terribly provoked & the Danger is, their setting up the Scotch and Welsh Collierys against us, from whom we got the Trade, & we must take care to nourish & not oppress it . . . . The very name of a Committee for Governing the Coal Trade must be Extinguished . . . . I am glad to tell you that the Trade is quite open'd again, & that you may lay down yo[r] own Rules for the Government of it."[162]

Maryport made rapid progress after the Act to improve the harbour had been obtained. A "census" of Maryport taken on Christmas Eve 1765 gives its population as 1,169, "besides 11 Families, found not the names or number".[163] According to Pennant it had 1,300 inhabitants in 1772.[164] By 1792 a "census" of Maryport taken by the overseers James Wheelwright and Daniel Osborne shows that its population had risen to 3,019.[165] Although records of the first coal shipments from Maryport do not exist the growth of the harbour duties suggests that coal exports increased quickly:[166]

Year	£	*s.*	*d.*	*Year*	£	*s.*	*d.*
1750	22	2	10	1758	207	2	2
1751	77	1	10	1759	215	15	0
1752	110	2	4	1760	239	0	10
1753	141	10	10	1761	245	6	0
1754	169	3	4	1762	337	0	0
1755	189	2	5	1763	467	1	2
1756	135	13	6	1764	354	1	4
1757	141	19	0				

The increase in coal shipments after 1756 resulted largely from the opening out of Broughton Colliery which between 1757 and 1760 shipped in successive years 3,836, 8,101, 9,990 and 6,748 waggons of coal.[167] By 1756 about sixty vessels were constantly employed in the coal trade, each making between ten and twelve voyages to Ireland and

Scotland in the course of a year. Early in 1758 Spedding found that sales of Whitehaven coal to Dublin were much below the quantities raised at the colliery and in his opinion, "A great deal of this is oweing to the setting up of Ellenfoot, where the Masters gain so well by the Smuggleing Trade that they care not how much they lower the price of Coals when it helps them to make a greater Number of Voyages".[168]

Shipmasters at Maryport were handicapped by the necessity to make two separate journeys to the Whitehaven Customs House, fourteen miles away, every voyage – one to enter and one to clear out. To reduce the expense and loss of time caused by these formalities fifty people (coal proprietors, ship owners and shipmasters) in 1756 signed a petition to the Treasury asking for customs clearance at Workington instead of Whitehaven.[169] The petition must have failed because not until 1841 was Maryport freed from its dependence upon the port of Whitehaven.

Another petition from Maryport (3 August 1768),[170] addressed to Sir James Lowther and signed by forty-one inhabitants, complained that the coal trade there was being adversely affected by the bitter feud which then existed between Lowther and Henry Curwen. Not only were they political enemies but they also clashed over the control of the harbour at Workington where Lowther owned the north bank and Curwen the south bank. On the death of John Christian in 1767 Curwen became the guardian of his young nephew John Christian and he used this position in his campaign against Lowther. The Maryport shipmasters and others who had signed this petition alleged that Curwen refused them coal unless they paid south side tonnage duties at Workington on all coal loaded on the north side. His next step, they said, was to deny coal to any vessel at Maryport that had lately taken any of Lowther's coal at Workington. This he was able to do since he now controlled the property of the Christian family which included the collieries at Broughton and Ewanrigg whose coal was shipped at Maryport. Hence there was little work for the Maryport vessels if they came under this ban. They were not allowed coal at Maryport and at Workington they had to meet the competition of the larger Whitehaven vessels in loading north side coal, in addition to the inconvenience of having always to go to Workington to find a cargo. These petitioners therefore, were trying to persuade Lowther to open a colliery at Flimby or Dearham in order to provide employment for their ships.

How much damage was done to the coal trade at Maryport by Curwen's actions is uncertain; there are few statistics available about coal shipments from Maryport before 1780 to enable any judgement to be made. The estrangement between Curwen and the Maryport shipmasters certainly did not last as long as Curwen's quarrel with Lowther. By 1770 Curwen was trying to help the coal trade by using his influence to free seamen from impressment. In that year the Maryport traders were sufficiently busy to draw up an agreement regulating the wages of seamen employed in the Irish coal trade. At a meeting (30 October 1770) twenty-four masters and owners of ships signed the following agreement,

> "Whereas the advance Wages demanded at present by the Seamen are thought to be ruinous and distructive the Coal Trade, as being more than can be afforded when Coals are selling at low prices, But the owners and Masters being desirous to give all due encouragement by permitting their Wages to advance with, and be governed by the price of Coals It is therefore proposed and agreed to give the following Wages for the different Voyages as undermentioned (That is to say) To Dublin and the ports between Rathline Isle and Tuskar, Two Guineas or the average price of Three Tons of Coals for each of those Voyages, at the option of the Seamen; To Cork and other ports to the

Westward of Tuskar, Three guineas or the average price of Eighteen Barrels of Coals for the Voyage at the seamen's option; To Waterford in the same proportion with Cork."

Any owner or master of a vessel hiring seamen contrary to these terms would not be allowed to load coal for at least six months.[171]

In 1766 34,733 tons of coal were said to have been shipped from Maryport.[172] Coal shipments increased in later years – 17,255 waggons of coal in 1780, 15,548 waggons in 1781 and 21,202 waggons in 1786.[173] A proposal was made in 1794 to link the Solway Firth and the North Sea by means of a canal from Newcastle to Maryport which would facilitate the export of Tyneside coal to Ireland and the import of Irish linen and agricultural products. At the same time the Cumberland coalfied would have benefited and Maryport would have become a great port. Reports on the proposed canal were published in 1795 and 1796 by Ralph Dodd, William Chapman and William Jessop. The scheme had to be abandoned owing to differences about the proposed route of the canal and revivals of the scheme in 1807 and 1817 yielded no results.[174]

In the latter half of the eighteenth century Whitehaven coal had to meet competition from the coal exported from the new harbours at Maryport and Harrington in addition to the continuing competition from Workington coal. Maryport, Workington and Harrington, as their pits were opened out, joined in the trade on a large and growing scale. Custom House returns give these statistics of coal shipments from the Cumberland ports:[175]

Year	*Whitehaven*	*Workington, Maryport and Harrington*	*Total*
	Waggons	*Waggons*	*Waggons*
1781	55,600	55,128	110,728
1782	57,392	54,960	112,352
1783	61,136	58,384	119,520
1784	59,680	63,412	123,092
1785	72,688	70,576	143,264
1786	64,224	66,336	130,560
1787	73,805	76,664	150,469
1788	79,062	75,483	154,545
1789	75,633	74,440	150,073
1790	67,208	74,344	141,552
1791	54,650	83,556	138,206
1792	55,972	78,740	134,712

Thus by 1790 Whitehaven's coal shipments were exceeded by the combined shipments of the other three Cumberland ports. At this time Whitehaven's share of the Irish market was only 28 per cent; in 1740 Whitehaven had supplied 70 per cent of the coal used in Ireland.[176] At the end of the century Cumberland still supplied the bulk of the coal imported into Ireland, but its relative importance had declined. In 1799 coal exports from Cumberland were only 65 per cent of the total exports of coal from England to Ireland, a decline caused largely by the growth of exports from Swansea and Liverpool.[177]

During the nine years 1782-90 annual average coal shipments from Whitehaven amounted to 65,965 waggons. In the next twelve years, 1791-1802, they dropped to 48,782 waggons largely because of the incompetence of Thomas Wyley, the manager of the Whitehaven collieries. The more energetic policies of his successors John Bateman and John Peile led to a marked improvement after 1802; annual average coal exports between 1803 and 1811 increased to 82,623 waggons and rose to 110,193 waggons in 1814.[178] At Workington annual coal shipments between 1791 and 1807 exceeded 40,000 waggons or 120,000 tons; after 1807 there was a steady decline until 1815 when only 23,378 waggons of coal were shipped.[179] From the ports north of Whitehaven coal shipments between 1809 and 1813 averaged 63,000 waggons per annum – 28,000 waggons from Workington, 19,000 from Harrington and 16,000 from Maryport. Out of this total J. C. Curwen's annual average shipments came to 55,000 waggons from his collieries at Workington, Harrington and Broughton.[180]

At this time the coal traders were hampered not only by inefficient marketing methods but also by the exigencies of war. Vessels were liable to attack by enemy privateers or members of their crews seized by the press-gang. During the American War seventy-two Whitehaven vessels were lost by July 1781[181] and before the war was over almost a hundred had been sunk or captured. Many were victims of American raiding squadrons that lay in wait off Dublin and Whitehaven itself was attacked 23 April 1778 by the American warship *Ranger*, commanded by John Paul Jones who in his youth had been apprenticed to a Whitehaven merchant and whose local knowledge would facilitate the attack. Very little damage was done although the harbour was crowded with shipping – one ship was set on fire and the guns guarding the port were spiked.[182] The *Cumberland Magazine or Whitehaven Monthly Miscellany* for July 1780 says, "During this month, several privateers have made their appearance in the North channel, and captured a great number of vessels, among which were the following belonging to this and neighbourting ports."

Vessel	*Port*	*Capt.*	*Ransomed for*
Ann & Francis	Whitehaven	Whitfield	£400
Fortune	Maryport	Pegan	£500
William & John	Whitehaven	Ramsay	500 guineas
Ann & Francis	Maryport	Borrowskill	400 guineas
Vine	Workington	Martindale	£500
Olive Branch	Whitehaven	Powe	£400
Harmony	Workington	Keys	£500
Concord	Workington	Wilson	£500

During the war years therefore, many Cumbrian vessels were heavily armed to resist attack, the Seaton Ironworks supplying the cannon.* Local captains obtained letters of marque and combined normal trade with attacks on enemy shipping. From the reports and advertisements in the *Cumberland Pacquet* it is clear that some Whitehaven merchants engaged in privateering. Samuel Martin, a prominent tobacco merchant, commissioned six vessels which between them carried 68 guns and 231 men. For him the war was disastrous – he lost ten vessels and became bankrupt after his estates in Virginia had been confiscated by the government of that state.[183]

* Messrs Spedding & Co. of Workington were given permission, "to send coastwise from thence to Whitehaven in the County of Cumberland on board any Vessel or Vessels that may offer for that purpose, 60 Tons of Iron Cannon and 60 Tons Ball for the use of Vessels fitting out at Whitehaven". (Whitehaven Custom House Letter Books, Board to Collector, 1 January 1780.)

The leading colliery owners continued to hold shares in shipping but were never able to dominate the fleets of colliers as is shown by their disputes with the masters over coal prices, harbour facilities and port dues. Humphrey Senhouse bought a sixteenth share in ten vessels between 1782 and 1789.[184] By 1804 J. C. Curwen had bought shares in a hundred vessels, most of which were employed in the coal trade. His profits from ships were always a small proportion of his income; in 1797 they amounted to £594 2s. 3d. and to £303 16s. 6d. in 1798 out of a total taxable income of £10,141 2s. 1d.[185] The Lowthers had greatly reduced their investment in shipping. In 1771 Sir James Lowther held shares in only six vessels which in that year earned a profit of £391 8s.[186]

Coal prices at Dublin and elsewhere in Ireland could fluctuate quickly and widely in accordance with changing market conditions. But at the Cumberland ports the decisive factor was the control of supply by the coal owners who did not compete with each other in terms of price when selling to the shipmasters. It was the coal owners who fixed the price of their coal at the ports, and price increases often led to friction with the masters of vessels whose profits were often meagre enough because of their lack of organization and business skills. In the latter half of the eighteenth century the price of coal to the ships remained remarkably steady, but at the turn of the century the price rose sharply. A waggon of coal shipped from Whitehaven cost:[187]

Year		*Year*	
1742-1802	10s.–10s. 6d.	1804-1805	17s.
1802 21 Aug.	13s.	1812 23 Feb.	18s. 6d.
1 Nov.	14s.		
1803 9 Jan.	15s.		
4 Dec.	17s.		

Price increases in Cumberland were usually initiated at Whitehaven and were immediately followed at the other Cumbrian ports. Sometimes, however, it was Workington which set the example. At Workington, Harrington and Maryport, as well as Whitehaven, coal owners suffered from the widespread practice of shipmasters bribing waggoners and banksmen with money or liquor to provide excess measure when loading vessels. To combat this abuse Curwen decided to raise the price of his coal from 10s. to 10s. 6d. per waggon from 8 September 1792.[188] He was visited by a delegation of local shipmasters soon after the proposed increase in price was announced and after listening to their protests Curwen reluctantly consented to fix the price at 10s. 4d. per waggon.[189] In January 1803 Curwen again increased the price of his coal at Workington and Lord Lowther followed his lead, although, since he disliked Curwen intensely, he was most anxious that it should be made clear that he was not acting in concert with him.[190]

When John Bateman attempted to raise the price at Whitehaven to 18s. per waggon in December 1805 he aroused a storm of opposition.[191] For three months the irate shipmasters refused to take on coal at the new price. Bateman found himself the target not only of abuse and scurrilous letters but also of physical violence and had to be escorted by a bodyguard to and from his home. A settlement was reached which was a defeat for the colliery owner – the price remained at 17s. per waggon, but the waggon was now to contain 26 instead of 24 Cumberland bushels, with the result that Lord Lowther received 15s. 7d. for the same amount of coal that had formerly brought in

17s. According to Bateman, "The difference between getting 18/- per Waggon for Standard Measure and 17/- with the increase of Measure to be given will amount to about £7,000 per year".[192] An attempt in 1808 to raise the price to 18s. 6d. per waggon led to a stop of trade for fourteen weeks and Bateman had again to give way.[193] Three years later he was able to regain some of the lost ground by re-establishing 24 bushels as the content of a waggon.[194] In February 1812 Lonsdale and Curwen raised the price of coal from 17s. to 18s. 6d. per waggon. Again there was a buyers' strike but this time the colliery owners were successful, the shipmasters having to accept the new price which remained in force until 1816.[195]

Some information concerning profits earned by Cumbrian vessels in the coal trade is available in the files of the *Cumberland Pacquet*. In the 1770's the *Laurel*, a brigantine of 170 tons, earned £481. 8s. in nineteen months, showing a net profit of £302. 8s. and the *Success* earned £415 in ten coal voyages.[196] The *Maria* of Maryport earned a profit of £2,032. 8s. 3d. between 1793 and 1796, all of it from the coal trade apart from £725. 6s. 5d. gained in trade with Quebec.[197] It would be wrong to assume that vast profits were always possible in the coal trade; there was too much competition for that. Profits derived from coal went largely to the colliery proprietor rather than to the trader. Hutchinson[198] commented in 1794, "The profits to the shipping are very uncertain, no capital of such extent makes so poor a return; it barely affords a living profit to the industrious and the careful. The shares are divided into small parts, and are generally taken to oblige some individual, rather than with the view of much advantage". It is interesting to note that the *Postlethwaite* of Maryport, which made a profit of £5,547. 5s. 8d. between 1797 and 1804, derived very little of this from the coal trade with Ireland. As a prison ship in Belfast it earned £1,016. 3s. 7d. and £3,919. 8s. when employed in the transport service.[199] Evidently government service was much more remunerative than normal trade. During the Napoleonic War, however, when selling prices in Dublin were high, the coal traders were able to earn profits substantially greater than they were normally able to obtain. In 1802, according to Bateman, "Coal ships are selling at Dublin at about 40/- per Ton, the whole expence now I am told is 23/- per ton to Dublin ready to deliver to the buyers, so they will have about 17/- per ton for Freight which is very great".[200]

Notes and References

1 *Gentleman's Magazine*, lxii (1802), Pt. I, 587. D/Lons/W, Estate Books.

2 D/Lec, 79, Records of Bransty estate. Forster View Book of John Barnes. Report on Bransty Estate 1775 in *Reports of the late John Smeaton, F.R.S.* (1812), iii, 396-7. D/Lons/W, Estate Books.

3 Fletcher, *op. cit.*, 309-310. *VCH*, 363, G. Jars, *Voyages Métallurgiques* (Lyons, 1774), i, 238-41.

4 D/Lons/W, Bateman's Report on Collieries 1781.

5 *Cumberland Pacquet*, 2, 9, 15, 22 Feb., 27 Sept. 1791. Hutchinson, *op. cit.*, ii, 88.

6 D/Lons/W, Colliery Abstracts. Bateman, 26 Sept. 1803.

7 D/Lons/W, Colliery Abstracts. *VCH*, 364.

8 Bateman, 22, 29 Aug. 1802.

9 D/Lons/W, Lowther to Bateman, 25 Aug. 1802.

10 Bateman, 23 Aug. 1802.

11 D/Lons/W, Lowther to Bateman, 3 Feb. 1804; Colliery Abstracts.

12 Buddle Report on Whitehaven Colliery, 25 Nov. 1809. *VCH*, 363. *Cumberland Pacquet*, 1 April 1806.

13 D/Lons/W, Lonsdale-Bateman correspondence, Mar.-Aug. 1811. Bateman, 23 Aug. 1811.

14 D/Lec 240, Records of Cleator Moor Colliery.
15 D/Lons/W, Peile to Lords Lonsdale, 28 April 1834.
16 Whitehaven Library, MS 48, Workington Colliery Book 1799-1853.
17 D/Cu, 4/218, Banklands Colliery papers.
18 D/Cu, Bacon to John Christian, 7 Jan. 1783. D/Sen, Copy of Agreement between John Christian and Bacon.
19 Hutchinson, ii, 140.
20 D/Lons/W, James Spedding to John Robinson, 25 Mar. 1770, to Lowther, 1 April 1770.
21 D/Cu, 4/219, 6/7; Chapel Bank Colliery papers.
22 D/Cu, General Ledger 1757-64.
23 D/Cu, Misc. Colliery Papers.
24 C. R. Hudleston, "An 18th Century Squire's Possessions", CW2, lvii, 128-33.
25 D/Cu, 6/11, 6/12; Workington Colliery Accounts.
26 Buddle, vol. i, Report dated 18 Sept. 1822. *VCH*, 366.
27 Whitehaven Library, MS 48, Workington Colliery Book 1799-1853.
28 D/Cu, 6/11, 6/12; Harrington Colliery Account Book 1784-89; Harrington Coals Shipped Books 1788-1815.
29 *VCH*, 370. D/Lons/W, Bateman Report on Collieries 1781.
30 D/Lec, Report of John Barnes, 2 Oct. 1773.
31 I. Fletcher, *op. cit.*, 297.
32 *VCH*, 370. Fletcher, *op. cit.*, 301.
33 *VCH*, 369, 373.
34 D/Lec, 116, Dean Colliery papers. Forster View Book of John Barnes.
35 D/Lec, 116, Dean Colliery papers.
36 Carlisle Record Office, R. W. Moore, MS "History of Whitehaven Collieries". *VCH*, 370-1.
37 *VCH*, 375.
38 D/Lons, Estate Papers.
39 D/Sen, Lease of Ellenborough Colliery.
40 D/Sen, Ellenborough Colliery accounts.
41 D/Sen, Ellenborough Colliery records.
42 D/Sen, Ellenborough Colliery records. *VCH*, 375.
43 Forster, View Book of John Barnes.
44 D/Cu, Broughton Colliery Lease 12 July 1755.
45 D/Cu, Broughton Colliery records.
46 Forster, View Book of John Barnes.
47 Hughes, *op. cit.*, 141. D/Lec, 65/8, Copy of agreement.
48 D/Cu, 6/2, Broughton Colliery records.
49 D/Lec, 65/11, Forster View Book of John Barnes.
50 D/Cu, 6/2, Misc. Colliery Papers, Colliery Accounts drawn up by executor of estate and by Charles Udale.
51 D/Cu, 6/2, Broughton Colliery records.
52 D/Cu, Broughton Colliery lease 1781. Hughes, *op. cit.*, 154-61.
53 Hughes, *op. cit.*, 150-2. D/Lec, 65/9, Report of John Peile 1831. D/Cu, Colliery leases.
54 D/Lons/W, Rental 1771. *VCH*, 376.
55 D/Lons/W, Bateman report on collieries 1781; Report of Flimby Park Colliery 1803; Misc. Colliery Papers, "Mr Lysons' Queries"; Legal Papers, Netherhall, 30 Sept. 1791. Carlisle Record Office, T/NCB/1 – 24·5 "Plan of Flimby Woodside Estate and Colliery, March 1816". Hartley family papers held by Messrs. Yelloly & Burnett, solicitors of Carlisle.
56 Forster, View Book of John Barnes. *VCH*, 377.
57 D/Lec, 41, 59, Forster View Book of John Barnes.
58 "Bishop Nicolson's Diaries", CW2, iv, 6, 36. Forster View Book of John Barnes. D/Lec, 90.
59 Dixon, *op. cit.*, 99-101. D. and S. Lysons, *Magna Britannia* (1816), IV, cxxii. Hughes, *op. cit.*, 171. J. Fisher, "Observations & Inquiries made upon and concerning the coal works at Whitehaven in . . . 1793", *Annual Register*, xxxvi (1794), 326-33.
60 Hughes, *op. cit.*, 202-3. *VCH*, 353. I. Fletcher, *op. cit.*, 307. Bateman, 8 Dec. 1806. R. Ayton and W. Daniell, *A Voyage Round* Great Britain (1814-15), ii, 150-1. Dixon, *op. cit.*, 100.

[61] *VCH*, 353.
[62] *VCH*, 354. Galloway, *op. cit.*, i, 353.
[63] R. W. Moore, "Whitehaven Collieries", *Trans. of the Federated Institute of Mining Engineers*, vii (1893-94), 630.
[64] D/Cu, Agreement between John Christian and Messrs Boulton & Watt, 1 July 1788.
[65] D/Cu, Colliery Account Book; Boulton and Watt to Udale, 18 June 1789; Udale to Boulton and Watt, 6 June 1789. (Copy of letter).
[66] D/Cu, Boulton and Watt to Udale, 18 June 1789; Account of cost of Elizabeth Pit engine. Birmingham Assay Office, Udale to Boulton & Watt, 17 Nov. 1790.
[67] Birmingham Reference Library, Boulton and Watt MSS, "Memorandums of Mr Curwen's Engines". D/Cu, 3/43. *VCH*, 355. Buddle, Vol. i, 27 Dec. 1814. D/Lons/W, Misc. Colliery Papers, "Mr Lysons' Queries".
[68] D/Lons/W, Misc. Colliery Papers, "Mr Lysons' Queries". *Cumberland Pacquet*, 1 April 1806. Birmingham Reference Library, Boulton and Watt MSS, "Mr Murdock's Memoranda Respecting Lord Lowther's Collieries at Whitehaven 1802". Buddle, i, 27 Dec. 1814. Galloway, *op. cit.*, i, 355. Birmingham Assay Office, Matthew Boulton – William Chippindall correspondence Dec. 1790.
[69] Birmingham Reference Library, Boulton and Watt MSS, "Mr Murdock's Memoranda Respecting Lord Lowther's Collieries at Whitehaven 1802". D/Lons/W, Collieries, List 2, 31-9. Hutchinson, ii, 140. Bateman, 5 Sept. 1802, 25 May 1803. Buddle, Vol. i, 27 Dec. 1814.
[70] Bateman, 20 Dec. 1802, 19 Dec. 1803. D/Cu, 3/43, Swinburn to Curr, 7 Oct. 1812, 14 Jan. 1814, 30 Jan. 1815.
[71] D. and S. Lysons, *op. cit.*, IV, cxxii.
[72] Ayton and Daniell, *op. cit.*, ii, 151.
[73] *Carlisle Patriot*, 27 Sept. 1828.
[74] Bateman, 4 June 1804.
[75] Ayton and Daniell, *op. cit.*, ii, 164.
[76] Whitehaven Library, MS44, Howgill Colliery payroll.
[77] D/Cu, 6/23, Register of men leaving the works 1806-12.
[78] D/Lons/W, Misc. Colliery Papers.
[79] D/Cu, 3/66.
[80] D/Cu, Book of Edmund Bowness.
[81] Bateman, 22 Aug. 1802, 5 Sept. 1802, 7 Feb. 1803.
[82] Bateman, 13, 27 June 1803, 4 June 1804.
[83] D/Lons/W, Lowther to Bateman, 10 May 1804.
[84] Bateman, 20 Oct. 1810.
[85] D/Lons/W, J. B. Longmire and John Peile to Bateman, 22 Oct. 1810.
[86] D/Lons/W, Lonsdale to Bateman, 23 Oct. 1810.
[87] Bateman, 6 Feb., 19 Mar., 17 May, 20 Aug. 1804.
[88] Bateman, 7 Feb. 1803.
[89] Bateman, 23 Sept. 1805.
[90] Bateman, 21 Mar. 1803.
[91] Bateman, 8 Sept. 1802.
[92] Bateman, 9 May 1808
[93] Bateman, 15, 21 Nov. 1802.
[94] Bateman, 31 Jan. 1803.
[95] Bateman, 7 Feb. 1803.
[96] Bateman, 25 Mar. 1811.
[97] Bateman, 1 April 1811.
[98] D/Lons/W, Colliery Account Book 1757.
[99] Brown Carlisle Spedding to Brown, 11 Oct. 1752.
[100] Jars, *op. cit.*, i, 241.
[101] Bateman, 23 May, 1804.
[102] Whitehaven Library, MS 44.
[103] Hughes, *op. cit.*, 139-40.
[104] D/Cu, Harrington Colliery Accounts.

[105] D/Cu, Bill Book of E. Bowness.
[106] D/Lons/W, J. B. Longmire and J. Peile to Bateman, 22 Oct. 1810
[107] D/Cu, Harrington Colliery Accounts.
[108] Buddle, vol. i, 27 Dec. 1814.
[109] D/Cu, Misc. Colliery Papers.
[110] *Carlisle Patriot*, 22 June 1822.
[111] *Cumberland Pacquet*, 23 Apr. 1788.
[112] Pennant, *op. cit.*, ii, 50.
[113] Bateman, 5 Dec. 1803.
[114] Bateman, 11 April 1803.
[115] *The Courier*, 13 Nov. 1819, quoted in *The West Cumberland News*, 23 Sept. 1950.
[116] D/Cu, Records of Colliers' Societies in Curwen collieries. Sir F. M. Eden, *The State of the Poor* (London, 1797), ii, 78-104.
[117] Dixon, *op. cit.*, 100-1.
[118] Whitehaven Library, MS44, Howgill Colliery payroll 1802.
[119] Ayton and Daniell, ii, 155, 159.
[120] *Ibid.*, ii, 155-6, 160.
[121] D/Cu, Indenturers of apprenticeship.
[122] Whitehaven Library, Notebook of John Bragg.
[123] Galloway, *op. cit.*, i, 405.
[124] Bateman, 30 Apr. 1810.
[125] Bateman, 17 Jan. 1803.
[126] Bateman, 24 Oct. 1802.
[127] D/Lec, Coroners' reports.
[128] Forster, View Book of John Barnes.
[129] D/Lons/W, Collieries 32.
[130] D/Lons/W, Colliery account book 1757.
[131] D/Lons/W, Colliery Abstracts.
[132] D/Lons/W, Bateman's report on collieries 1781.
[133] Watson, Vol. 92, Howgill Colliery Account, 17 May 1800.
[134] D/Cu, Harrington Colliery Account.
[135] Forster, View Book of John Barnes, Report on the Earl of Egremont's Collieries, 2 Oct. 1773.
[136] D/Lons/W, J. B. Longmire and J. Peile to Bateman, 22 Oct. 1810.
[137] Bateman, 6 June 1808.
[138] Jars, *op. cit.*, i, 179.
[139] *Ibid.*, i, 238.
[140] Pennant, *op. cit.*, ii, 50.
[141] D/Lons/W, Rental 1771.
[142] D/Lons/W, Colliery Abstracts.
[143] D/Sen, Ellenborough Colliery Accounts.
[144] Hughes, *op. cit.*, 177.
[145] *Ibid.*, 142.
[146] D/Cu, Udale's Colliery Account.
[147] Hudleston, "An 18th Century Squire's possessions", 130.
[148] D/Cu, Harrington Colliery Account Books.
[149] D/Cu, 5/158.
[150] Bateman, 23 May 1810.
[151] D/Cu, 3/43 Letter Book of William Swinburn.
[152] *Ibid.*
[153] D/Lec, 65/6-7.
[154] D/Lons/W, Collieries 11.
[155] *Report of Royal Commission on Coal in the United Kingdom* (1871), iii, 14.
[156] D/Lons/W, Letters, John Spedding to Sir James Lowther, 1 May 1757.
[157] *Report of Select Committee of the House of Commons into the Coal Trade* (1800), Appendix 42.

[158] J. Nicolson and R. Burn, *The History and Antiquities of the Counties of Westmorland and Cumberland* (1777), ii, 43.
[159] D/Lons/W, "Committee to Regulate the Coal Trade".
[160] *Ibid.*, D/Lons/W, Letters, John Spedding to Sir James Lowther, 9, 13 March 1757.
[161] D/Lons/W, Letters, John Spedding to Sir James Lowther, 12 Dec. 1756, 5, 26 Jan., 6 Feb 1757.
[162] D/Lons/W, Letters John Spedding to Sir James Lowther, 4 Feb. 1757.
[163] D/Ben, Broughton and Unerigg colliery book.
[164] Pennant, *op. cit.*, 52.
[165] Maryport Town Hall records.
[166] D/Cu, "Account of Harbour Duties at Ellenfoot since first Account".
[167] D/Sen, Broughton Colliery, Abstracts of Quarterly Accounts.
[168] D/Lons/W, Letters, John Spedding to Sir James Lowther, 12 Feb. 1758.
[169] D/Lons/W, Petition to Treasury 1756 concerning Ellenfoot coal trade.
[170] D/Sen, Harbour 12.
[171] D/Cu, 6/120, Agreement to regulate seamen's wages in the Irish coal trade.
[172] *Cumberland Pacquet*, 27 Oct. 1857 reporting a speech of H. Senhouse.
[173] Hughes, *op. cit.*, 162.
[174] J. Simmons, *Maryport and Carlisle Railway* (1947), 3.
[175] Quoted by Hutchinson, ii, 84.
[176] W. H. Makey, "The Place of Whitehaven in the Irish Coal Trade 1600-1750" (London University M.A. thesis, 1952), 258.
[177] Cullen, *op. cit.*, 84.
[178] D/Lons/W, John Peile's "Statement of the Export price and quantity of Whitehaven's Coal", Dec. 1842.
[179] D/Cu, Tonnage Day Books; Coal Shipments from Workington Books.
[180] D/Lons/W, Misc. Colliery Papers, "Mr Lysons' Queries".
[181] *Cumberland Pacquet*, 10 July 1781.
[182] Whitehaven Custom House Letter Books, 23 Apr. 1778. *Lloyd's Evening Post*, April 1778.
[183] Whitehaven Library, MS book entitled "Minutes of Proceedings had on behalf of the family of Martin".
[184] D/Sen, Shipping 8.
[185] D/Cu, 5/158, General Accounts and Ledgers.
[186] D/Lons/W, Rental, 1771.
[187] Forster View Book of Matthias Dunn. Bateman letters.
[188] D/Cu, Copy of Curwen's "Memorial" for raising the price of Coals.
[189] Hughes, *op. cit.*, 196-7.
[190] D/Lons/W, Lord Lowther to Bateman, 7 Jan. 1803.
[191] Bateman, 16 Dec. 1805.
[192] Bateman, 24 March 1806.
[193] D/Lons/W. Peile's "Statement of the Export price and quantity of Whitehaven's Coal".
[194] Bateman. 20 May 1811.
[195] D/Cu. 3/43 Letter Book of William Swinburn, Swinburn to Thomas Harriston, 4 May 1812, to William Renwick, 9 March 1812.
[196] *Cumberland Pacquet*, 8 Dec. 1774, 26 Jan. 1775.
[197] Information providced by the late Dr J. W. Crerar of Maryport.
[198] Hutchinson, ii, 140.
[199] Information provided by the late Dr J. W. Crerar
[200] Bateman, 5 Dec. 1802.

CHAPTER IV

1816-1850

WHITEHAVEN DISTRICT

The summary and enforced departure of John Bateman and his nephew J. B. Longmire in 1811 left the way open for John Peile, the viewer at Whingill Colliery, to take over as general manager of the Whitehaven Collieries. He had begun his career at Whitehaven in 1791 as a boy of fifteen and remained there until his retirement in 1847. During his period as manager, thanks to his energy and initiative, considerable progress was made. He sank Wreah and Castlerigg pits at Whingill, Countess Pit at Parton and at Croft, James and Saltom pits deepened the shafts from the Main Band to win coal from the Six Quarters seam. He also sank a new shaft at Duke Pit from the surface to the Six Quarters seam at 132 fathoms.[1] His greatest mining achievement was the sinking of Wellington Pit, begun in 1838, which reached the Main Band in 1843 and the Six Quarters seam in 1845. The cost of winning this pit until the end of 1847 amounted to £25,000. Coal was first raised from the Main Band 20 April 1843 and the pit soon proved highly productive and profitable. Between 1843 and 1847 the yearly output grew from 9,250 waggons to 34,630 waggons and annual profits from £1,117 to £11,687.[2] Unfortunately, an underground fire in 1847 proved very difficult to control and for months the workings near the area of the fire had to be sealed off.[3]

After being "totally lost" in 1776 work was resumed at Scalegill Colliery in 1806 when Bateman sank a new pit, Henry Pit, and installed a 30 h.p. atmospheric winding engine. This pit soon repaid the cost of winning and opening it out for by the end of 1806 this "little infant colliery" was making a profit of £30-£40 per week. Peile continued working the colliery which in 1814 employed twenty-six workpeople and made a profit of £2,052.[4]

Peile was responsible for several innovations – the construction of the Howgill surface incline to the Howgill staith (1813), the introduction of screens (1839) and the manufacture of coke and patent fuel at the Wellington and William pits; but the demand for coke was inadequate and the ovens were closed down soon after his retirement.[5]

Some idea of the work done in Peile's time and of the costs involved can be gained from his own "Rough Estimate of Extra Outlay at Whitehaven Colliery under John Peile as Agent from 1812 to 1841":[6]

				£
Howgill Staith	1813	Incline plane	say	500
	1839	Four new Iron Spouts &c.		2500
Croft Pit	1818	Second winning in sinking		2000
„ „		Yards Drifting		3000

Saltom	1819	Second Winning Sinking		2400
,,		,, ,, Drifting		3600
Duke Pit		New Shaft to the Six Quarters 132 fathoms		2500
,, ,,		Walling old shaft		500
Kells Pit		Upcast cased with Masonry		1000
New Scalegill		Sinking two Shafts	say	1000
Wellington		On account up to this time		20000
Parton Winning		Sinkings, Driftings, Roads		20000
James Pit		Walling Shaft &c.		1000
Wreah Pit		Winning &c.		500
Distington		Bottom Bank Dyon &c.		1000
Steam Engines		Nineteen new ones 600 h.p. @ £30		18000
Railways		Under and above ground	say	20000
Granary		Yard Sawmill &c.		5000
Bowthorn Colliery		Seven years rent at 200		1400
,, ,,		Boring		300
Patent Slip		Expenditure in erecting		2500
Saint Bees		School rent paid up to Midsummer Quarter 1841		22630
Rates Poor &c.		On Collieries since 1820, no		
£500 per annum		charge being made before		11000
		St Bees Colliery		500
		Bolton Bridge, Westmoreland		800
		Whitehaven Library		700
Rosehill Colliery		1841 Rent to G. W. Hartley Esq. this year	say at	200
				£144,530
		say £5000 per year		

It was during Peile's tenure of office that the Lowther family were compelled to rectify the injustice done to St Bees School. It will be recalled that in 1742 the school governors had leased to Sir James Lowther all their coal in the manor of St Bees for 867 years at an annual rent of £3 10s. In 1819 the Charity Commissioners inquired into this lease and their report (3 October 1819) led the Attorney General (26 February 1821) to initiate proceedings in Chancery to set aside the lease on the ground that Sir James Lowther and his steward were governors of the school when the lease was granted. The terms of the lease were obviously unfair to the school, for in 1742 a reasonable judgement could have been formed of the depth, quantity and quality of the coal. Further, the lease was granted for an unprecedented term, the annual rent was minimal and no tonnage rent was charged.

In 1825 Lord Lonsdale, who had inherited the lease, agreed to give up the old lease and to take a new one for a term of 42 years at a fixed rent of £300 per annum and a tonnage rent, to be determined by three arbitrators, was also to be paid. In addition, Lord Lonsdale had to pay tonnage rent on coal raised out of the school lands since the filing of the information, the rent to be the same as that fixed by the arbitrators. The three arbitrators, Matthias Dunn, John Buddle and Nicholas Wood, had thus to determine the amount of tonnage rent payable by Lord Lonsdale for coal raised between 25 February 1821 and 23 December 1826, when the new lease would take effect, and for coal raised since the latter date. Their investigations showed that a total of 679,794 tons had been raised out of the school royalty by October 1819 and that 111,750 waggons of 44 cwts. each had been raised between 26 February 1821 and 23 December 1826. The arbitrators, in a decision signed 3 June 1828, fixed the tonnage rent as one-eleventh of the selling price of Main Band coal and one-sixteenth of the selling price of Bannock Band coal. The case was finally settled by the payment of £13,280. 13s. 5d. to the governors of the school, out of which the expenses of the lawsuit had to be defrayed.[7]

Records of the output of coal in Cumberland before 1854 are scanty; in that year Robert Hunt of the Mining Record Office began the publication of mineral statistics, but these are not wholly reliable since they were based upon voluntary returns and estimates. The difficulty in assessing the output is further increased by the fact that before 1836 coal was sold by measure and not by weight – sometimes by the chaldron, and, more generally, by the waggon. Annual output figures for Whitehaven Colliery are not available for much of this period, but Peile has provided the following average output statistics:[8]

Years	*Howgill Colliery Annual average output Tons*	*Whingill Colliery Annual average output Tons*
1812-1818	116,739	109,886
1819-1825	114,191	107,205
1826-1832	127,041	89,090
1833-1839	117,572	97,782
1840-1842	132,182	73,095

From 1842 to 1847 the annual output of the two divisions of Whitehaven Colliery amounted to:[9]

Year	*Howgill Waggons*	*Whingill Waggons*
1842	56,908	52,042
1843	46,808	48,520
1844	54,349	53,302
1845	59,597	51,863
1846	65,114	53,223
1847	63,136	52,425

In 1827 Peile began the Parton Drift which was intended to drain a large tract of inland coal and in association with it began sinking Countess Pit at Parton in 1832. Development at Parton proved expensive; it cost about £20,000 to win Countess Pit and coal was not raised from it until the beginning of 1836. This pit was sunk to a depth of ninety-one fathoms, but the coal was only drawn up to the level of the Parton Drift, which was twenty-five fathoms deep at that point, and then conveyed to Parton Pit where it was raised ten fathoms to the surface. Although situated in a dislocated coalfield which was costly to work the Countess Pit from 1836 yielded satisfactory profits. In the eleven years 1836-46 annual average sales from the pit amounted to 24,220 waggons and the annual average profit to £6,434. 12s.[10]

Between 1841 and 1849 Peile was working coal at Rosehill Colliery, Moresby in a royalty leased from G. W. Hartley, the rent being one-tenth of the value of the coal raised.[11] Nearby he began sinking another pit, Moresby New Pit, in 1846 and, as at Countess Pit, he used the Parton Drift as an outlet for its coal. Moresby New Pit was abandoned in 1850. Countess Pit ceased production in 1863 and Parton Drift was given up in the same year.[12]

After a lapse of twenty years coal mining in Cleator Moor was resumed when Richard Barker and George Harrison took over an Egremont royalty and raised 1,754 tons of coal in the year ending 26 September 1836. They began sinking Whinny Hill Pit in 1839 and found Main Band coal, 9 feet thick, at a depth of 114 fathoms in 1843. They leased the colliery for a term of twenty-one years from 29 September 1841 but it was sub-leased in 1847 by the Whitehaven Hematite Co. to provide fuel for their Cleator Iron Works. The colliery produced 11,462 tons of coal in 1846 and in three successive years 1848-50 14,614, 8,022 and 6,236 tons.[13]

A new colliery at Cleator Moor began in 1846 when the Threapthwaite Coal Co. (Messrs Ainsworth & Hartley) sank Threapthwaite Colliery in a Lonsdale royalty. Here the Bannock and Main Band seams were worked until 1872.[14]

WORKINGTON DISTRICT

By 1816 Curwen had reduced the number of his pits at Workington to four – Lady, Union, Isabella and Church pits. He did try to sink more at Chapel Bank. Drainage, however, was a major difficulty; sinking of four pits had to be abandoned owing to a large feeder of salt water, the pumps being quite ineffective to remove it. Between 1818 and 1820 £2,855. 15s. 11d. was spent on sinking the new Buddle Pit, but to no avail, and not until 1837 did this pit draw coal. At this time pumps were kept going most of the day at the four working pits, and since Church Pit was approaching exhaustion pillar robbing was being undertaken there in order to remove the maximum amount of coal before the pit was abandoned. In March 1821 Church Pit was drowned out by a fall of roof bringing down a feeder of water and in the following year Union Pit ceased production because the drainage problem proved insoluble.[15]

Workington Colliery was now entering upon a prolonged period of decline from which it never recovered. Mining became more difficult and expensive, coal output was declining and matters were made worse by the post-war depression in the coal trade. Workington Colliery had shown a profit of £13,580 in 1806;[16] between 1818 and 1820 the annual average profit was only £1,072.[17] Curwen's investment in the colliery had left him heavily in debt – from 1819 to 1827 he owed over £120,000.[18] It was essential to increase output and reduce costs. John Buddle, the leading mining engineer of the time, was appointed in 1814 to advise on the management of the Curwen collieries in the hope that by his skill he could convert them into a prosperous undertaking. His many reports, covering the period 1814-1837, give a detailed description of their working and of the defects from which they suffered. He found that the Workington Colliery was not being efficiently administered and the daily output of 165 tons from four pits (August 1816) he regarded as much too small. He recommended:[19]

1. The suspension of work at Lady Pit, since it was the most expensive, and more work to be done at Union and Church Pits.

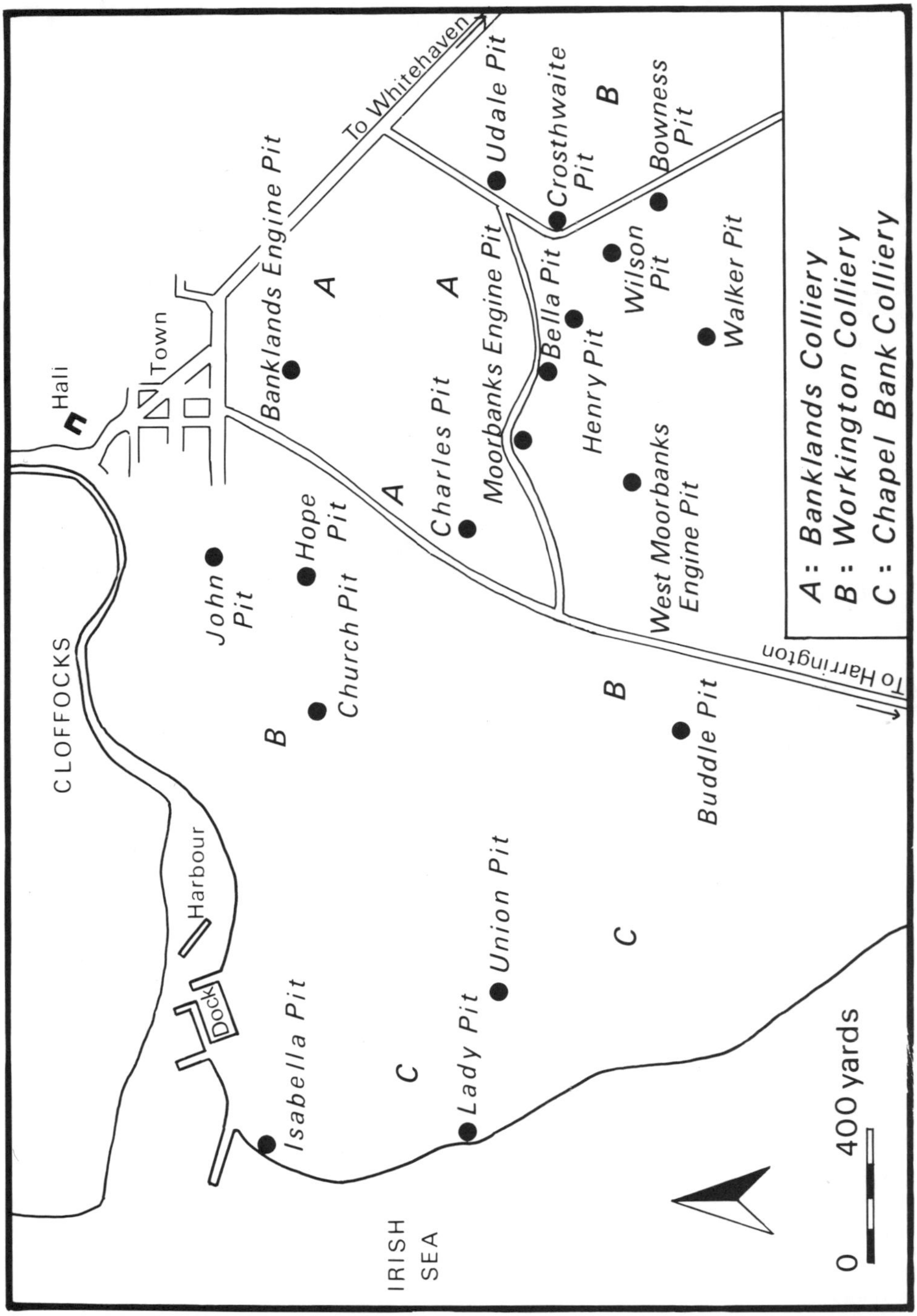

FIG. 3. – Workington Pits 1836

2. A reduction in the size of pillars at the Union Pit to eighteen yards long by eight yards broad which would increase output by 12% and still leave sufficiently strong pillars at a depth of thirty fathoms.

3. The sinking of Isabella Pit to the Main Band – estimated to take thirty weeks and to cost £1,370.

4. Church Pit should be allowed to drown up when the coal is exhausted in order to reduce pumping costs and to liberate the engine for use elsewhere.

5. After exhausting the Main Band at Union Pit the shaft there should be sunk deeper to win coal in the Four Foot seam.

6. Ventilation at Moorbanks should be greatly improved and at Union Pit a small furnace placed at the bottom of the pit should replace the lamp then in use.

His later reports criticised the poor condition of the rolley-way, the steepness of the descent into the stear, which distressed the horses drawing out full baskets, and the filthy condition of the stables. Some of his suggestions were acted upon. Union Pit was ventilated by a furnace instead of a lamp, the stables were cleaned up and by 1821 45·8% of the coal was being taken out of Union Pit where formerly 32·7% of the coal had been extracted. Pillar robbing followed at the other pits – by January 1821 the entire output of Church Pit came from the pillars; at Lady Pit most of the coal and at Isabella Pit half of the coal obtained was the result of pillar working.[20]

Workington Colliery produced 40,707 tons of coal in 1816 and 39,670 tons in 1829.[21] To increase output pillar robbing was the easiest and cheapest way and provided care was taken to ensure that the pillars remained strong enough to bear the weight of the roof there was no danger. Unfortunately, as events were to show, this care was not always taken and the continuation of pillar stripping as the working approached the sea floor led to catastrophe. On 8 November 1833 four men were drowned in Isabella Pit. The *Cumberland Pacquet*, 12 November 1833, says of this accident,

> "It appears that the workings in the Isabella Pit have lately been brought to the vicinity of the surface, and that near the spot where the colliers have lately been excavating, there was sixty years ago some old workings, which had been abandoned, as tradition reports, from the sea breaking into them. For some time, as we have been informed, the water was seen to ooze out of the coal as the workmen approached the old pit, but unfortunately those symptoms were not sufficiently attended to, and the consequence was that about 10 o'clock on Friday morning last they had approached so near the old workings south of Lady pit, that the water broke in upon the men employed in the Moorbank seam. The water rushed with astonishing velocity down to what is called the Low Bottom."

These deaths aroused considerable anger among the workmen. According to a contemporary writer, "The overman of the Isabella Pit has been dismissed at the request of the men who went in a body to Mr Curwen to demand it but the fault cannot properly rest on the poor man who simply obeyed orders at the same time informing his superiors of the danger".[22]

This accident was followed by the appointment of Matthias Dunn, an eminent Newcastle mining engineer, as director of the Curwen collieries on a three year contract from 1 January 1834 at a salary of £200 per annum plus expenses.[23] But the actual control of the pits was in the hands of a manager appointed by Dunn since the latter's function was essentially advisory. The first of these managers was Robert Rankin, formerly overman at Saltom, who proved unsatisfactory and was succeded by Ralph Coxon. It is clear that Coxon was under pressure to obtain the maximum output of coal and in his anxiety to please his employer took unnecessary risks which endangered the lives of the colliers. Heedless of the warning afforded by the 1833 accident the working of coal to the rise continued, little attention being paid to the large quantity of water making its way into the colliery. This was well known outside Workington. A letter, 12 September 1836, reported,[24]

> "It appears Mr Curwen's Colliery is under a considerable difficulty from the discharge of Water by working a Rise Barrier of Coal approaching near the Gravel & Surface – and being now above the power of the Pumping Engine – the Low Colliery is at present filling and they have been compelled to take the Horses out in that part. What may be the result is uncertain, if the Feeder is from the Gravel – and the Gravel connected with the Sea & particularly in Spring Tides which covers the land in this Quarter occasionally – it may be serious to the future working of the Colliery. Mr Dunn persevering in working this area of coal with a constant increase of water as they proceeded, has been the surprise of all acquainted with it."

By this time Dunn and Coxon were at loggerheads over the manner in which the colliery should be run. Mr Curwen however, delighted with the increased output at a time when coal prices were rising, supported Coxon and refused to renew Dunn's contract. He wrote to Dunn, 12 January 1837, "I hasten to state in answer to your letter of the 9th that Coxon informed me, the first time I saw him after your last visit, that if you continued to have the management of the collieries, he would not remain under you, he would rather break stones in the road, than do so, after what had passed between you. As I have every reason to be satisfied with his management as an Agent I had no alternative but to comply with the terms on which alone he would remain here, and this was the cause of my not renewing your engagement".[25] Coxon, anxious to get out as much coal as possible, not only persisted in working the coal to the rise but also began to cut into the pillars of coal supporting the roof. This was followed by falls of roof and incursions of sea water. Experienced miners, conscious of the danger, frequently warned Coxon of possible disaster and one under-agent at the colliery wrote to Dunn, "Unless some interference can be made, a very few days or weeks will most assuredly bring down the waters of the sea; and this opinion is now so generally expressed that men are leaving the colliery every day".[26] Higher wages had to be paid to encourage the men to continue working at Chapel Bank.[27]

At this time the principal workings at Lady Pit extended 1,500 yards under the sea and the Camperdown district, where most of the men were working, sloped towards the surface at the rate of one in three until the workings were only fifteen fathoms below the sea floor, only four fathoms of which were freestone and the rest loose sand and gravel. Disregarding the warnings frequently given to him Coxon carried on stripping the pillars. On Friday night, 28 July 1837, the sea broke through the roof of Lady Pit and in ninety minutes Lady, Union and Isabella pits, which were inter-connected, were flooded. This caused the deaths of twenty-seven people and twenty-eight horses and the loss of the colliery; but for the fact that it was the time for changing shifts the death roll would have been much higher. As it happened there were only fifty-seven men and boys in the mine at the time of the disaster and thirty of these escaped by making their way to the bearmouth.[28]

Coxon was immediately dismissed and was fortunate to escape being lynched by the infuriated people of Workington. This was an unnecessary and avoidable calamity. Throughout its history the colliery was under threat from the sea and the utmost care should have been taken at all times in working it. The 1833 accident afforded ample warning of the danger but this was disregarded and with criminal folly a most hazardous mode of working the colliery was pursued. The final tragedy was all the more deplorable since before it occurred it was a widely known probability and wiser men that Coxon had repeatedly cautioned him. According to Edward Stanley, agent to the Curwen family, "Coxon in my opinion was blamable but it was Mr Dunn who ordered him to take the Union Pit coal which let in such a deluge of Water that it drowned out the low bottom – Coxon had no resource then but to take coal from the Lady Pit for the purpose of keeping up the quanity, now he was blamable in not telling that he could only do it by losing the colliery".[29]

The loss of Chapel Bank Colliery threw some 300 men and boys out of work. Many of these were re-employed at Harrington Colliery and others were set to work sinking Buddle Pit which was begun in 1818 but had been abandoned for coal nearer the shore.[30] An attempt was made to re-open Moorbanks Engine Pit but was abandoned in 1840. The sinking of Jane Pit was begun in 1843, a task seriously hampered by the presence of water and gas; at sixty-five fathoms a blower of gas was encountered which was conducted by pipes to the surface where it burned for over two years.[31] Curwen's dissatisfaction is reflected in his vain attempt to let the colliery in 1845.[32] However, he continued with Jane Pit and in 1846 the Hamilton seam was reached at seventy-three fathoms, and another pit, Jackson pit, was also sunk. By 1847 the Workington Colliery was producing 20-22,000 tons of coal per annum.[33]

In the Harrington area an attempt was made in 1825 to win coal from the seashore by sinking the Micklam or Lowca pit to a depth of ninety fathoms. The attempt was unsuccessful owing to the volume of sea water which penetrated into the workings.[34] A report on Harrington Colliery in 1832 shows that it was then in a neglected state – the workings were old and in bad condition, the waggons and the wooden waggon-way, which ran to Harrington harbour nearly two miles away, were dilapidated and the harbour itself in need of considerable improvement. Although the colliery contained much coal it was of inferior quality.[35] By 1847 only two pits, the John and Hodgson Pits, were working, producing 25-30,000 tons per annum.[36]

Mining was resumed at Clifton in 1803 when Lord Lonsdale began opening out a colliery on the west side of Lostrigg Beck which he worked until 1815. Between 1811 and 1814 the annual output fell from 5,390 tons to 1,923 tons and the profit from £491. 5s. 10d. to £182. 17s. 6d. The colliery was leased by John Johnson from 1815 to 1822 and then by John Fletcher until its abandonment in 1829. In 1827 Thomas Westray sank Westray Pit in Cookson's royalty at Great Clifton, a royalty acquired by Lord Lonsdale in 1842 and worked by him until 1856. A total profit of £11,149. 5s. 6d. was earned by the colliery in the five years 1843-47.[37]

Greysouthen was producing some 10,000 waggons of coal annually in 1816.[38] Joseph

Harris & Co. worked a number of pits there – Harris, Sandriggs, Whitecrofts, Town End and Lamb Hill pits – with the following results:–[39]

Year	*Output*	*Profit*			*Loss*		
	Tons	£	s.	d.	£	s.	d.
1823		491	17	8			
1824	6,108	67	0	10¼			
1825	6,477				3	15	7¼
1826	13,071	494	13	5¾			
1827	11,409	244	1	4			
1828	14,241	1,093	5	1¼			

This firm opened out new workings at Nepgill (1837) and at John Pit (1838) which together comprised the Millbanks Colliery situated near Bridgefoot.[40] Joseph Birbeck and J. W. Fletcher began operations in Greysouthen in 1823 when they leased a Curwen royalty for nine years at a yearly rent of £20 and 1s. per ton above an annual output of 400 tons.[41] They also leased a royalty belonging to Lord Egremont where they worked the George and Hope pits; this royalty sold 10,348 tons of coal in 1832 and 9,189 tons in 1833. Mary Pit, sunk by Messrs Birbeck & Fletcher, was producing about 100 waggons of coal per week in 1842.[42]

Other small pits were in production in the Workington area. Between 1813 and 1824 J. C. Curwen was working coal at Branthwaite.[43] Dean Colliery was leased by Anthony Wild, a coal miner, and Henry James Johnston, a grocer, for fourteen years from 25 March 1816 at a yearly rent of 6d. per ton of coal up to 900 tons and 9d. per ton above 900 tons. The lease was renewed to Wild's widow and Johnston in 1834 for a further fourteen years at an annual rent of £37. 10s. and 9d. per ton above 1,000 tons. For most of the years 1819-48 this colliery's output was below 3,000 tons per annum and only in 1833 and 1834 did it produce more than 5,000 tons.[44]

In Distington Lord Lonsdale continued to work the Yard Band and Six Quarters seams at Bottom Bank colliery until April 1845 when the coal was exhausted and the pit laid in. He also sank Dyon Pit in Distington in 1815 which was worked until 1851; it was drowned in September 1840 and production was not resumed until the summer of the following year. Distington coal was carted to Harrington for shipment; in 1844 it was sold at 11s. 6d. per waggon, leaving a profit of 4s. 6d. per waggon. Between 1832 and 1847 Lonsdale's annual profits from his Distington pits ranged from £1,115 to £3,521.[45]

Boonwood Colliery, Distington, belonging to Allinson Crosthwaite was also drawing coal 1831-33 and the royalty was subseqently bought by Lord Lonsdale. At Gilgarran Captain Robertson Walker had a colliery and lime works in production from 1830 to 1838. His three pits were taken over by Ralph Tate in 1843 and worked by him until 1854 when lack of capital forced him to give up.[46] Ralph Cook's colliery at Camerton was leased in 1825 by John Fletcher who sank the Greengill pits to the Ten Quarters seam at 30 fathoms in 1829. He was succeeded by William Thorburn in 1840 who continued with the colliery until 1857. Although Thorburn was shipping over 20,000 tons of coal annually between 1842 and 1847 it cannot have been a very profitable venture for him. In 1842 he was selling coal at 6s. per ton, but cartage costs to Workington were almost 5s. per waggon and rent to the lessor amounted to 1s. 6d. per waggon.[47]

MARYPORT DISTRICT

The Christian family continued to work the Ewanrigg Colliery which in 1826 consisted of two pits, No. 7 and John Pits. No. 7 Pit closed in September 1826 and another pit, Thomson Pit, was sunk in the same year.[48] John Christian in 1831 was unable to let the colliery. Lord Lonsdale, on the advice of John Peile who had thoroughly examined the colliery and believed it to be exhausted, refused the opportunity of leasing it.[49] Christian therefore carried on the colliery himself until 1836. There was much friction between him and Mr Walker, the lessee of the neighbouring Broughton Colliery, the latter contemplating bringing an action for trespass against Christian who "has lately behaved unhandsomely to Mr Walker in preventing him the use of a level by which Broughton Colliery was drained, as well as the Waggon Way leave granted by the late Mr Curwen and Mr Walker is consequently anxious to return the incivility in any way he can".[50] Between 1836 and 1840 the colliery was worked by A. W. Hillary who ran into debt, the colliery's equipment having to be sold under a sheriff's writ.[51] He was succeeded by Joseph Harris who also leased Ellenborough Colliery for twenty-one years from 1 June 1848 and then worked both collieries together.[52] At Ellenborough he sank No. 2 shaft which was completed to the Ten Quarters seam at 100 fathoms in 1851.[53]

A Senhouse royalty at Risehow was leased by John Fletcher for twenty-one years from 11 November 1835 at an annual rent of £500 and 2s. 6d. per 61 cwts. over 244,000 cwts. Because his pit was heavily watered Fletcher had little success in winning the Main Band coal and in February 1843 he surrendered the lease owing the lessor £1,369. 18s. 2d. J. P. Senhouse then worked Risehow Colliery himself for the next fifteen months and when it closed in May 1844 sold its materials for £825. 6s. 3d.[54]

Broughton Colliery was taken over by John Christian (son of J. C. Curwen by his first marriage), but he proved to be an incapable manager and from 1811 to 1822 "Mr Curwen re-occupied the colliery for Mr Christian's benefit". During these years annual sales from the colliery ranged from 15,138 waggons in 1811 to 6,891 waggons in 1822.[55] Up to 1816 the colliery's five pits showed a profit, but this was followed by annual losses.[56] In 1823 the colliery was next leased by John and Thomas Walker for fourteen years at an annual rent of £688. 12s. and 10d. per ton over 20,658 tons of coal per year.[57] They had also to pay 1s. 7d. per waggon for wayleave from Broughton Moor to Maryport harbour and 100 guineas per annum for water passing from the colliery through the Ewanrigg level.[58] In the years 1824-36 the Walkers shipped annually between five and seven thousand waggons of coal from Broughton.[59] They worked the colliery until 1837 when it was taken over by Messrs Ross, Fletcher & Co. who continued with it until 1856. Messrs Ross & Co. did better than their predecessors; coal shipments increased annually, reaching 13,294 waggons in 1840 and 41,303 tons in 1846.[60]

In addition to their collieries at Flimby and Broughton the Walker family mined coal at Dearham. A Lonsdale royalty there was leased by John Walker (about 1820) who obtained coal from the Hope, William and Bell pits. His sons carried on and sank the Scott, George, Croft and Victoria pits, but their principal pits were the Lonsdale (sunk about 1830) and the Lowther (completed 1840). Between 1839 and 1847 annual coal shipments from this colliery exceeded 20,000 tons; in 1846 they amounted to 34,200 tons. Others besides the Walkers exploited coal in Dearham but not to the same extent. Ephraim Bell in 1820 had a pit near Bell Pit and in 1823 was working a small pit at

High Crosshow. In 1840 Messrs Ostle & Duglinson were getting coal from a thirty fathom pit near Townhead. Messrs Wood & Steel in 1842 sank Orchard Pit in a royalty belonging to Mrs Dykes where they mined Main Band coal at a depth of twenty-four fathoms. At that time their coal was carted to Maryport for shipment at a cost of 4s. 8d. per waggon; in later years it was moved to Maryport by the M. & C. Railway. They also sank John Pit which was in production between 1846 and 1850. Rail transport, available from 1843, accelerated the movement of a growing volume of export coal, 19,023 tons in 1843, 27,941 tons in 1845 and 33,962 tons in 1846.[61]

The small inland collieries met with little success in the early nineteenth century. Gilcrux Colliery in 1807 was full of water; Mr Grieve of Edinburgh advised the sinking of new pits and the driving of an undergrond level from the sea at Blue Dial to the colliery which would drain the colliery as well as carry the coal. In addition he surveyed the route of a projected waggon road from the colliery to the sea at Blue Dial. The proprietor, Mr Dykes, did not live to carry out either proposal. His widow in 1831 leased the colliery, which then had only one pit, Jane Pit, in production, to two Whitehaven colliery viewers, William Quayle and Williamson Peile (son of John Peile). The colliery was carried on by them and their widows until the expiration of the lease in 1852. The coal was mined in the Ten Quarters seam and from 1841 was sent by railway to Maryport for export. Between 1833 and 1840 annual coal shipments from Gilcrux ranged from 1,800 to 5,300 waggons, rising to 14-19,000 tons between 1842 and 1847.[62]

Nothing was done at Birkby for thirty years until in 1832 Messrs Tickle & Thompson leased the colliery for twenty-one years at an annual rent of £50 and 1s. per ton above 1,000 tons. They sank two new pits, one to a depth of forty-nine fathoms without finding coal that it would pay to work, and the other, which had two thin seams of coal, they worked for a short time until the losses they incurred compelled them to abandon the enterprise in August 1833.[63] They were succeeded in 1837 by Messrs Ross, Fletcher & Thompson who held the colliery until 1856, but they did not work any coal there. Between 1830 and 1838 Messrs Tickle & Son worked the Cannel and Metal Band seams at Dovenby from a pit sunk south of Row Beck Mill.[64]

Westmoor Colliery, Oughterside was leased by Brown, Hodgson & Co. in 1814 for fourteen years at a yearly rent of £200 and 1s. 3d. per ton for all coal over 2,400 tons, the lease being extended for a further seven years in 1816. The company laboured under difficulties. As early as 1818 it was stated that the Main Band coal, formerly worked by the Bolton Company, was exhausted and that competition from two new collieries at Arkleby was denying a market to Westmoor coal. It was alleged that Arkleby coal was sold at half the price charged by Brown, Hodgson & Co. largely because a royalty rent of 1s. 3d. per ton had to be paid on Westmoor coal compared with the 7d. per ton and 4½d. per ton paid by the lessees of the Arkleby collieries. Mining was made more expensive by the abundance of stone which interspersed the coal seams and in 1825 Messrs Brown & Co. gave up Westmoor Colliery after having lost more than £3,000.[65]

In 1825 Joseph Holliday obtained a licence from Lord Egremont to bore for coal in Westmoor and in the following year, in partnership with John Brown, began to work coal there. A pit was sunk near the north-west boundary about eighteen fathoms to a 27 inch seam. Little coal was found and that was badly intersected by dykes. John Kirkhaugh took over Westmoor Colliery from Holliday at Michaelmas 1828 having paid the latter's arrears of rent of £107. 3s. 8d. He leased the colliery for twenty-one years from 29 September 1830 at an annual rent of £20 and 6d. per ton of coal above 800 tons.[66] After a number of unsuccessful borings he sank a new pit forty-eight fathoms to the Yard Band; his efforts were hampered by many dykes, the inferior quality of the coal and by high transport costs, for until the building of the Maryport & Carlisle Railway his coal had to be carted to staithes at Allonby for shipment.[67] Between 1817 and 1837 output from this Egremont royalty never reached 5,000 tons per annum. After 1840 this figure was exceeded, rising to 11,520 tons in 1842. Production was further increased when Kirkhaugh in 1845 began working coal in a Lonsdale royalty at Oughterside, the coal being drawn up the shaft of Westmoor Colliery. Output from this royalty rose from 7,073 tons in 1846 to 14,523 tons in 1848.[68] Kirkhaugh must have incurred heavy losses during the early years of his tenancy of Westmoor but he continued with the colliery until 1850 when he was succeeded by the Aspatria Coal Co. (Messrs Thomas Westray, Isaac Fletcher and William Bragg). About 1830 Joseph Harris of Greysouthen was working the Yard Band and thirty inch seam at the old Domain Colliery at Oughterside from Nos. 1, 3 and Hall Pits. The colliery continued until 1861.[69]

Harris also began coalmining in the Aspatria district when he sank a pit at Plumbland in 1822 in a royalty belonging to Sir W. Lawson. There he worked the Yard Band up to the outcrop. Coal shipments from this colliery increased from 1,765 waggons in 1837 to 6,085 waggons in 1840. From their pits at Arkleby Messrs Drewry & Co. in 1826 and Mr Thornthwaite in 1836 were winning coal from the same seam.[70]

In 1831 the Bolton Company renewed their lease of Lord Egremont's coal in the manors of Wigton, Westward and Bolton for a further twenty-one years. At this time the company had three working pits – Engine and Limekiln Pits at Bolton Colliery and Crummock Pit near Weary Hall. A fourth pit was sunk at Weary Hall in 1833 which reached Main Band coal in June 1834. In the early 1830's the company paid no dividends because the coal seams at Bolton were nearly exhausted, Crummock colliery was working at a loss in 1833 and the new winning at Weary Hall was drowned out in 1835. Matters were made worse by competition from Harris's pit at Oughterside which obtained a great part of the market in the Wigton area and from another pit at Weary Hall controlled by Messrs Drewry.[71] The last pit was advertised for sale in 1841 "by order of the assignees of Rebecca Drewry, bankrupt", with eighteen years of the lease still unexpired.[72] Between 1814 and 1821 the Bolton Colliery raised over 30,000 tons of coal per annum. Thereafter output gradually declined, to 25,000 tons in 1828 and to 16,000 tons in 1837.[73] Despite their difficulties the company continued with the colliery until 1847 when they were succeeded by Messrs Addison and Paisley who leased Bolton Colliery for a term of twenty-one years from 25 March 1847 at an annual rent of £250 and a royalty rent of one-tenth of the selling price of all coal sold.[74]

As coal mining developed better harbour facilities were needed, particularly at Maryport where the harbour was completely controlled by the Senhouses, a family lacking both capital and business enterprise. Maryport harbour in 1815 was said to be "very small, and ill-suited to the wealth and commercial importance of the place. The channel is so narrow in parts, that two vessels can scarcely lie in it abreast; and the shipbuilders are obliged to launch their vessels into the water sideways . . . . Violent remonstrations have been made to the lord of the manor by the ship-owners concerning the inconvenience and insecurity of the harbour, but he has not attended to them, and some of them have consequently deserted the port".[75] Nothing could be done until 1833 when the grip of the Senhouses on the port was broken; the previous Acts relating to Maryport harbour were repealed and a new system of government was introduced. The lord of the manor became a trustee in perpetuity with power to nominate four other trustees, and a further eight trustees were to be elected by those townspeople who paid at least £6 per annum in rates or who owned shares in local shipping. This new body became responsible for the management of both town and harbour, and was empowered to levy rates on both, although the accounts for the town and harbour were to be kept separate.[76]

Humphrey Senhouse nominated Humphrey Senhouse the younger, the Rev. Anthony Dickson, James Pitcairn and Thomas Mark, and the first eight trustees elected by the townspeople were: William Ostle, John Walker, John Ritson, Kelsick Wood, William Chatterton, John Hodgson, Joseph Huddleston and Nicholas Ross.[77] This election is of some historical interest since the usual practice of voting by open poll was replaced by a form of ballot, possibly the first occasion in England when the ballot was employed in municipal elections. Unfortunately complete secrecy of voting was not observed on this occasion, candidates being allowed, by some error, to take the voting papers to the homes of the electors, where, doubtless, they guided the choice of the voters. But at the subsequent quinquennial elections complete electoral secrecy was enforced.[78]

The trustees immediately set to work to provide improved accommodation for shipping. In less than three weeks after their election they had resolved upon building a new dock on the south side of the harbour, and four months later they resolved to build a substantial bridge across the harbour.[79] By January 1837 the new dock (now the basin of the Elizabeth Dock) was completed; it covered an area of about two acres, but like the harbour, it afforded only tidal accommodation.

Railway building also assisted the rise of Maryport. Until the coming of the railway the northern section of the Cumberland coalfield remained largely undeveloped, cartage costs to the harbour at Maryport being too high to allow coal mining at a profit to be undertaken in that area. It was to increase the exploitation of the coal deposits north of Maryport that the Maryport and Carlisle Railway, the first public railway in the county, was called into existence. A company was formed in 1836 to construct a railway from Carlisle to Maryport, with George Stephenson as the engineer. Parliamentary powers were obtained in 1837 to build a railway from Maryport to join the Newcastle and Carlisle Railway (completed June 1838) at Bogfield.[80] The first section of the M. & C. Railway from Maryport to Arkleby Pits was opened 15 July 1840 and by the following February Aspatria coal from Kirkhaugh's new pit was being shipped at Maryport.[81] In February 1845 the whole line was open for traffic and in the next two years it was extended to Whitehaven by the Whitehaven Junction Railway. The early engines on the M. & C. Railway were supplied by the Lowca Foundry, and until the completion of the railway between Whitehaven and Maryport the locomotives were sent to Maryport by sea on rafts from the works.

By 1847 the ports on the coalfield were linked up by railways and the inland collieries could now develop quickly, particularly those in the neighbourhood of Aspatria which could not only ship their coal from Maryport but also share in the Carlisle market. As early as March 1844 the coal traffic at Maryport was stated to have nearly trebled since the construction of the M. & C. Railway.[82] The rise of Maryport as the leading port in

Cumberland was largely the work of the railway which, by providing easy access to the harbour from a number of pits, added greatly to the volume of coal which passed through the port; 140,473 tons of coal were shipped from Maryport in 1842 and 213,152 tons in 1846.[83] At the same time the revenues of the railway company were highly satisfactory, rising from £32,842. 15s. 10d. in 1847 to £51,038. 7s. 11d. in 1858;[84] between 1850 and 1922 the company paid an average dividend of over seven per cent, reaching a maximum of thirteen per cent in the 1870's.[85] The expansion of the coal trade led to an improvement in the status of the port; in 1838 ships were registered there for the first time and from 19 June 1841 Maryport became an independent port with bonding privileges, i.e. it was freed from the control formerly exercised by the Port of Whitehaven.[86]

With the opening of the M. & C. Railway the amount of coal carted to Maryport for shipment declined rapidly as the inland collieries took advantage of the quicker and cheaper facilites offered by rail transport:–[87]

Year	*Moved by Carts* *Tons*	*Moved by Rail* *Tons*
1842	80,752	59,721
1843	66,484	76,653
1844	61,243	107,466
1845	70,057	119,949
1846	72,411	134,865
1847	47,827	119,868

It may be noted that in 1846 5,876 tons and in 1847 10,923 tons of Newcastle coal (via Carlisle) were shipped from Maryport.

In this period coal from Ewanrigg and Broughton continued to be carted to Maryport. Other collieries, particularly those at Brayton, Gilcrux and Oughter side, benefited greatly from railway development. The port of Maryport now became available to them and the reduction in transport costs meant that their coal could compete effectively with Whitehaven coal in the export market. It was access to the Irish market, made possible by the railway link to Maryport, which brought about increased production at the inland collieries. Although his son was a partner in Gilcrux Colliery, John Peile realized that these changes, whilst helpful to his son and others, were injurious to his employer's welfare, "The promoters of Maryport & Carlisle Rail Way had the avowed object in view to injure Whitehaven, in which they have been successful. Mr Harris with Sir W. Lawson's coal paid Cartage before of 9/- the Waggon & now conveyed on Rail Way at 4/-. Gilcrux before 7/- the Waggon now 3/6. Other collieries are preparing to join it also".[88]

Other railways followed the M. & C. – the Whitehaven and Furness, the Whitehaven, Cleator and Egremont, and the Cockermouth and Workington – and these too benefited the coalfield as a whole by making it possible to move to the ports a growing output of coal. The Cockermouth and Workington Railway, opened in 1847, was especially useful in the growth of coal mining in the Camerton area.

Railway construction was generally beneficial to landowners. Not only did they sell tracts of land at higher than agricultural prices to companies building railways through their estates but they also profited from the increase in the value of their land that lay within a few miles of the railways. It was particulary beneficial to those whose estates contained minerals which could be sent quickly and cheaply by rail to the ports for shipment. Colliery "railways" with horse-drawn waggons had been long known in Cumberland and other coalfields and it was a logical step for colliery owners to promote railway companies using steam locomotives. Just as the Lowthers, Curwens and Senhouses aided port development as a stimulus to the coal trade so did they assist the construction of railways in West Cumberland and for the same reason; the ports were linked to the collieries and to each other. The Senhouse family and Sir Wilfred Lawson supported the building of the Maryport and Carlisle Railway. William, second Earl of Lonsdale (1787-1872), was responsible for much railway development south of Maryport. He was a close friend of George Stephenson, the railway engineer, and as soon as the M. & C. Railway was completed did his utmost to ensure its extension from Maryport to Whitehaven. Lonsdale was the leading promoter of the Whitehaven & Furness Junction Railway Co. which in 1847 began constructing a line from Whitehaven to join the existing Furness Railway at Broughton in 1850. Again, it was he who did most to bring about the construction of the Whitehaven, Cleator & Egremont Railway which was opened for goods traffic, largely iron ore and coal, in 1855.[89]

MINING METHODS

In the nineteenth century Cumberland colliery owners turned to the Newcastle district for guidance in the working of their mines, Newcastle mining engineers being appointed as consultants and local viewers being sent to Newcastle to observe mining methods in use there. John Buddle's system of panel working was adopted at Whitehaven, i.e. the workings were divided into districts (panels) enclosed within barriers of coal in order to restrict the spread of explosions and this, together with the use of dam-doors, helped to reduce the damage caused by explosions. Dam-doors, made of three or four inch planks, were suspended by hinges from the top and made to fit perfectly tight into a solid frame. These doors were normally kept open by a prop, but if an explosion took place the prop was removed, the door coming down to fit tightly into the frame; in this way the panel or division where the explosion occurred was sealed off from the rest of the mine, the exclusion of air causing the fire to die out.[90]

Until about 1870 mines in Cumberland were ventilated by underground furnaces. Buddle, when giving evidence to the Select Committee on Accidents in Mines in 1835, said, "I certainly must say, in speaking of the Whitehaven Colliery as a mine, I consider it to be the finest in England, both in point of thickness of the seams and in extent . . . I would futher add, in point of ventilation, both as to the system adopted and in all its details, it is one of the most completeley managed mines that I know of".[91]

The ventilating furnaces were large open fires about nine feet square, kept constantly burning on a grate set into a brick archway a short distance from the bottom of the upcast shaft and all the return air from the workings passed over it. These open furnaces were always a potential source of danger. If the return air drawn over the furnace contained too much gas disastrous explosions could take place. To overcome this hazard various kinds of mechanical ventilators were invented but the early types had little success. In 1814 a mechanical ventilator invented by William Swinburn, Curwen's agent, was tried out at Union Pit, Workington and proved to be of little value. At Whitehaven the first attempt at mechanical ventilation took place in 1840 when James Reed, the colliery engineer, erected a fan eight feet in diameter and four feet wide in the upcast shaft of Duke Pit.[92] Driven by a high pressure engine it produced 23,000 cubic feet of air per minute and at first John Peile was delighted with its performance.[93] It cannot have been very successful because it was discontinued after running for a few years and furnace ventilation was resumed.

In 1844 a Fourness fan, sometimes called a "rotary air drum" or "barrel churn ventilator", was installed at Wyndham Colliery, Cleator Moor. This consisted of a sheet iron drum five feet in diameter and twenty-seven inches broad which contained six vanes. Driven by power taken from the pumping engine it produced 13,500 cubic feet of air per minute compared with the 5,700 cubic feet per minute previously supplied by furnace ventilation.[94] But it was not until the 1870's, when the Guibal fan replaced furnace ventilation, that mechanical ventilation was generally successful in Cumberland.

In the first half of the nineteenth century corves continued to be used when winding coal out of the pits, but the flat hemp ropes which raised the corves were expensive and did not always wear well. An improvement came with the adoption of wire ropes (patented by Andrew Smith in 1835)[95] and some were installed at Whitehaven in 1842. John Peile reported to Lord Lonsdale, "I am contemplating an early visit to him (i.e. John Buddle) to gain some information before fixing the Rope Wheels and other matters at Wellington Pit. Wire ropes are being introduced and are likely to reduce the Hempen".[96] Four years earlier Peile had returned from another visit to Newcastle much impressed by the new cage and tub system in use there for winding coal out of the pits instead of the old method of using corves. This was much faster and safer than corves swinging about on the end of a rope and more coal could be raised up the shaft each day. The cage, capable of taking one or two or three tubs, was drawn up the pit with very little vibration since it ran between guides or conductors throughout the entire length of the shaft. Peile intended to adopt this system but there is no evidence to show that he did so.[97]

Fatal accidents occurred all too frequently at the Whitehaven Colliery. Fire-damp remained the chief killer and as long as naked lights were used underground the danger of explosion where gas was present could never be wholly excluded. The invention of the Davy safety lamp seemed to herald a new age of security in coal mining and in 1816 this lamp was tested at William Pit, Whitehaven, regarded by many as one of the most dangerous pits in the country. A local journalist described the test,[98]

> "Last week, Mr Murray, the Lecturer, descended William Pit, accompanied by Mr Peile, and put to the test Sir H. Davy's Safe Lamp – The wire gauze cylinder was, with the Lamp burning brightly, introduced into one of the blowers in a dangerous recess, where had they penetrated with a candle, their destruction would have been inevitable; the effect was grand beyond description. At first a blue flame was seen to cap the flame of the lamp, – then succeeded an increase of lambent flame, playing in the cylinder; and shortly after, the flame of the fire damp expanded, so as completely to fill the wire gauze. For some time, the flame of the lamp was seen through that of the fire damp, which became ultimately extinguished without explosion. Results more satisfactory were not to be wished . . . . The Lamp offers absolute security to the miner, when approaching where carburetted hydrogene is to be found, (no matter to what extent or quantity) and is a perfect index to his feet, how far to advance; even with so much precision as comparatively by a graduated scale; yet with the most perfect security."

After this test the safety lamp was generally adopted in Cumberland mines although candles continued to be used in pits where little gas was to be feared. Unfortunately the introduction of the safety lamp was followed by an increase rather than a decrease in mining accidents, for it led to the working of dangerous mines which could not have been worked without it and, by inducing an exaggerated sense of security, encouraged the winning of coal in districts which should have been left alone. It was not, however, simply a matter of greedy owners or managers, anxious to increase output, compelling men to take unnecessary risks. Far too often the miners themselves precipitated gas explosions by misusing their safety lamps.

Among the Whitehaven Colliery employees in the early nineteenth century was Taylor Swainson, a very capable engineer highly regarded by his contemporaries but, like his mother, an inveterate drunkard.[99] According to R. W. Moore,[100] Swainson in 1812 "tried his invention of the 'iron horse' or locomotive engine, on the Croft Pit waggon way. The engine answered admirably, but the track was not strong enough to bear the weight". This statement, repeated by many later writers, is inaccurate and apparently based upon the erroneous "Random Recollections" of Noble Steele, a Whitehaven printer. As has been recently shown,* a locomotive built to William Chapman's patent was ordered in 1816 by John Peile from Phineas Crowther of the Ouseburn Foundry, Newcastle. Parts of this engine were built at Whitehaven under Taylor Swainson's supervision and this may account for the local belief that Swainson invented the locomotive. The locomotive was tried on the Croft Pit waggon-way in 1817 but it broke up the cast iron rails and in the following year the locomotive experiment was abandoned and coal was again moved to the harbour by horse-drawn waggons.

Little progress was made in the conduct of Workington Colliery at this time and the reports of Matthias Dunn show how far methods employed there lagged behind the best mining practice. His report of December 1832 had many criticisms:–[101]

1. The daily output of 175 tons was much too small in proportion to the size and costs of the colliery.

2. Underground haulage was inefficient – the roads were wet, ill-constructed and needed to be straightened. The rails were of cast iron of the ledge pattern, long superseded elsewhere by the edge rail of malleable iron. (He found that 6 ft. lengths of underground rail, which cost 3s. 6d., weighed 75 lbs., whereas 6 ft. of improved malleable iron weighed only 30 lbs. and cost 2s. 6d. Allowing for a wastage of ten rails per day, the adoption of the new rails would save £120-£130 per annum).

3. Machinery and equipment above and below ground were clumsy, inefficient and old-fashioned, leading to excessive costs for repairs and to much wasted material.

4. Pumping costs were very heavy – often work had to be abandoned for weeks owing to the presence of water.

5. The coal, as good as that at Whitehaven, sold for 1s. per ton less because it was not properly screened.

6. Horses were expensive, twenty-five being needed in the only two workings and they were badly tended.

* P. Mulholland, "The First Railway Locomotive in Whitehaven", *Journal of the Cumbrian Railways Association* (1978). I am indebted to Mr Mulholland for providing this information.

Dunn urged that more attention should be paid to output and costs. He suggested that a good half-inch screen at every pit heap would be enable a better price to be obtained for the screened coal. Since it would be too expensive to change the rolley-ways to edge rails of malleable iron he recommended the adoption of bogies capable of carrying two baskets at a time instead of the single-basket bogies then in use; in this way more work could be done by the horses.

The lead formerly held by Cumberland collieries in mining practice had been lost by 1850. Matthias Dunn reported, "In the Cumberland district colliery improvements are considerably behind . . . with a single exception the collieries in Cumberland continue to work with baskets, notwithstanding the evident advantage to be derived from tubs, slides and cages".[102] At Whitehaven, apart from William Pit, no guides or conductors were used for corves moving in the shafts. As a result the baskets often tangled together as they went up and down and sometimes discharged their loads, which could include human beings, down the shaft.[103]

THE MINERS

In Cumberland, as in other mining areas, children continued to be employed underground until 1842. According to the Report of the Children's Employment Commission they began working in Cumberland sometimes as early as five or six years of age, more frequently at eight and generally at ten.[104] No correct estimate of the number of children working in Cumberland mines is possible, but they were stated to be not as numerous, even in proportion to the men, as in Yorkshire.[105] Out of a total of 1,300-1,400 employees in Lord Lonsdale's mines 200-250 were under eighteen years of age. At William Pit, Whitehaven fifty-four young persons under eighteen were employed underground who were divided into these age groups:[106]

Age	*No.*	*Age*	*No.*
7	1	13	4
9	2	14	10
10	4	15	5
11	7	16	4
12	8	17	7
		Near 18	2

In most collieries young children were employed as trappers – at Whitehaven there were 124 trappers aged between nine and fourteen.[107] At Whitehaven, where there were thick seams, coal could be drawn directly from the face to the shaft by horses so that children were not needed as trailers. But in the smaller inland collieries children were employed as trailers and also helped to fill and riddle the coal.[108] Williamson Peile, a colliery viewer at Whitehaven, when giving evidence to the Commission, stated,[109]

"In no one of Lord Lonsdale's collieries do the colliers themselves employ children – they are exclusively paid by us. Their work is first that of trapping; secondly that of coupling, which consists of linking and unlinking the trams or carriages, or greasing the axletrees of the carriages; and thirdly, of driving. The drivers have to hold back the carriages in descending, but I don't recollect any accident resulting from it. Girls do it very easily, and it is not hard work for boys . . . . It is only the empty baskets that they push back in this way, never full ones; with full ones they use a bit of stick put through the wheels, called burrs or breaks. We have no trailers. We carry the baskets completely to the foreheads, or place of working, with the horses in general."

An analysis of the ages of trappers and drivers at Whitehaven Colliery at this time gives these figures.[110]

Trappers Age	No.	*Drivers* Age	No.
9-10	13	12-13	3
10-11	13	13-14	15
11-12	36	14-15	18
12-13	27	15-16	33
13-14	19	16+	64
14+	16		
Total	124	Total	133

The drivers had to drive along the tram-ways the horses which pulled trams of baskets loaded with coal, several of which were hooked together. Their work was described by J. C. Symons,[111]

"The journeys along these tram-ways are greatly lengthened, owing to many of the large collieries being sub-marine. In the William Pit they have 500 acres under the sea, and the distance is 2 miles and a half from the shaft to the extreme part of the workings. There is a stable also under the sea in this immense pit for 45 horses. The shaft is 110 fathoms. A feature exists in this driving employment I have not hitherto seen, and which constitutes the chief labour of the occupation. To prevent the baskets from running down hill and falling on the heels of the horses, it is customary for the driver to place himself as a post between the foremost basket and the buttock of the horse. He places the left shoulder against the horse, the right foot on the rail of the tram, and the right hand on the top of the basket; the left leg being generally supported by the trace. When the train of corves is heavily laden, or the descent very steep, a pole is placed through the hind wheels of the trams, and thus it is in a measure dragged. Nevertheless the work is very toilsome, and, as will be seen by the evidence of the surgeon attending Earl Lonsdale's collieries, accidents sometimes occur by the foot slipping off, and getting struck by part of the wheel or axle. The leaning position in which they stand is not in itself, I think injurious; but the work strikes one as being palpably unnecessary, and as a barbarous preference of the human body for a mere mechanical process; in which shafts might be, and in some of the inland pits are, used instead."

In Cumberland pits children regularly worked twelve hours per day and often thirteen hours. At Thomas Westray's colliery at Little Clifton boys worked fourteen or fifteen hours per day and eighty hours per week all the year round. This was because of "the inefficient power of the drawing engine" and it was said that "this cannot be considered other than an accidental circumstance".[112] In the Lonsdale pits children were directly employed by the owners; at some of the inland collieries they were employed by the hewers, often their parents. As trappers they were paid 6d. per day and 1s. 3d.-2s. per day as drivers or trailers. Apart from their excessively long working day these children appear to have been fairly treated by their masters, whether colliery proprietors or coal miners, but their toil left them stunted in body and mind. They were reported to be "as ignorant as it is well possible to conceive children to be; nor are the lads from 13 to 18 years old one jot more informed; but, as far as I could ascertain, less so. It is not to be supposed that children confined for 12 hours in a coal-pit can have opportunities for any kind of education . . . the mental and moral destitution in which the great body of the collier children are growing up in Cumberland is fearfully great".[113] Illiteracy was widespread and to many religion something unknown. One eleven year old collier stated, "I don't go to church or chapel. I have never been taught to pray. I never heard of Jesus Christ". Another boy of the same age declared, "I don't know who Jesus Christ was. I never heard tell of God neither. I can't read the Testament. I was taught to say my prayers, and I say them. I don't know who to pray to".[114]

The employment of women underground in Cumberland had almost died out by 1842. Only at Whitehaven Colliery were women employed and there were twelve of them, their ages ranging from 34 to 72.[115] Even here no women had been taken on in the preceding fourteen years, those employed before being allowed to remain until they were too old to work.[116] Women, however, remained until recent times to work as surface hands.

Danger, discomfort and physical hardship prevailed in the mines, but in one important respect the miners were better off than many other manual workers – they were better paid. According to the *Cumberland Pacquet* of 16 August 1831 Whitehaven hewers then earned 3s. 4d. to 5s. 6d. per day; the latter figure is doubtful, for it would be earned by very few, and this newspaper, which consistently served the Lonsdale interest, was anxious to show that the miners had no justification for striking. More reliable sources stated that wages were good for men and children; in 1842 adult hewers earned 18s. to 25s. per week, trappers 6d., drivers 1s. 8d. and engine-men 2s. 2d. per day.[117] Deductions for picks, lamps, shovels and other tools supplied to the men amounted to 1s. 6d. per week. Colliers were assured of medical attention for themselves and their families on the payment of 4d. per month, single men paying 2d. per month; if injured at work they received 5s. per week until they were fit enough to resume employment.[118] It was the custom to allow the widow of a man killed in a pit accident to have a house rent-free and free fuel.[119] One Whitehaven witness stated in 1842, "My father had a free house, and was a long-time ailing, and supported by the employers, and his widow has the free house now".[120] At Cookson Colliery, Little Clifton hewers earned as much as 30s. per week, but their high wages did them little good since they drank heavily and had often to absent themselves from work in order to recover from their drunken bouts.[121]

At that time night work was customary in the mines, twelve hour shifts being worked in the Lonsdale collieries, the men working day and night shifts in alternate weeks. In a few collieries eight hour shifts were worked, but in none of them were regular hours assigned for meals.[122]

Life outside the pits was dreary and unpleasant for all miners, young and old, male and female. Since much of the coalfield is rural many mining families lived in small, ugly and often isolated villages consisting of a few rows of terraced cottages squalidly huddled together. Living conditions were much worse in the filthy, overcrowded and disease-ridden slums of the towns, particularly Whitehaven. In a scathing report to the General Board of Health Robert Rawlinson describes in detail the sanitary condition of

Whitehaven in 1849.[123] The 1,825 inhabitants of the Mount Pleasant area were "without any form of privy accommodation or any regular supply of water" and many of their dwellings were "dreary, black and loathsome". The New Houses built in 1788 for the use of Lord Lonsdale's mineworkers had nothing to recommend them,

> "They are built on the side of a hill and form three rows or streets, the roofs being, in many instances, level with the roadway in front of the houses behind, and the roofs of the highest run full against the hill side. There are no sewers or drains, and consequently the roads and houses are damp. On the front row there are 77 tenements and 5 ash-pits; on the middle row 111 tenements and 9 ash-pits; on the back row 78 tenements and 7 ash-pits; total number of houses 266; total number of ash-pits 21. There is not a single privy belonging to the whole property . . . . The water supply is very inadequate. It is not uncommon to see 20 women waiting at the stand-pipe for water. In the summer months this frequently fails and the inhabitants are then obliged to fetch their water for domestic use more than a mile; or they resort to any nearer place if it can be obtained, even when of inferior quality . . . . Pig-sties and stagnant water in contact with the houses are common. These houses are very seldom clear of fever."

Two years later another official report on these miners' dwellings was much more mildly phrased, its understatement obscuring the facts of the situation, "Buildings which, when erected many years ago, were probably thought, together with the spaces about them, to fulfil every purpose for which they were designed, now excite remark by defective arrangements for cleanliness and comfort".[124]

Coarse and brutal in their habits the miners led harsh, monotonous lives that bore few traces of education or religion. Their chief diversions took the form of heavy drinking, cock fighting, hound trailing and the traditional sport, wrestling. Few facilities existed to lighten the darkness of their drab existence and to many of their contemporaries the miners seemed little better than animals.

In the 1840's railway construction and the development of the iron industry in Cumberland led to a shortage of mineworkers, dissatisfied colliers seeking better-paid employment outside the pits. According to John Peile there was,

> "quite a ferment with the Labouring class. The Demand is in excess of the supply and a general clamour for increased Wages. We have lost several for refusing . . . but obliged to comply in many cases, and the mischief is, in giving way in one instance brings all others upon us . . . . I believe our Scale of Wages generally were less than any other Works in the district . . . . The Railway Contractors & us have an understanding to decline taking each other's people which will be so far useful."[125]

He refused to increase the wages of all categories of mine workers, believing it better policy to lose some of them than to accede to their demands for higher wages; but found himself, "obliged to advance some that could not be done without".[126] The agreement with the railway contractors helped to ease the situation, for the miners on leaving Whitehaven Colliery found that work on the railways was not available to them.*

To strengthen their demands for higher wages the miners attempted to form trade unions, but in the first half of the nineteenth century the employers were almost invariably too strong and were able to smash the unions. Cumberland did not escape the trade union fever which swept the country in the early 1830's and the Whitehaven

* Peile to Lord Lonsdale, 26 May 1845, "We are very accommodating to the Railway Contractors here, and they refuse to employ any of our people on being apprized which in part tends to check leaving us".

miners were particualrly susceptible. Peile, strongly anti-unionist in outlook, was most anxious to prevent the establishment of a miners' union at the colliery and, alert to the danger spreading from other areas, kept himself informed of the situation in distant collieries. In January 1831, when there was no labour trouble in Whitehaven, he wrote,[127]

> "I have cautioned the officers to be careful in employing any strange Haggers or trampers. I think our Pitmen never joined any of the Union Societies at other places, and have long been too poor to contribute to such Funds, and thus, we may be secure against their receiving aid from other places in case of a revolt being attempted. Our men have no real cause of complaint, their greatest evil is such a disposition for Idleness & disinclination to Work, and their present poverty is our only security."

Five weeks later he reported,[128]

> "We have had another letter from the Coal Owners in Flint, with a copy of the Rules of the Associated Societies, which they secretly obtained, the whole is completely organized for carrying on the combination ag[t] the proprietors and held as Lodges No. 1 being at Bolton near Manchester as headquarters but connected with other Lodges over all the mining districts in Lancashire, Shropshire, Wales &c. The system is yet in its infancy, but regular contributions are to be paid to the Stock Fund and in a little time the Men will be Masters. The letter states to save property and life they had given way to the first demand by advancing the wages through[t] fully $\frac{1}{6}$th part or nearly 17 per cent and were apprehensive of further demands."

In April 1831 collieries in the Newcastle area were closed following a strike by the pitmen to obtain better conditions. This example was not lost on the Whitehaven miners who, in a petition to their employer, complained of their long hours, fortnightly pays and being paid on Saturday morning instead of on market day.[129] Peile had no intention of meekly accepting the miners' claims and asserted, "We are quite satisfied nothing but firmness will do on our part, whatever is done for them must be voluntary, one step of concession would lead to many and end in overthrow". Trouble-makers he proposed to get rid of at the earliest opportunity.[130] When the men began to form a trade union Peile summoned its most active members and informed them that a trade union would never be allowed at Whitehaven and unless they signed a declaration against it they must seek employment elsewhere. They refused and were immediately dismissed.[131] Delegates from the Lancashire collieries arrived in Whitehaven, presumably to help in forming a miners' union. Peile decided that a direct confrontation with workmen was the best policy and accordingly ordered three or four men from each pit to meet him and again made it clear that no union would be permitted and unless the workmen signed a declaration renouncing the union "the Colliery workings thro'out would be laid in".[132] The declaration referred to by Peile stated,

> "We the undersigned Pitmen & others employed in and about the Earl of Lonsdale's Collieries at Whitehaven, do humbly disclaim having any connexion with the Union or any other Secret Society now forming or attempting to be formed for the purpose of combination amongst the Pitmen & others in the said Employ, and that we are not and will not become party or partners to the said Association or any such Combination so long as we remain employed in the said Collieries."

The men refused to sign the declaration and a three weeks' strike ensued. Guards stationed at the entrance to the colliery were badly beaten up by strikers and the houses of non-unionists were attacked. Peile informed the local magistrates of the situation and they immediately summoned troops, then stationed at Cockermouth, to Whitehaven and swore in 110 of Lonsdale's men as special constables to maintain order. To the forces of law and order the local Catholic priest, Father Gregory Holden, added his considerable weight by exhorting his congregation to repudiate the union entirely. By the end of August the strike was over and the union dissolved.[133] Whilst the struggle was going on Peile had not wavered in the slightest in his determination to break the union, nor had he shrunk from conflict with his labour force. In his own words, "if a Battle is to be fought in this part, we must do it . . . . My own opinion is, we sh[d] make a direct stand, be the consequence what it may".[134] Lacking the means to bear the strain of a prolonged strike the miners had to return to work on the employer's terms; their povery constituted the strongest weapon in the hands of the colliery owner.

Twelve years later the same trouble arose when P. M. Brophy, a delegate of the recently formed Miners' Association, tried to persuade the Cumberland miners to form a union. Peile immediately imposed a test requiring the Whitehaven miners to renounce all connection with the union, threatening the dismissal of those who did not comply. This threat, together with the dismissal of unionists, precipitated a six weeks' strike which was joined by colliers in the Maryport district. Attacks by unionists on non-unionists again resulted in the movement of troops to Whitehaven and the justices, in a warning printed in the local newspapers, promised protection to miners who continued to work and asserted, "no-one has a Right to dictate to another what Wages he ought to receive; still less to endeavour by Intimidation and Threats to prevent him from working for any Master, on any Terms he thinks fit. And if any Union has any such Objects as these in view, it becomes an Illegal Combination to impede the due course of honest Industry, and all Persons assisting in it become liable to severe Punishment by law".[135] But combinations of employers were apparently different, for all the coal owners joined together to refuse employment to union members.[136] The Miners' Association tried to help the strikers; all members in Northumberland and Durham paid a 6d. levy and the Lancashire colliers a 3d. levy.[137] But again the miners had to submit, first at Maryport and then at Whitehaven, the men signing a declaration that they had abandoned their union before being allowed to resume work. Those who refused to sign were denied employment in the pits.[138] One journalist, commenting upon this strike, expressed the wish, "that on all future occasions of a similar nature, the pawnbrokers and provision shops would unite . . . in refusing to give either money or goods on credit. Strikes would then be of shorter duration, and the privations of the families of the turn outs would be infinitely less extended".[139] The failure of the strike was a severe blow to the trade union movement in Cumberland. Although the *Cumberland Pacquet* of 9 April 1844 reported that a Whitehaven delegate, representing 120 members, had attended a meeting in Glasgow of the Miners' Association of Great Britain and Ireland, by the autumn of that year the county branch of that Association had ceased to exist.[140] Disheartened by their defeat the Cumberland miners turned away from trade unionism and for nearly thirty years it was impossible to persuade them to help themselves by forming a trade union. To many of them it seemed that any attempt to organize another miners' union would inevitably lead to another struggle with their employers which was certain to collapse and which would inflict great hardship upon themselves and their families.

Some of the Cumberland miners soon showed that they had abandoned trade union principles. In April 1844 the pitmen of Northumberland and Durham went on strike and they remained on strike for about twenty weeks. Many of them who lived in tied cottages were evicted from their homes by their employers. It might have been expected that the miners of Cumberland would support such a strike, more especially because in the previous year the colliers of Northumberland and Durham had given financial help to the Cumberland miners when they were on strike. This did not happen. Not only did the Cumberland miners give no help to the pitmen in the north east they also joined in to break the strike. The strike ended largely because the colliery owners were able to import strike breakers from other areas and one of these was Cumberland.[141]

The colliery's grim record of accidents did not abate; all too often lives were squandered needlessly through the carelessness of the workers themselves. The death of a girl trapper in 1819 is laconically reported by a colliery manager, "Jane Grave, Trapper, fell down Croft Pit sliped by the side of the Basket when getting in no person saw her – Boaden & Thompson was in the Basket when she fell down – very little mangled – every person about the place went home on account of it".[142] Miners were responsible for many of the accidents by tampering with their lamps, unscrewing the tops to obtain more light or to light their pipes, or simply preferring to use candles. In January 1819 five persons were killed by an explosion of fire-damp at Kells Pit; the fire-damp "appeared to have fired at John Fox's candle as he was formost & had his Lamp Case upon his back and his Lamp in it".[143] Despite this warning twenty more people were killed in October of the same year in the same pit and for the same reason – the men preferred to use candles although safety lamps were provided.[144] In an attempt to prevent a recurrence of such disasters caused by the foolhardy behaviour of the men clear regulations were issued to the Whitehaven miners:[145]

> "Notice is hereby given That if any Hewer or other Workman, employed in the Whitehaven Coal-Pits where any Danger can exist from Fire Damp, shall neglect to approach his Work with the SAFETY LAMP, or shall use CANDLES in doing so, he shall (for every offence) forfeit the Sum of Five Shillings.
>
> And when Orders are given to continued the Use of the Lamp, during the whole time of the Working, any Person, who shall remove the SAFETY CYLINDER, or use Candles in the Workings, shall also forfeit Five Shillings . . . . The Overmen, of the respective Districts in the Collieries, are responsible for the Miners being compelled to use the Lamp, and not to use Candles, except in those Parts where it is ascertained that no Danger will be occasioned by removing the Safety Cylinder, or using Candles."

It was a long time before the Whitehaven miners learned their lesson as the following list of accidents, caused by open lamps or the use of candles, shows:

Year	*Pit*	*No. Killed*	*Year*	*Pit*	*No. Killed*
1821	William	12	1834	Saltom	1
1823	William	32	1839	William	23
1824	Saltom	3	1842	Duke	1
1828	Croft	6	1844	Duke	11
1831	Croft	23	1847	Croft	4

Such accidents were not confined to Whitehaven; they occurred at other Cumberland collieries and the evidence given at the inquests following these accidents affords ample proof of the almost incredible carelessness and stupidity which then prevailed in the mines. The Report of the Children's Employment Commission, 1842, had acid comments to make upon the William Pit accident in 1839,[146]

> "In this case not only the same absurdity repeated of letting the men down into the pit before the "waster" had completed his survey, as produced the catastrophe of the John Pit* – the lesson afforded by the latter being wholly disregarded – but there was the further absurdity of the boy carrying a Davy lamp in the one hand for the sake of safety, and a naked lighted candle in the other, as if for the sake of danger.
>
> Beyond this there can be no conceivable thoughtlessness and folly; and when such management is allowed in the mines of two of the most opulent coal-proprietors in the kingdom, we cease to wonder at anything that may take place in mines worked by men equally without capital and science."

Men, women and children died from other causes – from falls of roof, suffocation by choke-damp, falling down the shaft, or being crushed by waggons. In 1826 nine men were suffocated by smoke from a fire originating in an underground engine room at William Pit.[147] Peile suspected that the Croft Pit explosion in 1831 was the result of arson since there was much discontent in that colliery at that time and commented, "if the misguided men committed the act of wilfully setting the place on Fire, of which we feel suspicious, the sacrifice of their lives may be considered an act of providence". He was unable to find any proof that the accident had been deliberately contrived by the miners, "but great proofs of their carelessness and neglect of orders, in not having the safety tops on their lamps, as most of the lamps were found without them".[148]

An attempt was made in 1835 to compile statistics of mining accidents in the preceding twenty-five years and the returns, although incomplete, show that 140 lives had been lost in Cumberland collieries during that period.[149] My own search of contemporary records shows that for the years 1810-34 inclusive at least 187 persons were killed in Cumberland mines. Deaths from fire-damp explosions caused by the use of open lamps or candles at collieries other than Whitehaven were:

Year	*Colliery*	*No. killed*
1825	Moorbanks Pit, Workington	3
1829	Isabella Pit, Workington	1
1831	Lady Pit, Workington	1
1833†	Lady Pit, Workington	13
1833	John Pit, Harrington	1
1834	John Pit, Harrington	1
1835	Isabella Pit, Workington	6
1838	John Pit, Harrington	40
1841	Countess Pit, Moresby	1
1842	Cleator Moor	1
1843	Countess Pit, Moresby	2
1845	Wyndham Pit, Cleator Moor	2
1847	Cleator Moor	1
1848	Whinny Hill Pit, Cleator Moor	30
1849	Threapthwaite Colliery	1

* 24 Oct. 1838 an explosion of fire-damp at John Pit, Harrington, which belonged to Mr Curwen, killed 40 people. This explosion was caused by the overman taking a lighted candle along a main road into the pit.

† According to Moore (*VCH* 367), "Workington Colliery until 1837 had been free from any very serious catastrophe". How many death constitute a "very serious catastrophe"?

In addition to the twenty-seven killed in the Chapel Bank disaster in 1837 and those listed above at least thirty-seven more persons were killed at Workington Colliery between 1816 and 1867. Of these eighteen were killed by falls of roof, four were drowned and four fell down the pit shaft.[150]

Following the accident at Workington in 1833 a local writer asserted, "To imagine that any other cause than the folly and neglect of workmen, in nine cases out of ten, produces these accidents, is preposterous, and it mostly happens that the unthinking instrument is himself a sacrifice to his own negligence".[151] The enquiry into the John Pit disaster in 1838 revealed that the pit was badly managed, its ventilation inadequate and that much slackness prevailed among the workmen. Witnesses stated that working with candles was quite common and one went as far as to say, "Never a hagger worked with a Lamp". This explosion had been caused by the overman, Harrison Kay, a drunken and headstrong man, taking a lighted candle along a main road into the pit.[152] Ten years later thirty people were killed at Whinny Hill Pit, Cleator Moor because some of the men had been foolish enough to remove the tops of their safety lamps in order to light their pipes; this was a common practice in this pit despite standing orders against it.[153]

According to the report of the Children's Employment Commission, 1842 "Accidents from ropes breaking are almost unheard of in Cumberland; nevertheless the ropes used are no larger than in Yorkshire, the pits are douible the depth; and the weight drawn is quite as heavy, and often heavier. This can be alone attributed to the system of drawing the baskets up bratticed shafts without any conductors at all. There is nothing to catch or jerk the ropes, to which the accidents are chiefly referrible, which occur so very frequently in Yorkshire".[154] This statement is inaccurate. In 1806 two men riding the basket down the shaft of North Pit, Whitehaven were killed when the rope broke and they fell to the bottom of the pit.[155] Three men were killed at Hodgson Pit, Harrington 28 December 1832 in the same way. This accident angered local miners who smashed the house windows of Curwen's agent. Anonymous letters to the editor of the *Cumberland Pacquet* alleged that the ropes in use were old and defective and known to be dangerous. One of these letters declared, "There can be no doubt Mr Tate, the overlooker and others higher in place were the sole cause of the mischief by not having sufficient materials for the purpose. The rope was an old one, and was so when put up and they were told by the men that it was not safe a long time before the accident occurred and it had broken three or four times within a month before. I hope you will not let them escape but expose such infamous conduct as it ought to be exposed". The editor of the *Pacquet* passed on the anonymous letters to the Curwen family and at the same time expressed his disbelief in the accusations of culpable negligence which they contained.[156] Very few local journalists would then have contemplated launching a newspaper campaign on behalf of miners against colliery proprietors. Moreover, the *Cumberland Pacquet* was virtually controlled by the Lowthers, the most powerful of Cumberland coal owners, and editorial policy was governed by the necessity to propagate the views of masters rather than men.

COSTS AND PROFITS

Higher wages contributed to the rise of mining costs in the early nineteenth century. John Buddle in 1817 calculated the cost of working and delivering a 44 cwt. waggon of coal to the ships from the individual pits of the Workington Colliery and the results of his calculations were:[157]

	Union		*Lady*		*Isabella*		*Church*	
	s.	*d.*	*s.*	*d.*	*s.*	*d.*	*s.*	*d.*
Working and delivering to ships	10	3	12	7	9	6	7	9¼
Difference between Pit & Shipping ton, one sixth.	1	8½	2	1	1	7	1	3½
Engine Coal & Machine	1	0½		2	1	4¼		3¼
Total	13	0	14	10	12	5¼	9	4

At that time a waggon of coal for shipment was sold for 16s. An analysis of the costs at Workington Colliery in 1832, based on pay-sheets Nos. 7, 8, 9 and 11, gives the following average figures:[158]

	Lady Pit Main Band			*Isabella Pit Moorbanks*			*Isabella Pit, Main & Little Bands*		
	£	*s.*	*d.*	£	*s.*	*d.*	£	*s.*	*d.*
Hewing	60	2	2	44	11	6	84	16	0
Overmanship	9	2	0	7	3	0	6	6	4
Horsekeeping	2	18	0	1	4	0	5	0	0
Driving & Trapping	15	9	0	6	16	4	10	13	0
Staithers	3	8	0	2	5	10	2	19	3
Bankmen	5	2	0	1	16	6	6	4	0
Heap expenses	2	6	6		14	6	3	3	0
Engine	4	12	8	4	4	0	12	0	0
Basket maker	8	8	0	3	5	10	9	3	0
Hookers & Drivers	2	17	0		3	4	6	15	0
Yard expenses	62	19	0	24	16	0	54	16	0
Horses Hire	20	2	0	4	16	0	12	16	0
Shift Work		—		11	4	0	6	1	0
Engine Coal Leading	4	11	0		—			—	
Sundries	10	3	0		—		3	3	0
Total	£212	0	0	£113	0	0	£223	0	0
Workings in Scores		185			79			198	

Yard expenses (i.e. the cost of the labour and materials of blacksmiths and carpenters) were particularly heavy and confirmed Dunn's assertion that the machinery and equipment were antiquated and inefficient. Horses, too, were expensive – they were hired to the colliery at 16s. per week each and they still had to be fed and drivers employed. Both at Workington and Harrington the leading of coal was the work of contractors who provided their own men and horses. They were paid 5d. per waggon from Isabella Pit and 5½d. per waggon from Lady Pit; coal from John Pit, Harrington was carried to Harrington harbour at a cost of 1s. per waggon.*[159]

* In 1814 15,391 waggons of coal were shipped from Harrington Colliery and for this coal horse costs amounted to £897 16s. 2d. or 1s. 2d. per waggon. (Buddle MSS. Vol. I, 27 Dec. 1814.)

At Whitehaven costs continued to rise. The annual average working cost per ton of coal at the pithead was:–[160]

Year	Howgill Colliery s.	Howgill Colliery d.	Whingill Colliery s.	Whingill Colliery d.
1812-1818	4	9	4	0½
1819-1825	4	9	4	2½
1826-1832	4	9½	4	4¾
1833-1839	4	6	4	8
1840-1842	5	3	1	10

Working costs varied widely from pit to pit. A comparison of the working costs per waggon of coal at four pits in 1841 gives these figures:–[161]

Pit	Cost per Waggon s.	Cost per Waggon d.
Saltom	9	6
Croft	8	0
Wilson	6	4
Duke	4	10

Costs and profits per waggon also varied from one colliery to another:–[162]

Year	Howgill				Whingill				Parton			
	Cost		Profit		Cost		Profit		Cost		Profit	
	s.	d.	s.	d.	s.	d.	s.	d.	s.	d.	s.	d.
1842	12	3½	4	2½	12	0¾	4	5¼	11	7¾	4	10¼
1843	12	3	2	5	10	5	4	3	10	7	4	1
1844	11	4	3	5½	9	9½	4	10½	10	11	4	1
1845	11	10	4	3	11	1½	4	10½	10	7½	5	2
1846	12	0	7	0	10	6½	8	7½	12	4½	6	8½
1847	13	1½	6	9	11	6	8	0	12	10	7	2

The increase in profits after 1844 is a reflection of the sharp rise in the price of coal, not of increased efficiency.

The Lowther family in the first half of the nineteenth century enjoyed a substantial income from their collieries. In the thirty-two years 1812-43 the Whitehaven collieries earned a total profit of £1,146,313, an average of £35,822 per annum. This income was subject to wide variations, declining sharply from the 1814 peak of £58,687 to £11,361 in 1843.

Profits of Howgill and Whingill Colliers[163]

Year	£	s.	d.	Year	£	s.	d.
1816	51,575	12	6	1836	29,069	14	9½
1817	45,909	7	4¾	1837	30,895	10	4½
1818	30,231	4	7¼	1838	22,927	0	3
1819	35,523	13	3¾	1839	20,461	17	10
1822	30,790	14	11	1842	15,564	9	7½
1823	28,484	11	7	1843	11,361	5	11½
				1844	16,311	13	1½
1832	27,318	14	3½	1845	19,516	19	7½
1833	27,630	10	5½	1846	36,307	18	11
1834	21,526	9	6	1847	33,002	6	7
1835	25,779	6	5½				

The post-war fall in profits may be attributed to the reduction in the price of export coal, lower coal shipments, the increased capacity of the waggon and the growth of development costs. Between 1816 and 1824 Whitehaven coal was sold to the ships at 16s. per waggon and at 15s. in 1834-5; between 1812 and 1815 the price was 18s. 6d. per waggon, over 100,000 waggons of coal being shipped at this price in 1814 and 1815. Coal shipments from the Whitehaven Colliery fell to 76,660 waggons in 1818 and it was not until 1836 that they again reached 100,000 waggons. To increase their profits the shipowners of Whitehaven established a selling agency in Dublin towards the end of 1826 and its work in keeping up prices aroused much opposition there. As a result coal exports from Whitehaven fell from 99,859 waggons in 1826 to 84,436 waggons in 1827 and to 82,436 waggons in the following year with adverse effects upon Lowther profits.*[164] In the next decade profits continued to decline. According to John Peile, writing in 1835, since 1815 the price of export coal had been reduced from 18s. 6d. to 15s. per waggon and the measure increased (equal to 2s. per waggon). These together equalled 5s. 6d. per waggon and upon the gross sales would amount to about £30,000 per annum.[165] Colliery profits were further diminished by the defalcations of a colliery viewer, William Quayle. Peile discovered that Quayle had embezzled £3,770 from the proceeds of coal sales. When confronted by Peile Quayle was unable to clear himself and to avoid arrest committed suicide by throwing himself down the shaft of James Pit, 12 July 1837. This did not end the matter. Peile, together with Lonsdale's lawyers, exerted a great deal of pressure on Quayle's widow to compel her to refund the money stolen by her husband.[166]

After 1837 profits from Whitehaven Colliery fell year by year, amounting to only £11,361 in 1843. A severe depression in the coal trade injured both colliery owners and coal traders and at Whitehaven in 1843 short-time working became necessary because of the fall in the demand for coal. A six weeks' strike at the colliery in the same year made matters worse. To attract buyers the price of coal had to be reduced from 16s. 6d. to 15s. per waggon. This was followed by a period of recovery, profits rising to £36,307 in

* Figures of Whitehaven Colliery profits 1824-31 are not available but the total profits of *all* Lowther collieries fell from £39,347 in 1826 to £29,751 in 1828. (R. W. Moore, MS. "History of Whitehaven Collieries").

1846, more than three times the 1843 figure and the highest annual profit since 1817. A revival of the coal trade, accompanied by large increases in the price of coal, accounted for this sudden change of fortune. From 15s. per waggon in 1843 the price of coal rose to 20s. in 1845 and to 21s. per waggon in 1846. The new Wellington Pit made a valuable contribution to Lord Lonsdale's income, earning a total profit of £34,164. 17s. 8d. in the years 1843-47; it had, therefore, more than repaid the £25,000 spent in winning it.[167]

Other collieries added to the Lowther revenues. In the eleven years 1836-46 the Countess Pit at Parton produced a total profit of £70,783. 9s. 7d., an average of £6,434. 12s. per annum; annual average coal sales amounted to 24,220 waggons, each waggon costing 11s. 1½d. and yielding a profit of 5s. 4d. Between 1832 and 1847 annual profits of the Distington Colliery ranged from £1,100 to £3,500 and Clifton Colliery in the years 1843-47 earned a total profit of over £11,000. From their collieries at Whitehaven, Parton, Distington and Clifton and from a lime works at Overend the Lowther family derived the following income:–[168]

Year	£	*s.*	*d.*	*Year*	£	*s.*	*d.*
1816	52,995	5	4¼	1836	38,368	19	3
1821	34,239	15	5½	1841	22,883	8	3½
1826	39,347	18	4¾	1846	48,356	17	11
1831	25,913	14	7				

For many years the Curwen collieries had earned good profits; in 1806 as much as £20,000. Thereafter profits declined and J. C. Curwen, like other colliery owners, was badly hit by the post-war trade depression, by the fall in prices and coal sales. For an undertaking of its size Workington Colliery after the Napoleonic War yielded small profits:–

Year	*Colliery*	*Profit* £	*Colliery*	*Loss* £
1818	Moorbanks	1,966	Chapel Bank	1,573
1819	Moorbanks	1,058		
	Chapel Bank	2,467		
1820	Moorbanks	488		
	Chapel Bank	1,666		

The colliery's average annual profit between 1818 and 1820 was therefore £2,024. But the above figures disregard the sum of £2,855. 15s. 11d. spent on the new Buddle Pit which if included in the working costs would reduce the average annual profit to £1,072.[169] Profits at the colliery amounted to £3,342. 18s. 7½d. in 1821 and £2,805. 12s. 8d. in 1822. In the latter year Harrington Colliery produced a profit of £1,549. 8s. 2½d. and Broughton Colliery £215. 12s. 8¾d.[170] Clearly coal mining did not always yield unreasonably high profits and the return to capital was sometimes slight. It was a business that involved much risk and, to quote Buddle, "Perhaps no stronger Proof of the great Risk can be adduced than that it is a property which cannot be insured either against Fire, Water, or any other Accident. We have never been able to effect an insurance on a Coal Work". In addition it was impossible to borrow money on collieries without collateral security almost equal in amount to the money borrowed.[171]

Increasing costs and declining profits compelled Curwen to borrow money on a large scale from a wide variety of individuals. It is reasonable to suppose that the expenditure involved in sinking Isabella Pit was a major cause of his indebtedness. His financial position in his last years may be summarized in this way:[172]

Year	*Money Owed to J. C. Curwen*			*Money Owed by J. C. Curwen*		
	£	s.	d.	£	s.	d.
1818	14,649	13	9½	118,042	4	7
1819	16,003	17	10	120,522	4	1½
1820	15,749	19	5	120,567	15	7½
1821	21,020	15	8	124,302	7	6
1822	20,836	14	6½	124,710	7	10
1823	16,007	15	10½	124,111	3	1½
1824	14,827	9	1½	124,796	17	2½
1825	15,202	12	3½	126,818	16	11½
1826	17,304	17	9½	126,763	13	7½
1827	16,996	12	5½	122,606	18	1
1828	16,579	12	8	118,334	9	0

Thus the combined resources of the Christian and Curwen estates had proved inadequate to finance the collieries and the injection of borrowed capital could not stop the rot. The high hopes and prosperity of Curwen's early years gave way to falling output, smaller profits and a large debt burden and the collieries themselves were badly in need of a thorough overhaul.

The loss of the Chapel Bank colliery in 1837 was a grievous blow to the Curwen family. For a very short time the dangerous policy of pillar robbing to increase the colliery's output at lower costs seemed to be succeeding. Curwen's coal shipments from Workington went up from 21,086 waggons in 1835 to 32,089 in the following year. But this success was temporary and led to the destruction of the colliery. As a result only 2,460 waggons of coal were shipped by Curwen from Workington in 1838.[173] After the Chapel Bank disaster the profits of the Curwen collieries slumped badly. For the five years 1837-1841 Harrington Colliery earned an average annual profit of £2,277 and Workington Colliery an average annual loss of £1,969. For the two collieries together the annual average profit was no more than £308.[174] It is not surprising that Henry Curwen in 1839 was trying to borrow £40,000 on the security of his estate.

In this period the coal belonging to Lord Egremont continued to be worked by others; but his income compared favourably with that of the Curwens and was obtained without capital expenditure or risk to himself. In 1832 and 1833 his coal rents were:[175]

Colliery	*1832*			*1833*		
	£	s.	d.	£	s.	d.
Bolton	1,063	12	1	928	3	4
Aspatria	84	3	6	67	10	0
Broughton	1,034	8	3	1,039	10	9
Greysouthen	323	7	9¾	287	3	6½
Dean	59	17	2	199	2	3
Total	2,565	8	9¾	2,521	9	10½

THE COAL TRADE

Complete statistics of the Cumberland coal trade between 1815 and 1850 are not available but the *Report of the Royal Commission on Coal in the United Kingdom* (1871) provides figures of coal shipments from the "Port" of Whitehaven (i.e. shipments not only from Whitehaven itself but also from Workington, Harrington and Maryport) for the years 1819-28, 1833 and 1835-41. After Maryport became an independent port in 1841 separate statistics of its shipments were collected and these together with those from the "Port" of Whitehaven give the total of coal exports from Cumberland. These are given below; the figures for 1829 and 1834 are taken from Parliamentary papers. John Peile in his colliery reports assembled statistics of Cumberland coal exports, based upon Custom House returns, for the years 1831-40 and 1842-7 and these may be compared with those in the 1871 Report.

Cumberland Coal Shipments[176]

Year	*1871 Report* *Tons*	*Peile Reports* *Waggons*
1819	256,036	
1820	253,588	
1821	197,436	
1822	268,084	
1823	297,629	
1824	297,230	
1825	287,468	
1826	295,935	
1827	307,346	
1828	277,413	
1829	349,144	
1830		
1831		161,402
1832		173,097
1833	363,528	162,849
1834	372,645	165,531
1835	373,704	174,731
1836	423,504	182,975
1837	437,683	188,790
1838	422,819	177,828
1839	461,804	180,340
1840	447,697	190,076
		Tons
1841	474,848	
1842	465,647	451,734
1843	445,783	451,648
1844	475,837	484,499
1845	533,030	533,298
1846	547,256	532,835
1847	507,604	489,135
1848	482,836	
1849	530,650	
1850	544,624	

Official papers do not provide figures of coal shipments from the individual Cumberland ports until they were separated for customs purposes. Thus figures for Maryport are not available before 1842 or for Workington before 1850; The Workington statistics also include shipments from Harrington. Fortunately, John Peile's records are more informative than the official returns, giving details of coal shipments from the various Cumberland ports between 1815 and 1847 and since these do not exist elsewhere they are given here:

Coal Shipments from:–[177]

Year	*Whitehaven Waggons*	*Workington Waggons*	*Maryport Waggons*	*Harrington Waggons*
1816	92,097			
1817	93,926			
1818	76,660			
1819	87,174			
1820	91,531			
1821	87,024			
1822	90,335			
1823	93,169			
1824	87,070			
1825	90,732			
1826	99,859			
1827	84,436			
1828	82,436			
1829	87,471			
1830	94,416			
1831	90,908	34,702	29,466	6,326
1832	96,434	37,254	30,969	8,440
1833	93,600	31,951	27,595	9,703
1834	94,279	33,402	28,855	8,995
1835	98,247	39,350	27,646	9,488
1836	101,019	46,944	22,799	12,213
1837	105,904	29,571	31,694	21,621
1838	94,846	25,560	35,789	21,633
1839	94,665	27,485	44,170	21,197
1840	97,684	23,148	49,359	19,885
	Tons	*Tons*	*Tons*	*Tons*
1842	205,518	71,731	140,473	34,012
1843	205,800	73,954	143,147	28,757
1844	215,182	74,209	168,709	26,399
1845	231,893	80,094	190,006	31,305
1846	220,870	70,430	213,152	28,383
1847	217,576	60,021	183,384	28,154

Whitehaven remained the leading port in Cumberland for coal exports and until 1840 its coal shipments exceeded those of the other three ports put together. This was not the case afterwards for its supremacy was being increasingly threatened by competition from Maryport whose exports were greatly enlarged by the coal moved there from the inland collieries by the new Maryport & Carlisle Railway. In the last years of the Napoleonic war there was a boom in the coal trade with shipments from Whitehaven reaching record heights, 110,193 waggons in 1814 and 107,100 waggons in 1815. This was followed by a trade depression. A reduction in the price of coal at the Cumberland ports after the war brought little advantage to the shipmasters for the trade was now entering a period of decline. In the summer of 1816 Whitehaven coal was reduced from 18s. 6d. to 16s. per waggon. This substantial reduction in price was followed by a strike of the Whitehaven seamen engaged in the coal trade. They objected to proposed wage cuts and for three months nearly a hundred vessels were laid up, but in late October the seamen had to accept the reduced wages and return to work.[178] At Workington the coal traders were also suffering. William Swinburn, J. C. Curwen's agent, informed John Buddle (17 May 1817), "The coal trade had never since I knew it a worse aspect – the shipments have indeed not been so bad lately as the profits to the ship owners . . . . The principal cause seems to me to be from the poverty and distress of the Irish – compelling them to use only (perhaps) half the quantity of what they formerly consumed & the same quantity being exported from this coast causes such an overplus in the market as to leave no profits to the carriers".[179] In 1818 only 76,660 waggons of coal were exported from Whitehaven. Nevertheless, large quantities of Cumberland coal were still being sold in Ireland. In the year ended 5 January 1819 Ireland imported 323,335 chaldrons of coal, Winchester measure, from Great Britain; of this quantity 155,934 chaldrons or 48·22 per cent, came from the Port of Whitehaven,* this being the largest amount sent from any British port. All the Scottish ports together shipped 76,324 chaldrons, Liverpool 41,117 chaldrons and Swansea 26,225 chaldrons.[180]

Local newspapers suggest that 1823 was a particularly bad year for the coal trade which was then said to be even more depressed than agriculture, "It will no longer bear insurance; no new vessels are embarked in it".[181] A later issue of the same newspaper stated that there were only 50-60 vessels at Whitehaven shipping coal and that, "No new vessels have been built for the coal trade for many years . . . . Last year there were 90 vessels in the coal trade".[182] The editor of the *Whitehaven Gazette* believed it was a flourishing trade for the colliery owner who continued to charge the same high price for his coal, but a declining trade for the ship owner who was receiving a much lower price in Ireland than formerly and had to pay higher duties and port charges.[183] Ship owners were also harmed by the malpractices of the Whitehaven waggoners who "are merely paid for filling and the less quantity they give, the less labour they have to perform. This is an abuse of long standing, to rectify which, masters of vessels are obliged to fee the waggoners to the amount of from 15s. to 20s. each cargo. If they do not comply with this imposition the waggoners give short quantity".[184]

In the year ended 5 January 1826 Great Britain exported 695,028 tons of coal to Ireland of which 265,121 tons or 38·14 per cent came from the Port of Whitehaven. Thus in seven years Cumberland's share of the Irish market had fallen by ten per cent. Coal shipments from some of the other ports were:[185]

Port	*Tons*	*Port*	*Tons*
Gloucester	5,667	Chester	5,557
Cardiff	11,644	Liverpool	61,932
Newport	96,668	Preston	8,114
Swansea	43,659	Glasgow	18,051
Llanelly	20,339	Irvine	105,716
Milford	1,463	Air	47,334

* Shipments from the "Port of Whitehaven" must include shipments from Workington, Harrington and Maryport as well as from Whitehaven.

Clearly between 1818 and 1825 Cumberland coal was having to meet increasing competition from Scottish and Welsh coal in the Irish market. Whitehaven coal continued to dominate the Dublin market, but by 1830 it had been supplanted in Cork and Waterford by Welsh coal. According to one witness, "There is some sent from Wales to Dublin; but there is none can compete with the Whitehaven for use in houses. Whitehaven coal used to supply Cork and Waterford; but the Welsh coal has beaten the Whitehaven coal almost entirely out of those markets".[186] Peile explains the loss of these markets,

> "When we had so large a share of the Trade to those Ports the Welsh collieries were scarcely known to export any Coal, but as these Collieries extended and being so much more contiguous, the reduced expense of sailing the Welsh vessels compared to ours & the Brokers or Merchants at these ports in Ireland, connecting themselves with the Welsh shipowners, a double Interest was created against us, as well being able to undersell the Cumberland vessels, and thus combined to drive us out of the market. At the time I allude to, Workington & Harrington also shipped largely to these ports, and are now equally excluded with ourselves . . . . But altho' we lost the down channel Trade it has been compensated by the increased Sale to Dublin & other Ports."[187]

By this time the Whitehaven shipowners had learned the benefits of organization; the profits derived from their trade within Dublin resulted from a policy of price-fixing and controlled sales. To ensure high prices for their coal they established a selling agency in Dulin in 1826 – they asserted, of course, that unorganized sales led to price-cutting and total loss of profit. Two agents were appointed to reside in Dublin to whom all cargoes were consigned and who were responsible for their disposal, the expenses of the agency being met by a levy on the amount realized by the sale of each cargo.[188] One Dublin merchant declared that this "conspiracy" enabled them to obtain £20,000 a year more for their coal than they would otherwise have received. He further alleged that any attempt to undercut the price fixed for the coal resulted in vessels being delayed at Whitehaven the next time they tried to load coal. These restrictions applied to the Whitehaven-Dublin trade only, Whitehaven coal destined for other ports being free from such price-fixing since the Whitehaven shipowners were not in a position to dictate prices elsewhere. Thus Whitehaven coal cost 16s. per ton in Dublin, but only 12s. per ton in Belfast where it competed with Scottish coal.[189]

Peile had little love for the Whitehaven shipowners and less for their Dublin agency and it was with reluctance that he co-operated with them. He realized that the scheme was designed to benefit the shipowners rather than the colliery owner and that any odium generated in Dublin by the maintenance of high prices could easily be transferred from the coal traders to the Lowthers and their collieries. The Dublin agents incurred much unpopularity and this was reflected in some reduction in the Whitehaven coal trade in Dublin. The reasons for this decline are listed in a letter to Peile from his brother:[190]

1. The greatest of all is "scarcity of money" and selling prices nearly alike in winter and summer. There is no incentive to lay up coal stocks in summer.

2. Government contracts nearly always specify best Whitehaven coal. But some factors mix inferior coal (from Workington, Harrington etc.) with Whitehaven coal.

3. Workington and Harrington coal cost 2s. 6d. per ton cheaper than Whitehaven coal and therefore there is less demand for Whitehaven coal.

4. The foolish pertinacity of the Whitehaven Committee keeping to 15s. per ton created an increased prejudice to the agency system and lessened our consumption.

In February 1834 Peile wrote to Lord Lonsdale, "The Mas^{rs} of the vessels are we think keeping the price in Dublin too high, unfair to the consumers & detrimental to us by encouraging the sale of the inferior Coals". He added that he had informed the chairman of the Whitehaven Committee to regulate the coal trade "that we could not sanction any thing like imposition on the citizens of Dublin, or submit to sacrifice ourselves for their exclusive benefit".[191] Later in the same year he refused the committee's request that ship-masters who disobeyed the regulations should be delayed in port for fourteen instead of seven days.[192]

Relations between Peile and the Committee deteriorated. By 1840 Peile had become tired of the Committee which for years had been growing "more irksome" to him; the shipowners profited but hardly the Lowther family which earned undeserved unpopularity for "supporting a monopoly". In May 1840 Peile, against the wishes of the Whitehaven Committee, reduced the price of coal by one shilling per waggon and also announced his intention of withdrawing entirely from the system of trade regulation.[193] At last (31 December 1840) Peile carried out his threat and severed his connection with the coal agency.

At Whitehaven the coal trade in 1841 and 1842 was said to have yielded no profit to those who had invested in ships and many shipowners had suffered great loss. Peile reported in December 1842, "The Whitehaven Coal vessels are double the number wanted, are extravagantly managed in Wages and Dietary and the Captains & the Men are in League to resist any reduction". Irish vessels, however, "on low wages & diet" were regularly increasing. A depression in the iron industry had restricted exports of iron ore and this, together with the decline in the coal trade, had brought shipbuilding at Whitehaven almost to a standstill.[194]

Although the shipowners had devised the agency system to control coal prices in Dublin it proved impossible to enforce complete price unformity. Too often the shipmasters, anxious to dipose of their cargoes, were ready to undersell each other.[195] Nor were the two Dublin agents, Adamson and Tomlinson, wholly satisfactory. They were unable to co-operate in their work and both were late in handing over the proceeds of coal sales. In 1831 Tomlinson was appointed sole agent; sixteen years later he turned out to be an embezzler. In exchange for the money the shipmasters had obtained from coal sales and which they had entrusted to him, Tomlinson gave them worthless letters of credit. His frauds did much harm to many shipowners and coal factors.[196]

In 1815 a waggon of coal exported from Whitehaven cost 18s. 6d.; thereafter the price fluctuated:–[197]

Year				
1816	28 July	17s.	5 Aug.	16s.
1825	12 Sept.	17s.		
1828	11 Feb.	16s.	8 Dec.	17s.
1829	22 June	16s.		
1834	7 Apr.	15s.		
1836	1 Jan.	16s. 6d.	17 Oct.	17s. 6d.
1837	2 Jan.	18s. 6d.		
1838	Apr.	17s. 6d.	May	16s. 6d.
	Oct.	17s. 6d.		

Year		
1839	May	16s. 6d.
	Oct.	17s. 6d.
1840	11 May	16s. 6d.
1843	6 Mar.	15s.
1844	2 Dec.	16s.
1845	1 Sept.	17s.
	2 Oct.	18s.
	1 Dec.	20s.
1846	26 Oct.	21s.
1848	10 Apr.	19s.

The capacity of the waggon was not constant. In the eighteenth century the Whitehaven waggon held about 40 cwts. which increased to 42 cwts. in 1814. An Act of Parliament compelled the sale of coal from 1 January 1836 by weight and not by measure and the weight of coal per Whitehaven waggon was then fixed at 45 cwts. In 1841 this was raised to 48 cwts. and to 50 cwts. in 1870.

As at Whitehaven, coal shipments from Workington suffered from the fluctuations in the coal trade, but the general trend was downwards. In 1836, however, 46,944 waggons of coal were exported from Workington, the highest figure since 1805; of this total 32,089 waggons came from Curwen's Workington Colliery. The loss of his Chapel Bank Colliery in the following year ended any chance of such quantitities being ever exported again; in the years 1837-40 coal shipments from Workington were less than 30,000 waggons per annum. Curwen's shipments from Workington fell to 2,460 waggons in 1838 and two years later amounted to only 2,681 waggons.[198]

For many years coal shipments from Maryport lagged behind those of Whitehaven and Workington. In 1816 only 6,169 waggons of coal were exported from Maryport,[199] but in later years the coal trade there expanded considerably, close to 50,000 waggons of coal being shipped in 1840. By 1837 Maryport was exporting more coal than Workington and by 1846 was approaching the level of Whitehaven exports. A major reason for the growth of coal shipments in the 1840's was the Maryport & Carlisle Railway. The owners of inland collieries such as Brayton, Gilcrux and Oughterside were quick to take advantage of the cheaper and faster transport provided by the M. & C. railway. As early as 1841 W. Peile of Gilcrux Colliery was moving his coal by rail to Maryport and by 1844 over 100,000 tons of coal a year were being carried by the railway to the port for shipment.[200]

Co-operation among the colliery owners in Cumberland was closest when coal prices were fixed. Thus when Henry Curwen reduced the price of Workington coal in 1834 Whitehaven coal was reduced by one shilling per waggon, a reduction which would cost Whitehaven Colliery about £5,000 per annum. To make up for this Peile lowered the wages of all employees by 6d. to 1s. per week.[201] An increase in the price of Whitehaven coal in 1836 led Curwen to write, "In consequence of a communication from Mr Peile, that they had advanced their coals another shilling we have done the same & as we give our former weight I believe our coals are yet the cheapest on the coast".[202] When the sale of coal by weight became compulsory Benjamin Thompson, Curwen's agent, wrote to Peile, "If you will be so good as to let me know what you intend doing in the Coal Trade as to the selling Coals by weight & the price I will endeavour to cooperate as far as I can". A month later he wrote, "Since my last to you we have finished the weighing Machine & we find the weights of the waggons much exceeding what was expected. Mr Curwen has fixed the standard at Fifty two Cwt. to the waggon which we are now giving. It began yesterday. The price is the same as before".[203] The difference in the capacity of the waggon at the Cumberland ports proved embarrassing to Whitehaven as Peile pointed out to Mr Curwen in 1841,[204]

> "We have been under the necessity of giving an increase in Weight in consequence of such excess being given at the northern ports (particularly Maryport) and with less price with some. Your Ship Mas^{rs} have been thus enabled to so far undersell and make voyages as nearly to bring our Trade to a stand-still whilst defending the whole, a course of things you must agree with me could not remain and must eventually involve the whole in greater difficulty.

Our weight before was 45 Cwt. to the Waggon & I believe the delivery by the Ships was regular 45 to 45½ Cwt. – whilst all to the North exceed 50 & some very much higher – our increase is to 48 *Net* (no extra allow^e) weighing every waggon. I trust you will not find it requisite at present to make any change.

I have had a conversation with an extensive shipper on the benefit that would arise to us all by abolishing the Term Waggon and adopting the *Ton* a definite quantity and each Shipper to regulate his price per Ton according to the Quality. This would give a steady character to the Trade and by adhering to the Net Ton do away with deceit towards each other. If such a measure meets your approbation I shall be glad to assist in bringing it about."

By 1842 the colliery owners north of Whitehaven were meeting regularly to fix the price of coal to be charged by individual collieries.[205] A letter from Thomas Westray to J. P. Senhouse states, "I am requested to forward to you a reduced Scale of prices upon Coal exported from the undermentioned Collieries fixed upon and agreed to at a special meeting of Coal proprietors held yesterday at the Green Dragon Inn Workington, in consequence of Ld. Lonsdale's having reduced the export price of his Coal at Whitehaven 1/6d. a Waggon".[206] In 1844 when the shipping price of Whitehaven coal was increased by 1s. per waggon the increase was adopted at all the Cumberland ports. Peile, in a letter dealing with this matter, informed his employer, "I wrote to Mr Curwen on Saturday (advising as we had done on former occasions) of the intended advance, intimating that it would be adopted by the other shippers and hoped he would concur in it". He estimated that this rise in price would add £5,000 per annum to Lord Lonsdale's profits and although some advance in wages might be necessary he hoped it would be trifling and every reasonable resistance to it would be made.[207]

A reduction in the price of Whitehaven coal in 1848 led to a meeting of the other Cumberland colliery owners at Workington where they agreed to the following scale of prices per ton to be charged by members of their association:–

Wood & Co.,	Dearham	7s. 8d.	Mr Harris	Ewanrigg	7s. 6d.
Messrs Walker	Dearham	7s. 8d.	Mr Harris	Greysouthen	7s. 8d.
Mr Harris	Brayton	8s. 2d.	Mr Fletcher	,,	7s. 8d.
Mr Kirkhaugh	Oughterside	8s. 2d.	Lord Lonsdale	Clifton	7s. 8d.
Mr Peile	Gilcrux	8s. 2d.			

These prices could not be changed except by giving notice to the association. The price of Whitehaven coal at this time was 7s. 11d. per ton. A fortnight later these prices were further reduced by 10d. per ton.[208]

Notes and References

1 D/Lons/W, Peile reports and letters.
2 D/Lons/W, Peile report on Whitehaven Collieries 1847. *VCH*, 364.
3 Peile, 30, 31 May, 20 July, 4 Oct. 1847.
4 D/Lons/W, Bateman's Report on Collieries 1781; Unlisted papers, "Mr Lysons' Queries". Moore, MS "History of Whitehaven Collieries". Bateman, 22 Dec. 1806.
5 R. W. Moore, "Whitehaven Collieries", *Trans. of the Federated Institute of Mining Engineers*, vii (1893-94), 632-4.
6 D/Lons/W, Collieries 46.

[7] Forster, View Book of Matthias Dunn. Watson, Vol. 92 D11-70. M. Dunn, *Winning and Working of Collieries* (Newcastle, 1848), 328-31.
[8] D/Lons/W, Collieries 46.
[9] D/Lons/W, Peile reports on Whitehaven Collieries 1842-7.
[10] D/Lons/W, Peile reports on Whitehaven Collieries 1846. *VCH*, 372.
[11] D/Lons/W, Royalty Accounts 1841-9.
[12] *VCH*, 372.
[13] D/Lec, 240, Cleator Moor Colliery papers.
[14] *VCH*, 366.
[15] Buddle, vol. i, 27 Dec. 1814, 18 Sept. 1822.
[16] D/Cu, 3/43, Letter book of William Swinburn 1805-17.
[17] Buddle, vol. i, 18 Sept. 1822.
[18] D/Cu, Account of loans to and from J. C. Curwen.
[19] Buddle, vol. i, Aug. 1816.
[20] Buddle, vol. i, 20 Oct. 1817, 9 Jan. 1821.
[21] *VCH*, 367.
[22] D/Lons/W, Letters W. Peile to Lord Lonsdale, 18 Nov. 1833.
[23] D/Cu, 3/45, Letter book of B. Thompson, Thompson to Dunn, 9 April 1836.
[24] Peile, 12 Sept. 1836
[25] D/Cu, 3/45, Letter book of B. Thompson.
[26] M. Dunn, *op. cit.*, 230-2.
[27] Children's Employment Commission 1842, *First Report*, App. Pt. I, 310 – evidence of the Rev. Henry Curwen.
[28] *Cumberland Pacquet*, 1 Aug. 1837.
[29] D/Cu, 3/83, E. Stanley to E. S. Curwen, 29 April 1838.
[30] *Cumberland Pacquet*, 8 Aug. 1837.
[31] *VCH*, 367, Mannix and Whellan, *History and Directory of Cumberland* (Beverley, 1847), 423.
[32] *Cumberland Pacquet*, 3 June 1845.
[33] *VCH*, 367, Mannix and Whellan, *op. cit.*, 423.
[34] Forster, View Book of M. Dunn.
[35] *Ibid.*
[36] Mannix and Whellan, *op. cit.*, 335.
[37] D/Lons/W, Moore MS, "History of Whitehaven Collieries"; Peile reports on Lonsdale collieries 1842-7. *VCH*, 368-9.
[38] Lysons, *op. cit.*, IV, cxxiv.
[39] Carlisle Record Office, DX/425.
[40] *VCH*, 370.
[41] D/Cu, Colliery lease.
[42] D/Lec, Peile reports on Lord Egremont's collieries 1834. *VCH*, 370.
[43] *VCH*, 372.
[44] D/Lec, 116, Dean Colliery papers.
[45] D/Lons/W, Peile reports on Lonsdale collieries; Moore's MS, "History of Whitehaven Collieries".
[46] *VCH*, 371.
[47] D/Lons/W, Peile reports on Lonsdale collieries. *VCH*, 373.
[48] D/Cu, Unerigg Colliery Accounts 1826-7.
[49] D/Lec, Peile to W. Tyler, 29 Mar., 20 June 1831.
[50] D/Lec, Peile to Tyler, 21 Jan. 1831.
[51] Peile, 7 June 1841.
[52] D/Sen, Netherhall 19/234.
[53] *VCH*, 376.
[54] D/Sen, Netherhall 19/125, 19/173.
[55] D/Ben, Christian colliery records, Broughton Colliery Sales 1811-22.
[56] D/Cu, Misc. Colliery papers, "Mr Christian's profits of Broughton Colliery".
[57] D/Lec, 65/7, Broughton Colliery papers.
[58] D/Cu, Misc. Colliery papers, Settlement of wayleave by John Buddle and John Peile 1823.

[59] D/Lec, 65/7, Broughton Colliery papers.
[60] *VCH*, 376. D/Lons/W, Custom House returns quoted by Peile in his reports of coal shipments 1831-47.
[61] *Ibid.*
[62] *VCH*, 377. D/Lons/W, Custom House returns quoted by Peile in his reports of coal shipments 1831-47.
[63] D/Lec, 65/9, Birkby Colliery papers.
[64] *VCH*, 377-8.
[65] D/Lec, 41, Westmoor Colliery papers.
[66] D/Lec, 41, 59 Westmoor Colliery papers.
[67] *VCH*, 378. Peile's report on Lord Egremont's collieries 2 Mar. 1831.
[68] D/Lec, 41, Westmoor Colliery papers.
[69] *VCH*, 378.
[70] *VCH*, 378. D/Lons/W, Custom House returns quoted by Peile in his reports of coal shipments.
[71] D/Lec, 89, 90 Bolton Colliery papers; Peile reports on Lord Egremont's collieries 1831-4.
[72] *Cumberland Pacquet*, 19 Jan. 1841.
[73] D/Lec, 89, Bolton Colliery papers.
[74] D/Lec, Dixon report 1860.
[75] Ayton and Daniell, *op. cit.*, ii, 166.
[76] W. Whellan, *History and Topography of Cumberland and Westmoreland* (Pontefract, 1860), 321.
[77] Maryport Harbour records, Minutes of meeting, 8 Oct. 1833.
[78] J. W. Crerar, "Maryport and the Ballot", *West Cumberland News*, 5 Aug. 1939.
[79] Maryport Harbour records, Minutes of meetings, 23 Oct. 1833, 13 Feb. 1834.
[80] J. Simmons, *op. cit.*, 1-9.
[81] *Carlisle Patriot*, 13 Feb. 1841.
[82] *Carlisle Patriot*, 23 Mar. 1844.
[83] D/Lons/W, Custom House returns quoted by Peile in his reports of coal shipments.
[84] Whellan, *op. cit.*, 322.
[85] Simmons, *op. cit.*, 1.
[86] Maryport Custom House records.
[87] D/Lons/W, Custom House returns quoted by Peile.
[88] D/Lons/W, Peile's report on Whitehaven Collieries 1842.
[89] W. Mc. G. Gradon, *Furness Railway: Its Rise and Development, 1846-1923* (Manchester, 1946), 12-31.
[90] *Report of Select Committee on Accidents in Mines, 1835*, 160.
[91] *Ibid.*
[92] *VCH*, 356.
[93] Peile, 21 Dec. 1840.
[94] Galloway, *op. cit.*, ii, 295.
[95] J. H. Clapham, *Economic History of Modern Britain* (Cambridge, 1931), i, 437.
[96] Peile, 7 June 1842.
[97] Peile, 9 April 1838.
[98] *Cumberland Pacquet*, 9 April 1816.
[99] Bateman, 24 Jan. 1803.
[100] *VCH*, 364.
[101] Forster, View Book of M. Dunn, Dec. 1832.
[102] M. Dunn, *Report on Coal Mines, 1851*, 4.
[103] *VCH*, 356.
[104] Children's Employment Commission, 1842, *First Report*, 16.
[105] *Ibid.*, *First Report*, Appendix Pt. I, 299.
[106] *Ibid.*, *First Report*, Appendix Pt. I, 876.
[107] D/Lons/W, Misc. Colliery Papers, paper dated 1842 listing numbers of different categories of colliery employees.
[108] Children's Employment Commission 1842, *First Report*, 87.
[109] *Ibid.*, *First Report*, Appendix Pt. I, 305.
[110] D/Lons/W, Misc. Colliery Papers, paper dated 1842.
[111] Children's Employment Commission 1842, *First Report*, 87.
[112] *Ibid.*, *First Report*, 109, Appendix Pt. I, 881.

[113] *Ibid.*, *First Report*, Appendix Pt. I, 302.
[114] *Ibid.*, *First Report*, Appendix Pt. I, 303, 308.
[115] D/Lons/W, Misc.Colliery Papers.
[116] Children's Employment Commission 1842, *First Report*, Appendix. Pt. I, 876.
[117] *Ibid.*, *First Report*, Appendix Pt. I, 303.
[118] *Cumberland Pacquet*, 5 Sept. 1843.
[119] *Report of House of Lords' Select Committee on the State of the Coal Trade 1829*, 48.
[120] Children's Employment Commission 1842, *First Report*, Appendix Pt. I, 304.
[121] *Ibid.*, *First Report*, Appendix Pt. I, 300.
[122] *Ibid.*, *First Report*, 115-16, 120-1.
[123] R. Rawlinson, *Report to the General Board of Health on . . . the Sanitary Condition of the Inhabitants of the Town of Whitehaven* (1849), 13-17.
[124] *Report of the Commissioner on the State of the Population in the Mining Districts* (1851), 25.
[125] Peile, 21 April 1845.
[126] Peile, 19 May 1845.
[127] Peile, 31 Jan. 1831.
[128] Peile, 7 Mar. 1831.
[129] Peile, 24 May 1831.
[130] Piele, 13 June 1831.
[131] Peile, 28 July 1831.
[132] Peile, 6 Aug. 1831.
[133] Peile, 12-29 Aug. 1831.
[134] Peile, 28 July 1831.
[135] *Cumberland Pacquet*, 5 Sept. 1843.
[136] *Carlisle Patriot*, 7 Oct. 1843.
[137] R. Challinor and B. Ripley, *The Miners' Association* (London, 1968), 157.
[138] *Cumberland Pacquet*, 5 Sept., 30 Oct. 1843.
[139] *Carlisle Patriot*, 14 Oct. 1843.
[140] Challinor and Ripley, *op. cit.*, 158.
[141] A. R. Griffin, *The British Coalmining Industry* (Buxton, 1977), 88.
[142] D/Lons/W, Henry Jackson's Journal, 12 Jan. 1819.
[143] *Ibid.*, 27 Jan. 1819.
[144] *Cumberland Pacquet*, 2 Nov. 1819.
[145] D/Lons/W, "Regulations in the Whitehaven Collieries", 13 Nov. 1819.
[146] Children's Employment Commission 1842, *First Report*, 146.
[147] *VCH*, 364.
[148] Peile, 5, 26 Mar. 1832.
[149] *Report of Select Committee on Accidents in Mines 1835*, iv.
[150] D/Lec, Reports of Coroners' inquests. Local newspapers.
[151] *Cumberland Pacquet*, 24 Sept. 1833.
[152] *Whitehaven Herald*, 27 Oct., 3 Nov. 1838.
[153] *Cumberland Pacquet*, 31 Oct. 1848.
[154] *First Report*, Appendix Pt. I, 302.
[155] Bateman, 14 April 1806.
[156] D/Cu, Misc. colliery papers. *Cumberland Pacquet*, 1, 8 Jan. 1833.
[157] Buddle, vol. i, 20 Oct. 1817.
[158] Whitehaven Library MS 48, Workington Colliery Book 1799-1853, Forster, View Book of Matthias Dunn.
[159] Forster, View Book of Matthias Dunn.
[160] D/Lons/W, Collieries 6.
[161] D/Lons/W, Peile report on Whitehaven Collieries 1842.
[162] D/Lons/W, Peile reports on Whitehaven Collieries 1842-47.
[163] D/Lons/W, Collieries 46; Peile reports on Whitehaven Collieries; Moore, MS "History of Whitehaven Collieries".
[164] D/Lons/W, Peile reports on Whitehaven Collieries.
[165] Peile, 2 Mar. 1835.

[166] Peile, 5, 11, 12 July, 29 Aug. 1837.
[167] D/Lons/W, Peile reports on Whitehaven Collieries.
[168] *Ibid.*
[169] Buddle, vol. i, 18 Sept. 1822.
[170] D/Cu, Letter book of William Swinburn 1818-24.
[171] *Report of House of Lords' Select Committee on the State of the Coal Trade 1829*, 50, 66.
[172] D/Cu, 5/23, Account Book of J. C. Curwen.
[173] D/Lons/W, Custom House returns quoted by Peile in his reports of coal shipments.
[174] Peile, 23 Jan. 1843.
[175] D/Lec, Peile reports on Lord Egremont's Collieries 1834.
[176] *Report of Royal Commission on Coal in the United Kingdom 1871*, iii, Appendix, 71, 86. D/Lons/W, Custom House returns quoted by Peile in his reports of coal shipments 1831-47.
[177] D/Lons/W, Peile's "Statement of the Export Price and Quantity of Whitehaven Coal" Dec. 1842; Custom House returns quoted by Peile.
[178] *Carlisle Patriot*, 2 Nov. 1816.
[179] D/Cu, 3/43, Letter book of William Swinburn.
[180] Parliamentary paper 1819 (67), xvi, 333.
[181] *Whitehaven Gazette*, 24 Mar. 1823.
[182] *Ibid.*, 9 June 1823.
[183] *Ibid.*, 7 July 1823.
[184] *Ibid.*, 16 June 1823.
[185] Parliamentary paper 1826 (155), xxii, 147.
[186] *Report of Select Committee of the House of Lords on the State of the Coal Trade, 1830*, 228.
[187] Peile, 20 Aug. 1832.
[188] *Cumberland Pacquet*, 19 Dec. 1826.
[189] *Report of Select Committee on Manufacturers, Commerce and Shipping 1833*, Q2341, evidence of Stephen F. Dickson.
[190] Peile, 1 July 1833.
[191] Peile, 10 Feb. 1834.
[192] Peile, 15 Sept. 1834.
[193] Peile, 11 May 1840.
[194] D/Lons/W, Peile report on Whitehaven Collieries 1842.
[195] D/Lons/W, Letters John Tomlinson to Peile, 18 June 1844; Peile to Lonsdale, 17 May 1847.
[196] Peile, 7-21 June 1847.
[197] D/Lons/W, Peile reports on Whitehaven Collieries.
[198] D/Lons/W, Custom House returns quoted by Peile in his reports of coal shipments.
[199] Hughes, *op. cit.*, 366.
[200] D/Lons/W, Custom House returns quoted by Peile.
[201] Peile, 30 Mar., 21, 28 April 1834.
[202] D/Cu, 3/45, Letter book of Benjamin Thompson, H. Curwen to M. Dunn, 3 Nov. 1836.
[203] *Ibid.*, Thompson to Peile, 2 Dec. 1835, 3 Jan. 1836.
[204] D/Cu, 3/82, Peile to E. S. Curwen, 14 Jan. 1841.
[205] D/Lons/W, Peile report on Whitehaven Collieries 1842.
[206] D/Sen, Netherhall 19/173, Westray to Senhouse, 8 Mar. 1843.
[207] Peile, 2, 9 Dec. 1844.
[208] D/Lons/W, Misc. Colliery Papers.

CHAPTER V

1851-1914

WHITEHAVEN DISTRICT

Coal mining in Cumberland was carried on with increasing vigour after 1850 and although mining was confined for the most part to districts already known it was more productive. Workings were driven deeper and further seawards, the yield was greatly augmented and manpower in the mines grew steadily year by year.

John Peile retired in 1847 and was succeeded as principal colliery agent at Whitehaven by his senior assistant Peter Bourne who held the post for twenty years. At that time the colliery consisted of seven working pits – Saltom, Wellington, Croft and Wilson pits at Howgill and William, North and Wreah pits at Whingill – and produced 224,985 tons of coal. Wilson Pit ceased drawing coal in 1847 and Saltom Pit in the following year. An important development took place in 1862 when Main Band coal was proved in the St. Bees School royalty.[1]

Mining operations received a severe setback in 1863 when there was an extensive underground fire at Wellington Pit. The fire spread rapidly, causing much damage and on the advice of T. E. Forster it was decided to exclude all air from the fire by hermetically sealing Duke Pit and flooding the workings in Wellington Pit, the sea being let in by means of a drift, fitted with dams and sluices, driven from the shore. Not until 1866 was the fire completely extinguished and another six years passed before the water was removed and work there resumed.[2]

Bourne resigned in 1867 and the management of the colliery was taken over by T. E. Forster, G. B. Forster and T. G. Hurst, a team of experts who introduced many improvements in mining methods. They were responsible for the sinking of Henry Pit (1870-2) to a depth of 155 fathoms. This pit worked a large area both landward and seaward until 1891 when it was abandoned. Further progress resulted from the efforts of R. F. Martin who was appointed chief viewer in 1874. During his three years in office he did much to modernize the colliery by improving the ventilation of the pits and the underground haulage and winding of coal.[3]

When the lease of the St Bees School land expired it was renewed for a term of 42 years from 1 January 1869 at a certain rent of £500 per annum together with a tonnage rent of 6d. per ton on all coal raised. In addition there was a wayleave rent of 2s. per ton of twenty-two waggons, each waggon containing 50 cwts.[4] For many years the Lowther family had been working undersea coal at Saltom, which was Crown property, and afterwards at William and Wellington Pits, free of royalty. This happy position for the Lowthers came to an end in 1860 when the Crown put forward a claim to the minerals lying below low-water mark and for the next twenty years the submarine coal was mined under a Crown lease. By this lease (dated 27 August) the Crown received 2d. per ton for large coal and 1d. per ton for small coal raised from these undersea workings. Up to 1880 the average annual royalty paid to the Crown amounted to £1,757. 8s. 5d.[5] By an agreement (2 February 1880) the lease was renewed and Lord Lonsdale was given the option of buying the land in question, which extended ten miles seaward and was thirteen miles wide, for £50,000. Lord Lonsdale bought the Crown minerals, the Lonsdale Settled Estates Act being passed in 1880 to enable the trustees of the settled estates to raise the purchase money by leasing any of the mines and selling any of the plant and machinery.[6]

Whitehaven Colliery, which had been worked by the Lowther family for over two centuries, was leased in 1888 to the Whitehaven Colliery Company which consisted of Sir James Bain, his sons J. R. Bain and J. D. Bain and J. S. Simpson who then controlled the colliery and ironworks at Harrington. The decision to lease the colliery was prompted by the rapid and alarming fall in profits; a certain revenue from a leaseholder seemed preferable to the variable and uncertain return derived from the continuation of coal mining. For a term of 31 years from 11 August 1888 the new company took over Henry, William, Wellington, Croft, James, Duke, Saltom and Kells pits together with all their equipment, seventy-two coke ovens at Bransty and 578 cottages. For the pits, including a sub-lease of St Bees School coal, the certain annual rent was £10,000 with further annual payments of £3,000 for the fixed plant and coke ovens and £1,014 for 139 cottages. After the first five years a further rent, equal to their rateable value, was payable in respect of the remaining 439 cottages. The tonnage rent was 8d. per ton on all coal raised when the selling price of round coal was less than 9s. per ton; above 9s. one-tenth of the excess was to be added to the tonnage rent.[7]

This transfer of the biggest colliery in Cumberland aroused some fears in the minds of other royalty and colliery owners. A report[8] on the Leconfield estates says,

> "The depression of trade which has brought about serious losses on the Coast has, in the Coal trade, been minimized by the management of the Lowther collieries, who are the biggest producers, holding out for prices. This enabled them and other local owners, including of course, many of their own Lessees, to secure better prices than would have been possible with the unrestricted competition of the Lowther pits. It is generally feared by the Coal Owners that Messrs. Bain & Co. will, to secure sale for a largely increased output, compete most keenly in all their markets, with most serious results . . . if our surmise be correct they will be able to make the competition of some of our thin seam Royalties almost impracticable."

These fears proved to be groundless; Messrs Bain & Co. did not attempt to drive their smaller competitors out of business by underselling them. Instead they co-operated with them, particularly in resisting wage increases. Others hoped that the transfer of the colliery might lead to desirable changes.

> "Perhaps some of the restrictions which hamper the working of collieries held from the Lowther Estate Trust may be modified. They are now prevented selling in Whitehaven or shipping at the port, which makes the price of Whitehaven coal in Whitehaven about the same as it is at Belfast. This is equivalent to a protective duty of say 2s. 6d. a ton, or supposing the annual consumption in Whitehaven is 40,000 tons, a tax upon the town of about £5,000 a year. If the restriction were removed competition would soon relieve the consumers."[9]

For twenty-five years the company worked the colliery successfully, introducing new and improved methods and increasing output. But in its early years the new modes of working and management caused difficulties with the men and led to a resurgence of trade unionism at Whitehaven and to strikes. In December 1888 it was said that, "The chief thing about the collieries here is the constant discharge of old hands & the introduction of Scotchmen."[10] Between 1900 and 1902 the company sank Ladysmith Pit to a depth of 1,090 feet alongside of Croft Pit[11] which ceased to draw coal in 1903 after being in production for 129 years. By this time only the Main Band was being worked and apart from the St Bees School royalty the entire output came from undersea coal.[12]

By 1913 the lessees were anxious to surrender their lease of the colliery. For some time they had been unwilling to undertake exploratory work, partly because the lease was due to expire in 1919 and partly because only one of the original partners in the Whitehaven Colliery Company still survived and for many years he had taken no active part in the working or business of the colliery. The reduction in the colliery's output after 1909 worried Lord Lonsdale's agents who feared that the colliery would deteriorate. It was obviously in the interest of the Lowther family that the colliery should be acquired by a new company prepared to adopt a more vigorous and active policy. If the colliery continued to decline not only would tonnage rents diminish but also it might prove difficult to obtain new lessees.[13] Accordingly in 1913 the existing lease was surrendered and a new company, the Whitehaven Colliery Co. Ltd., took over the colliery. The new lease was for a term of fifty years from 11 August 1913, the company paying a certain annual rent of £20,000 and tonnage rents of 8d. for Main Band coal, 6d. for Bannock Band and 4d. for coal from other seams when the average selling price did not exceed 9s. 9d. per ton; above 9s. 9d the tonnage rents were to be increased by one-tenth of the excess over 9s. 9d. per ton. In addition the company sub-leased the St Bees School land for 30⅓ years at an annual certain rent of £666. 13s. 4d. and paying the same tonnage rents as in the main lease.[14]

Output at the Whitehaven Colliery increased steadily during the last forty years of its direct control by the Lowther family, from 224,985 tons in 1847 to a record total of 417,039 tons in 1886. After some setbacks in its early years the Whitehaven Colliery Company succeeded in increasing production beyond what it had been when the Lowthers controlled the colliery:-[15]

Year ending August	*Undersea Coal Tons*	*Cwts.*	*St Bees School Royalty Tons*	*Cwts.*	*Total Tons*
1890	321,573		5,251		326,824
1895	476,786		17,216		494,002
1900	497,813		50,773		548,586
1905	462,774		–		462,774
1910	352,889	17	215,982	19	568,872
1913	376,526		141,301		517,827

The Moresby Coal Co. Ltd. leased a Lonsdale royalty in Moresby for a term of sixty years from 25 March 1877 and in 1879 sank Walkmill Pit. This colliery, for many years highly profitable to both lessor and lessee, was worked continuously until 1961. Another Lonsdale royalty in Moresby was taken over by Sir James Bain & Co. in 1881 and worked by them, in conjunction with their colliery at Harrington, until 5 January 1891.[16]

At Cleator Moor the Whitehaven Hematite Iron Co. continued to win coal for their ironworks. The company sank a new pit, Hope Pit, which reached the Main Band in 1861 at a depth of 120 fathoms and cost £10,000-£12,000 to sink. No. 2 Pit, begun in 1853, was continued to the Six Quarters Seam in 1861. Whinny Hill Pit was exhausted in 1863 and used thereafter for pumping purposes. The entire output of the colliery went to the company's iron furnaces. It was inferior in quality and although it was washed in an attempt to improve the coke made from it the firm still had to mix their coke in the blast furnaces with about 75 per cent of the best Newcastle coke. Mining was difficult and expensive and hampered by many faults and much inflammable gas. Not surprisingly the company lost many thousands of pounds and refused to renew the lease of the colliery which closed in 1871.[17]

Thomas Ainsworth, who was largely responsible for the formation of the Whitehaven Hematite Iron Co., leased Bowthorn Colliery, a Lonsdale royalty to the north of the Leconfield royalty at Cleator Moor, for 14 years from 31 December 1858 at an annual rent of £20.[18] In 1869 this colliery was taken over by the Whitehaven Hematite Iron Co. and worked by them until December 1894 when production stopped and the colliery was finally abandoned in 1897.[19] Threapthwaite Colliery, controlled by Messrs. Ainsworth and Hartley, earned a total profit of £14,056 between 1846 and 1865 and remained in production until 1872.[20] Nearby John Stirling sank Montreal Pit in 1862. It was unique in that it was both a coal and iron mine, both minerals being drawn up the same shaft. Coal was produced here until 1918 and in the early twentieth century it was the only coal mine at work in the Cleator Moor coalfield.[21]

Coal mining was carried on in other villages near Whitehaven. The Venture Pit in the township of Whillimoor, sunk in 1852 by Thomas Hinde and leased to Joseph Ward and others in 1860, was discontinued in 1865. The Moorside royalty in Whillimoor was worked by W. Baird & Co. between 1874 and 1880 and then by A. Johnston until 1899.[22] In Arlecdon William Irving leased a Lonsdale royalty in 1860 and sank Asby Pit 53 fathoms to the Main Band. He was succeeded by J. H. Woolcock and J. Huntrods in 1892 and they were followed by A. Johnston and J. S. Peile in 1899. Asby Colliery produced 283,312 tons of coal between 1 January 1871 and 5 July 1890; after 1900 output declined rapidly and the colliery closed in 1908.[23]

At Frizington Messrs. Musgrave, Gibson & Co. sank the Boghole Pit to a depth of 48 fathoms in 1861 and mined coal until 1878 when the colliery was closed. Their lease, for 21 years from 30 June 1861 at an annual rent of £192, stipulated that no coal was to be sold in the borough of Whitehaven but was allowed to pass through to other places.[24] A small pit at Lamplugh, owned by R. Hannay & Co., produced coal between 1872 and 1879.[25]

WORKINGTON DISTRICT

In this period the Curwen collieries at Workington were rapidly becoming exhausted. In 1851 the output of the Jane and Buddle pits amounted to 20,000 tons.[26] By 1853 all the solid coal in both the Buddle and Jackson pits had been wrought out and most of the pillars had been robbed and in April of that year Buddle Pit had to be abandoned owing to flooding, continuous pumping for a hundred days having had no effect.[27] Hope

Pit, Ellerbeck, was sunk in 1854 to the Four Foot Seam at a depth of thirty fathoms. It proved unprofitable and after four years it was given up.[28] Jane Pit was the only pit in production at Workington Colliery in 1860 and its output was diminishing. The decline of the colliery is reflected in the dwindling coal shipments, from 10,356 waggons in 1849 to 1,905 in 1861.[29]

J. C. Curwen's successors inherited not only a heavy burden of debt which made it impossible to finance necessary improvements to the colliery but also some of the intractable problems inseparable from what is essentially a robber industry – the exhaustion of seams, increased depths, longer distances from the shaft to the working face and the obsolescence of pit equipment. It must have been a relief to the Curwen family when William Irving leased Workington Colliery in 1864 at a certain rent of £300 per annum together with tonnage rents of 6d. per ton for round coal and 1d. per ton for small coal.[30] He sank Annie Pit to the Hamilton seam at 72 fathoms. This pit was badly laid out, its ventilation inadequate and its working costs too high. Despite a very adverse report on Annie Pit (31 August 1870) from W. Peile, H. K. Spark of Darlington leased the colliery for 31 years from 1 April 1871 at an annual rent of £600 and 5d. per ton for over-workings. He had little success and his interest in the colliery was acquired in 1874 by J. H. Garbutt, a coal and iron merchant, who had insufficient capital to work the colliery. Pumping at Jane Pit was stopped in the summer of 1875 and as a result Annie Pit was drowned and production at Workington Colliery ceased. In the following year Garbutt went bankrupt.[31]

Harrington Colliery was no more successful than Workington Colliery. Coal shipments showed a similar decline, falling from 9,174 waggons in 1846 to 1,391 in 1861.[32] In 1865 Messrs. Bain, Blair and Paterson of the Harrington Ironworks took over the colliery. On the death of Robert Blair the lease was surrendered and a new one granted for twenty-seven years from 1 July 1871 at an annual rent of £1,000. Tonnage rents were: coal 5d., ironstone 1s., fireclay 8d., limestone and freestone 2d. The firm had also to pay 1s. 6d. for every thousand bricks made and double agricultural rent for all farm land used.[33] Ten pits were sunk here, although one of them was the old Micklam Pit and re-opened as No. 5 Pit. These pits had mixed fortunes. No. 1 was not completed because of flooding, No. 2 worked a small area of Four Foot coal, No. 3 (sunk in 1867) and No. 4 (sunk in 1871) both worked coal until 1879. No. 6 (sunk about 1871) and No. 8 (sunk in 1872) ceased in 1874.[34] At No. 8 Pit Main Band coal was worked at a depth of eleven fathoms. A report on this pit by R. W. Moore (21 March 1874) stated, "The Coal which Messrs. Bain & Paterson are working is in fact the Barrier which should be left to protect the other & greater portion of the Colliery. Nos. 6 & 7 Pits are to the dip of No. 8 and if the latter was inundated it might endanger the two former pits." Two months later the sea broke in and the pit was lost. In Moore's words, "The water broke in at 8 a.m. on Saturday morning last when all the men were in. They, however, escaped by the ladder which is placed up the side of the pit. The workings were filled in an hour."[35] No. 7 Pit produced coal between 1874 and 1901. In 1901 only three pits were working and in 1905 only one, No. 9, which had been sunk in 1880.[36] No. 10 Pit was sunk in 1910-11 and remained in operation until 1968.

By a Crown lease* H. F. Curwen obtained the right to raise minerals from the foreshore

* Although the Crown lease to Curwen is dated 24 January 1878 Bain & Co. were working the undersea coal in 1877.

and bed of the sea at Harrington. This right was transferred to Messrs. Bain & Co. who began to work the undersea coal in 1877. Between June 1865 and June 1867 the company raised 21,588 tons of coal and paid £360. 11s. 8d. in royalties. In later years output from Harrington Colliery and from the undersea coal amounted to:-[37]

Year	*Harrington Colliery Tons*	*Undersea Coal Tons*	*Total Tons*
1870	19,296		19,296
1875	78,681		78,681
1880	45,000	19,745	64,745
1885	46,308	13,554	59,862
1890	46,778	51,613	98,391
1895	48,477	54,255	102,732

In an attempt to reduce the excessive competition and uneconomic duplication of works in the Cumberland iron and steel industry the Workington Iron & Steel Co. Ltd. was established in 1909. This was an amalgamation of the Moss Bay Hematite Iron & Steel Co. Ltd., the Workington Iron Co. Ltd., the Harrington Iron & Coal Co. Ltd. and the Cumberland properties of Messrs. Cammell, Laird & Co.[38] Harrington Colliery was controlled by the new company which leased the Curwen royalty at Harrington and Workington (13 November 1911) and sub-leased the Harrington undersea coal (28 June 1909) for a term of sixty years from 1 July 1908.[39] The company sank No. 10 Pit which began to raise coal in January 1912 and built additional coke ovens to provide coke for its blast furnaces.

The Curwen royalty in Little Clifton was leased in 1852 to Isaac and William Fletcher who sank a pit near Crossbarrow forty fathoms to the Main Band, and in 1854 sank the Harry Gill Pit in John Cookson's royalty to the same seam. Their success led the Earl of Lonsdale to sink Lowther Pit half a mile westward, which reached the Main Band in 1855 at a depth of only thirty fathoms. A dispute about the Earl's title to the royalty under certain lands in Great and Little Clifton was ended by Lord Lonsdale's purchase of these estates and he thereby became the owner of nearly all the land in both townships. In 1856 he leased all his royalty in Great and Little Clifton to Messrs. Fletcher who completed the Lowther Pit and in 1861 sank the William Pit.[40] It was stated in 1860 that "the Workington Harbour and the Cockermouth and Workington railway are both chiefly dependent for their revenue upon the Clifton Colliery."[41]

Messrs. Fletcher took a further lease of this royalty for a term of 21 years from March 1867 at an annual rent of £3,500 and tonnage rents of 1s. per ton for Main Band Coal and 10d. per ton for Ten Quarters coal.[42] In 1873 the certain rent was reduced to £2,500 per annum and in the same year the Clifton Colliery was sub-leased by the West Cumberland Iron Co. in which the Fletchers had a large interest. Within a year the company completed a new brick works and erected sixty coke ovens at Lowther Pit.[43] With the growth of iron manufacture in the district the demand for both coal and labour increased. G. B. Forster reported in 1873 that, "All places in Cumberland are at present short of colliers, but increased exertions ought to be made to get more men so that a greater output of coal may be reached."[44] Since additional men could only be attracted by the provision of good houses he recommended that the firm should build miners' cottages; his advice was followed and forty were built by the end of the next year. As a result output went up and in 1874 the Lowther, Clifton and William Pits were yielding 1,750 tons per week.[45] The Westray (or Clifton) Pit was closed in 1875 and Lowther Pit abandoned in 1885. Two years later the Allerdale Coal Co. took over Clifton Colliery for a term of 31 years, paying an annual rent of £2,000 and tonnage rents of 6d. for Main Band Coal, 5d. for Ten Quarters coal and 3d. for coal from other seams. This firm continued with the colliery until 1947.[46]

In 1853 Messrs. Fletcher & Co. were working coal in a Leconfield royalty at their Bridgefoot Colliery, near Greysouthen and in that year raised 1,953 tons.[47] This royalty was divided into two sections and leased respectively to Harris & Co. of Greysouthen in 1856 and to John Harris of Darlington in 1859. Both these sections were sub-leased in 1860 to Messrs. I. & W. Fletcher who completed Melgramfitz Pit in 1863 after spending about £12,000 in sinking it. This pit was taken over by the West Cumberland Iron & Steel Co. Ltd. in March 1874 and worked by them until June 1880 when the Main Band having been exhausted work was suspended.[48] Harris & Co. carried on their Millbanks Colliery until 1874. Account books give the following outputs at Millbanks and Linefitts:[49]

Year	*Millbanks Colliery Tons*	*Linefitts Colliery Tons*
1864	1,291	23,476
1865	2,940	23,789
1866	3,288	18,953
1867	4,570	16,991
1868	3,412	15,938
1869	5,958	10,818
1870	11,745	
1871	10,999	

Another pit was opened in 1874 by Messrs. Kenyon & Campbell. This New Banks Pit was worked for ten years but since that time no coal was raised at Greysouthen.[50]

Mining was resumed at Winscales in 1873 when the Rev. A. F. Curwen and Messrs. A. B. Were and R. Blair leased the Curwen royalty and sank Wythemoor Pit. The lease was for thirty-one years at an annual rent of £100 in the first year, rising to £1,000 in the sixth and later years and a tonnage rent of 8d. The dead rent was subsequently reduced to £700 per annum.[51] The first two partners withdrew leaving the unfortunate Blair, who had no previous experience of mining, to manage the colliery after 1875. He proved to be extremely incompetent and was dismissed in 1878 by his wife's uncle, Dr. A. F. Richmond, the new owner of the colliery. Richmond carried on himself until 1886 when the pit, on which he had spent £22,000, was closed. Its failure resulted from mismanagement, the high tonnage rent and inadequate local demand.[52] In 1904 the Wythemoor Colliery Co. leased the same royalty[53] and worked the Two Foot seam at a depth of 16 fathoms. Four years late the company leased an adjoining Leconfield royalty[54] and raised coal from both royalties. A new pit was sunk (1909-11) and coal was won from the Two Foot seam, the Yard seam at 21 fathoms and the Four Foot seam at 54 fathoms. This company had no more success than its predecessors and went into liquidation in 1913.[55]

In the Distington area Ralph Tate worked a colliery at Commonside in 1859 and leased Haycastle Colliery from 1863 to 1872, raising coal from the Three Foot seam from a depth of thirteen fathoms. A company headed by Messrs. Mackenzie and Main re-opened some old Yard Band pits at Gilgarran in 1872, but having little success abandoned operations in 1875. No more coal was mined in Distington until 1880 when the Moresby Coal Co. completed Oatlands Pit in a Lonsdale royalty* to a depth of 108 fathoms.[56] This colliery proved productive, the average annual output in the ten years 1911-20 amounting to 82,712 tons.[57]

Dean Moor Colliery was leased by John Percival for 21 years from 25 March 1848 at a certain rent of £37. 10s. per annum and one-twelfth of the value of the coal sold at the pithead. He, and later his widow, worked the colliery until May 1868. Coal was mined at a depth of thirty-two yards for local sale and in 1860 was raised to the surface by a horse gin. In that year a new pit was sunk and a 13 h.p. steam engine was erected to drain the mine and draw coal up the shaft. The eight or nine hewers employed had still to trail the coal to the shaft along a road less than three feet high. Mrs. Percival was succeeded by George Grievson & Son who began operations in May 1869 and continued until 1880 when William Summerson leased the colliery. After his death his family formed the Dean Moor Colliery Co. Ltd. to run the colliery.[58]

This small, isolated colliery produced little coal – between 1854 and 1879 it never raised 7,000 tons in any year – and was hard to manage. Coal had to be carted for a mile to Wright Green station on the Whitehaven, Cleator & Egremont Railway.[59] Because there were no miners' cottages within more than a mile of the pit was difficult to secure an adequate labour supply. High transport costs were an additional burden; it cost about 1s. 3d. per ton to trail the coal from the face to the shaft and another 1s. 3d. per ton to move it from the pithead to the railway. The lessees showed little enterprise and in 1891 were said "to work from hand to mouth and to pick out the Coal where it can be worked without pumping or without their being stopped by Faults."[60] In 1896 they wished to abandon the colliery but a reduction in the tonnage rent persuaded them to carry on. They also worked an adjoining royalty belonging to Mr. Brookbank in the early years of the twentieth century and a small Lonsdale royalty from 1913.[61]

The Camerton Colliery was taken over in 1857 by Messrs. Cook & Co. who were succeeded by Joseph Mulcaster in 1873 and by Dr. R. S. Mutch in 1878.[62] An adjoining Lonsdale royalty at Seaton was leased by Mutch for twenty-one years from 23 June 1883 at an annual rent of £25 and tonnage rents of 4½d. for coal and 4d. for fireclay. This lease was renewed for a further twenty-one years from 24 June 1904 by Messrs. Mutch and Clark, later the Camerton Coal & Firebrick Co. Ltd. This royalty produced little coal; between 1906 and 1909 the output was less than 1,000 tons per annum and from June 1909 to December 1912 no coal was raised. No. 1 Pit ceased production in 1903 and No. 2 Pit was abandoned in 1911.[63]

* The company leased this royalty for 60 years from 25 March 1877 at a certain rent of £100 in the first two years rising to £1,800 per annum from the fifth year onwards. The annual rent was reduced to £1,500 from 1881.

From 1840 coal and fireclay were worked continuously at Moorhouse Guards, Seaton, by successive lessees. Thomas Fletcher & Co. leased this Lonsdale royalty in 1891 at a yearly rent of £55 and tonnage rents of 6d. for coal and fireclay. Between 1899 and 1913 the Seaton Firebrick Co. worked both coal and fireclay here. They were succeeded by the Seaton Colliery Co. which held the royalty for the next two years. From this colliery about 3,000 tons of coal were raised each year.[64]

About 1850 Messrs. Mulcaster & Co. sank No. 1 Pit, Seaton Moor to the Ten Quarters seam at 32 fathoms. They took over a Lonsdale royalty at Seaton Moor in 1854 for a term of 21 years at an annual certain rent of £1,000 and a royalty rent of 6d. – 1s. per ton. They opened Nos. 2 and 3 Pits where they worked the Ten Quarters, Metal and Rattler seams until 1869. In the following year Seaton Moor Colliery came into the possession of Robert Wilson.[65]

In 1861 Mulcaster & Co. re-opened and enlarged two old pits at St. Helens and continued them to the Cannel and Metal Band; these were Nos. 1 and 2 Pits, St. Helens Colliery. They leased the colliery in 1870 and began sinking No. 3 Pit, Siddick in 1877 which their successors, the St Helens Colliery & Brick Works Co. Ltd., completed to a depth of 170 fathoms in 1889. Nos. 1 and 2 Pits which had been closed in 1887 were re-opened and enlarged in 1901 to win coal over the faults that had stopped the former lessees.[66] A valuable tract of coal was added in 1902 when Lord Lonsdale leased an area of undersea coal from the Crown and immediately sub-leased it to the company on the same terms on which he had acquired it.* A battery of forty bye-product coke ovens made substantial profits and more than made up the losses suffered in some years by mining operations.[67]

MARYPORT DISTRICT

The Harris family continued with the Ellenborough Colliery, renewing their lease for a term of twenty-one years from 1 October 1871 at a certain rent of £800 per annum and a tonnage rent of 6d. per ton on all coal raised. Serious losses compelled the family trustees to give up the colliery in October 1878 which was then taken over by the Maryport Hematite Iron Co. for a term of fifty years on the same conditions as their predecessors.[68] The company made new haulage roads and installed improved haulage machinery, built fifty coke ovens and erected a new washing plant. From the two working pits coal was used to provide coke for the firm's blast furnaces.[69] Despite its efforts the company went into liquidation in 1891 and the colliery was then closed. It was re-opened in 1894 by Messrs. Graves & Semple (the Ellenborough Colliery Co. Ltd.)[70] and worked by them until 1914 when it was finally closed.

At Broughton Moor the Leconfield royalty was worked by Messrs. Ross, Fletcher & Co. until 1857. In three successive years, 1854-1856, the colliery sold 48,832, 44,548 and 38,036 tons of coal. Most of this was shipped at Maryport, some 2½ miles away, to

* Terms of St Helens Colliery lease: 21 years from 1 January 1870. Certain rent £1,000 per annum, tonnage rents 9d. – 1s. per ton. This lease was renewed in 1885 and again in 1903. Terms of Crown lease: 60 years from 1 January 1902. Certain rent £500 per annum for the first three years and then £1,000 per annum for the remainder of the term. Tonnage rents 1½ – 4d.

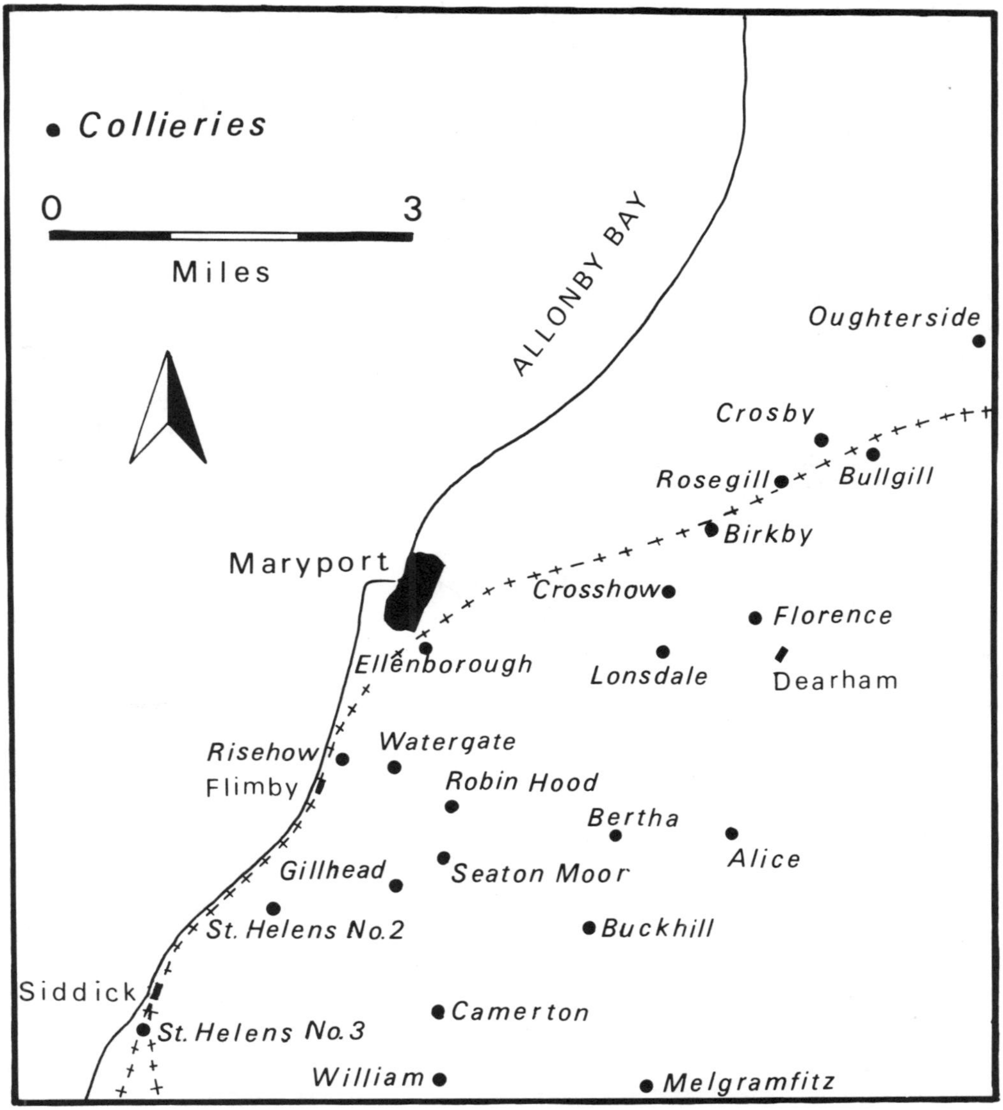

FIG. 4 – Coal Mines near Maryport.

which it was carted at a cost of 4s. 8d. – 5s. 8d. per waggon, a heavy item in total costs. John Harris of Darlington succeeded Ross, Fletcher & Co. in 1859 and in 1860 sank Bertha Pit, the two working pits, Henry and Mary Pits, being nearly exhausted. He also completed a private railway from the colliery to the M. & C. Railway at Dearham Bridge in March 1860 at a cost of £6,000 and thereby greatly reduced his transport costs. In the same year a portion of the royalty was sub-let to Messrs. I. & W. Fletcher who worked the Broughton coal from a shaft in Harris's freehold in Greysouthen. In 1862 the colliery produced 79,000 tons of coal.[71]

The Leconfield royalty at Broughton Moor was divided into three sections which were leased respectively to: (1) I. & W. Fletcher and E. Waugh, (2) J. & T. Walker and (3) Robert Wilson.* In March 1874 the section leased by Messrs. Fletcher, which included the Melgramfitz Colliery, was sub-let to the West Cumberland Iron Co. who completed the sinking of Buckhill Pit (begun 1873) and built a branch line to connect the pit with the Cockermouth & Workington Railway at Ribton.[72] At Buckhill coal was raised from the Main, Yard and Little Main seams from both the Leconfield royalty and an adjacent Lonsdale royalty. The company handed back the Buckhill and Melgramfitz collieries to Messrs. Fletcher & Waugh in March 1887 who sold off the plant at Melgramfitz and continued working Buckhill Pit for the next year. They were succeeded at Buckhill in 1888 by the Allerdale Coal Co. Ltd. who mined coal there continuously until the pit closed in 1932. The new lessees applied themselves vigorously to development and improvement. Thirty beehive coke ovens and new coal washing and crushing machinery were erected in 1890. Twenty years later the company installed a bye-product coking plant at their Clifton Colliery and the small coal from Buckhill was sent there to be coked.[73]

Robert Wilson, whose northern portion of the Broughton Moor royalty included Bertha Pit, sank the No. 2 or Bertha New Pit near the Old Bertha Pit. In the new pit coal was worked from the Little Main, Six Quarters and Three Quarters seams; in the old pit coal was obtained by robbing the pillars. The colliery in 1893 became part of the Flimby & Broughton Moor Coal & Firebrick Co. Ltd., formed by the Wilson family to control their collieries.[74] High labour costs made profitable working almost impossible; in 1894 the labour charge on coal sold for £30,000 came to £26,000 and there was a loss of over £800 without allowing for interest or depreciation.[75] To reduce wage costs different kinds of coal-cutting machinery were tried between 1898 and 1910 but they achieved little. In 1907 the colliery lost over £5,000. Despite this loss bye-product ovens were erected at the company's Watergate Colliery, Flimby in the following year. The ovens proved a success; the company, however, was unable to pay the contractors who erected them or to pay the rents due to Lord Leconfield. Shortage of money meant that the underground workings were neglected and the colliery generally was in a bad condition.[76] Rent concessions by the lessor helped a little but the firm's financial difficulties continued and in 1913 it was said that, "The Minimum Wage Act seems to be the great feature in rendering the working of this Coal expensive and impracticable at any margin of profit."[77] High costs were combined with low productivity, the output falling from 78,644 tons in 1907 to 27,955 tons in 1914 and it was probably this continuing

* All three leases were for 31 years from 1 July 1870. Annual certain rent £1,650, £200 and £150 respectively. Tonnage rent one-seventh of the selling price of coal.

decline in output which compelled the lessees in 1914 to give a year's notice of their intention to abandon the lease.[78]

The Walker family continued to mine coal at Dearham. More powerful pumping machinery enabled them in 1852 to re-open the Lonsdale Pit which had been flooded in the 1830's. Coal was obtained from the Ten Quarters, Kennel and Yard seams at Lonsdale Pit and from the Kennel and Metal seams at Lowther Pit. In 1871 conditions at the latter pit were bad – there was much water, the roads had been allowed to deteriorate, mining was hampered by the abundance of metal in the seams and the diminished ouput was insufficient even to pay the men's wages.[79] Nevertheless Messrs. J. & T. Walker renewed their lease of this Lonsdale royalty and added to it part of the Leconfield royalty at Broughton Moor which included Outfields near Dearham. These measures did not improve the situation and in 1877 Dearham Colliery was sold to John Osmaston of Derby for £60,000.[80] He carried on with the colliery until 17 November 1884 when it was taken over by the Cumberland Union Bank.[81] A new pit was sunk at Outfields about 1,800 yards south of Lonsdale Pit and worked in conjunction with the Dearham pits. The bank, however, was more concerned with quick returns than planned development, "The Lease being in the hands of a Bank no expenditure is incurred in developing the minerals properly, the cheapest and most available Coal is being worked and the more expensive Coal is left unworked. No operations are being carried out for winning the dip Coal."[82] Matters were made worse in 1891 by an influx of water from the neighbouring colliery at Dovenby. A declining output, high working costs, a large proportion of colliery consumption and waste and heavy railway charges made profit impossible; in 1892 there was a loss of 1s. 5·72d. on every ton of coal raised.[83] Two years later the bank gave up both Dearham Colliery and the Leconfield royalty at Outfields.

The Outfields royalty was leased to Matthew Walton (the former manager of the colliery) and John Williamson. In 1896-7 Outfields Colliery was troubled by water, much of it coming from Dearham Colliery and it was therefore decided to abandon the old pit and sink a new one, Alice Pit, in Outfields in 1898. Coal was worked in the Main, Yard and Ten Quarters seams, not only in the Leconfield royalty but also in several adjoining royalties belonging to Mr Walker, Mr Adair and Mr Dykes. In 1906 the colliery was amalgamated with the Flimby and Broughton Moor collieries belonging to Messrs. Wilson and operated as part of that undertaking. Output from the Leconfield royalty declined sharply from 36,905 tons in 1907 to 3,921 tons in 1914. To help the company the terms of the lease were modified in 1914, the certain rent being reduced from £250 to £50 per annum and the minimum tonnage rent on coal from 5d. to 3d.[84]

After the closure of Dearham Colliery in 1894 the miners determined to provide employment for themselves by embarking upon a scheme of co-operative mining. Two companies consisting largely of local workmen were formed in 1894, each leasing a Lonsdale royalty in Dearham. The Dearham Colliery Co. re-opened old shafts at Crosshow and the Townhead Colliery Co. re-opened an old pit at Townhead, at the Row Beck end of the village, in 1895. Whilst they were in production output at Crosshow ranged from 8,000 to 32,000 tons per annum and at Townhead from 2,400 to 4,200 tons per annum. It proved impossible for these working miners to succeed against the competition of wealthy coalowners who were themselves at this time having to struggle for a share in a highly competitive market. Mining problems, particularly drainage, insufficient capital and probably a lack of business skill obliged both companies to go into liquidation in 1903.[85]

Small quantities of coal were raised in other Lonsdale royalties at Dearham by Messrs. Walton, Burns and Saint between 1894 and 1901 and by the Row Moor Colliery Co. Ltd. between 1905 and 1909.[86] Coal and fireclay together were worked in Dearham from the thin seams below the Yard Band by W. Tickle & Sons (1866-77) and by Messrs. Steele & Co. (1894-1901).[87]

After establishing the Birkby Brickworks in 1862 William Tickle also worked coal and fireclay in nearby Leconfield royalty from adits in both banks of the Ellen. Production was on a small scale and ceased in 1877, the lease and the works falling into the hands of the City & District Bank, Carlisle from whom they were acquired by Matthew Watson and John Croudace in 1881. They formed the Birkby Firebrick Co. Ltd. in which they held two-thirds of the shares. The Birkby works were next sold in 1883 to the Birkby Coal & Brick Co. Ltd. which sank Mother Pit to the Little Main seam at 12 fathoms. It proved expensive to mine the thin seams of poor quality coal and by 1885 the company had to resort to robbing the pillars. The losses made by working coal were offset to some extent by the profits from brickmaking. Coal production stopped in August 1887 and, apart from a few tons of coal obtained at the beginning of 1888, was not resumed until 1890, this time by the Cumberland Coal & Brick Co. Ltd. which had taken over the works in that year. Mining at Birkby ceased in 1893 and was not revived until the Second World War.[88]

Gilcrux Colliery was leased by Edward Steel in 1852 who continued with the Jane Pit and two years later sank the Eliza Pit. John Steel and William Miller sub-leased the colliery in 1860 and completed the sinking of Ellen Pit (begun in 1859), near Bullgill station, to the Lickbank seam at a depth of 101 fathoms in 1862. In 1866 the colliery was leased to the Gilcrux Colliery Co. who sub-let it two years later to the Crosby Colliery Co. This company, for which William Mulcaster was chiefly responsible, began in 1854 and worked the Gilcrux pits in conjunction with their own colliery which consisted of No. 1 and Rosegill Pits. Both collieries were controlled by the Lonsdale Hematite Iron & Steel Co. Ltd. from 1883 until 1896 when they came into the possession of the Bullgill Colliery Co. Ltd. Crosby No. 1 Pit was abandoned in 1893 and Rosegill Pit was closed in 1896. Gilcrux Colliery continued to produce coal after the closure of the Crosby pits and was next taken over by the Bullgill Coal Co. Ltd. in 1901. Coal was worked at the Ellen Pit until 1910 when the colliery was closed.[89]

A little coal was mined at Dovenby in the latter half of the nineteenth century. About 1853 Messrs. Steel & Miller sank a pit near Dovenby Close to a depth of 25 fathoms and worked it until 1860. Another pit was sunk in 1872 by Messrs. Harris & Carlton and continued by others until 1895 when the colliery was laid in. Messrs. Wood & Co., the last owners of the colliery, found that the coal was rapidly becoming exhausted and mining hampered by excessive water. In 1891 there was a large influx of water from the colliery into the adjoining Dearham Colliery for which they had to agree to pay compensation. Messrs. Wood became bankrupt and since that time no coal has been raised at Dovenby.[90]

In 1850 Robert Wilson and John Harris of Darlington leased the Curwen royalty at Flimby where they sank the John and Mary Elizabeth (or Risehow) Pits and in 1855 the Robin Hood Pit. Wilson and Isaac Bass took over an adjoining Lonsdale royalty in 1855 and sank Watergate Pit (1866) and Moor Pit (1873). Flimby Colliery was not very profitable – between 1853 and its closure in 1860 the Mary Elizabeth Pit made a loss of £703. 8s. 8d. and made no contribution towards the repayment of the large amount of capital invested in it. Robin Hood Pit earned a total profit of £1,081. 11s. 5d. between 1860 and 1867 and John Pit £2,579. 4s. between 1853 and 1859.[91] Robert Wilson himself took control of Flimby Colliery, renewing the Lonsdale lease in 1869 and the Curwen lease in 1871 and leasing in addition a Senhouse royalty in Flimby in 1875. He and his sons worked the Flimby pits until 1893 when the collieries they owned were formed into the Flimby & Broughton Moor Coal & Firebrick Co. Ltd.[92]

Another pit was sunk at Gillhead, Flimby in 1867 by Messrs. Lucock & Carlton. This was taken over in 1878 by Henry Graves and worked by him and his son until 1897 when the Gillhead Coal & Firebrick Co. succeeded to the colliery. There they obtained coal, fireclay and ganister until 1917 when the colliery passed to the Workington Iron and Steel Co.[93]

In the 1850's the Aspatria Coal Co. found it impossible to carry on their Westmoor Colliery at a profit since the coal was rapidly becoming exhausted. Their two pits, Bank End Pit and West End Pit, ceased working in 1856 and 1857 respectively, and the lease was relinquished. A portion of this royalty was leased by John Harris of Greysouthen in 1859.[94] Harris, in 1850, had begun the Brayton Domain Colliery when he sank Nos. 1 and 2 Pits in the Brayton royalty belonging to Sir Wilfrid Lawson and worked them for twenty years.[95] He and his successors mined coal at Brayton from both the Leconfield and Lawson royalties which were contiguous. Other sinkings followed – No. 3 Pit (1868-70) which reached coal at 89 fathoms and was in production until 1902 and No. 4 Pit (1888-92), near Brayton railway station, which was sunk to the Yard Band at 92 fathoms. Between 1907 and 1910 No. 5 Pit was sunk 171 fathoms to the Yard Band where the coal was four feet thick.[96]

At Oughterside a Lonsdale royalty was leased in 1902 by Joseph James & Co. (later the Oughterside Coal Co. Ltd.) but no coal was raised until 1906. In the following year the company leased an adjoining Leconfield royalty and worked both royalties together, getting coal in the Ten Quarters and Yard seams. A battery of forty bye-product coke ovens was erected close to the pithead in 1913.[97]

To the east of Brayton No. 3 Pit Allhallows Colliery was sunk near Mealsgate station in 1874 by Messrs. I. and W. Fletcher to the 30 Inch seam at 63 fathoms. Two years later, after further sinking, Yard Band coal five feet thick was reached at a depth of 105 fathoms from the surface. This colliery, which comprised royalties 490 acres in extent belonging to Sir Wilfrid Lawson, George Moore and others,[98] was enlarged in 1886 by the addition of 221 acres of a Leconfield royalty in the manor of Allhallows.[99] In 1887 the colliery passed to the Allerdale Coal Co. Ltd. which soon found itself in difficulty. The colliery was costly to work and because the coal face was a long way from the shaft expensive rope haulage was needed to bring the coal to the shaft bottom. In the Carlisle area Allhallows coal had to compete with coal from the Brayton Domain Colliery which enjoyed favourable railway rates, paying about 4d. per ton less than Allhallows coal. Similarly Allhallows coal moved to Maryport by rail had to pay higher transport charges than Brayton coal. A high percentage of absenteeism and a lengthy strike in 1892 added to the company's problems. There was a loss of £2,600 at Allhallows in 1895; during the previous five years the Allerdale Coal Co. had lost about £24,000 from its Allhallows,

Buckhill and Clifton collieries. In the early years of the twentieth century the company steadily lost money and a strike in 1908 which lasted thirteen weeks was a very heavy blow. Lord Leconfield advanced £2,000 to the company in 1909 in an attempt to ease its financial difficulties but continuing losses compelled the company to surrender its lease a year later. The colliery was taken over immediately by the Allhallows Collieries Ltd.[100] The Allerdale Coal Co., however, retained the Brayton Knowe Pit which had been sunk in 1902 and added to it a small Leconfield royalty of about 18 acres in 1913. Here Yard Band coal was worked until 1917 although in the last two years of its life mining was confined to robbing the pillars.[101]

Bolton Colliery under Messrs. Addison & Paisley, like many of the inland collieries, did not flourish at this time. Knapethorn Pit closed in 1855 and Wyndham Pit in 1856. A new pit, Annie Pit, which was begun in 1854 reached coal in January 1856 but was not very successful. Another pit was sunk to the Main Band at Mealsgate in 1859. The colliery, having been extensively worked for over a century, was approaching exhaustion and its output between 1854 and 1860 was so small that royalties did not equal the dead rent of £250 per annum. The situation was made worse by competition from Messrs. Drewry & Co's adjoining Priestcroft Colliery (near Mealsgate) and from better and cheaper coal brought into the Wigton area, formerly dependent upon Bolton coal, by rail from collieries at Gilcrux, Dearham and Crosby.[102] Priestcroft Colliery closed in 1863[103] and Bolton Colliery in 1869. Bolton Colliery remained unlet until the Maryport Iron Co. leased it for a term of twenty-six years from 1 October 1872 at an annual rent of £1,000 and a tonnage rent of 9½d. on all coal raised.[104] Most of the output was sold locally and the remainder carted to Mealsgate station and then sent by railway to the firm's works at Maryport. This company had made a bad bargain for it had taken over an old colliery, hampered by many faults, whose output did not justify the high annual rent that had to be paid. Mining by the company at Bolton ceased in September 1886.[105] However, Joseph Faulder, the former manager of the colliery, with the help of a £200 loan from Lord Leconfield, worked the pillars left at Bolton Colliery from 1887 to 1897.* He had to pay a rent of £120 per annum and a royalty rent of one-fifth to one-seventh of the selling price of the coal.[106]

During this period both the output and manpower of the Cumberland coalfield increased notably. Between 1854 and 1914 coal output rose from 887,000 tons to 2,360,510 tons and the number of employees from 3,579 to 11,422. For the first time in its history the county, in 1894, produced more than two million tons of coal and in 1914 output reached a then record figure. Only once, in 1927, has this production level been surpassed.

Cumberland Coal 1854-1914[107]

Year	*Output Tons*	*No. of Employees*	*Year*	*Output Tons*	*No. of Employees*
1854	887,000	3,579	1884	1,707,582	6,480
1864	1,380,795		1894	2,058,867	8,282
1874	1,036,274	5,678	1904	2,119,382	8,967
			1914	2,360,510	11,422

* R. W. Moore (*VCH* 379) is mistaken in saying that coal was last worked in Bolton in 1874 by the Maryport Iron Company.

In some years more than half of the coal raised in Cumberland came from Lonsdale coal royalties. Output from these royalties amounted to:[108]

Year	*Output Tons*	*Year*	*Output Tons*
1911	1,192,662	1914	1,310,402
1912	1,179,091	1915	1,209,021
1913	1,316,114	1919	1,089,101

The biggest producer of all the Lonsdale royalties was Whitehaven Colliery:[109]

Year	*Whitehaven Output Tons*	*Cumberland Output Tons*	*Whitehaven Output as % of Cumb. output*
1866	255,505	1,490,481	17·14
1876	281,968	1,392,046	20·25
1886	417,039	1,788,981	23·31
1896	495,657	1,931,771	25·65
1900	536,493	2,022,327	26·52
1912	513,724	2,133,563	24·07

But the expansion of the industry did not keep pace with the development of coal mining in the United Kingdom generally. In 1854 Cumberland's production was 1·37 per cent of the national output (it never reached 2 per cent after 1854) and from 1899 onwards it was less than 1 per cent. Although output increased after 1850 mining operations in Cumberland were hampered by many difficulties and these were reflected in the declining output per person employed. Between 1874 and 1914 the annual output per man in Cumberland was always considerably less than the national average. In 1881 the annual output per person employed in Cumberland was 286·5 tons and from that date it declined until in 1914 it was only 206·6 tons; for the United Kingdom the figures were 311 tons and 234 tons respectively.[110] However, the figures of the annual output per man can be misleading since the number of persons employed is the number of persons "on the colliery books", whether regularly employed or not, i.e. these figures do not take into account short-time working and clearly the yearly output will vary with the total number of days worked during the year. Much more reliable figures are those of "output per manshift worked" but except for the single month of June 1914 these statistics are not available until after the first world war.[111]

In the latter half of the nineteenth century the growth of the iron trade stimulated the coal industry although until the early twentieth century iron production in Cumberland was dependent upon large imports of Durham coke* since much of the local coal was unsuitable for coking. However, an increasing proportion of Cumberland coal was used in local ironworks:[112]

* In 1907 five-sixths of the coke used in Cumberland ironworks came from the east coast. (*Report of Committee on Eight Hour Day*, 1907, Q 18394).

Year	*Coal used in Iron Furnaces Tons*	*As % of Cumberland Coal Ouptut*
1854	36,000	4·05
1855	49,628	6·10
1856	64,628	7·07
1857	53,000	5·60
1858	78,450	8·50
1859	152,728	14·65
1866	409,029	27·40
1873	500,000	28·61
1877	695,091	45·85
1878	550,000	39·60

Coke production in Cumberland increased as the iron industry developed. In 1873 the West Cumberland Iron Co. built sixty coke ovens at Lowther Pit, Clifton.[113] Eight years later coke ovens were erected on the foreshore near William Pit, Whitehaven.[114] Others followed – at Buckhill (1889-90), Watergate (1907-08), Clifton (1910-11), Oughterside (1913) and Harrington No. 10 (1911-13).[115] All the coke ovens in Cumberland were of the beehive pattern until 1894 when twenty-four Coppée ovens were constructed at St Helens Colliery, Siddick and these produced coke in much less time than the beehive ovens. The Coppée ovens ceased production in 1908 and no more ovens of this type were erected in the county. In 1908 the Koppers bye-product coke ovens were introduced at St Helens and Flimby Collieries. Coke output increased quickly in the years preceding the outbreak of the first world war:[116]

Year	*No. of Ovens*	*Coke Output (Tons)*		*Total Output*	*Value £*
		At Gas Works	*At Coke Ovens*		
1905	688			204,620	128,159
1906	562			177,894	123,186
1908	584	23,859	160,907	184,766	125,818
1909	555	24,519	234,825	259,344	186,445
1910	582	24,478	230,813	255,291	203,419
1911	664	25,828	309,716	335,544	254,897
1912	477	27,836	444,930	472,766	453,235
1913	509	31,266	506,374	537,640	500,475
1914	383	31,084	449,460	480,544	361,042

The progress of the coke industry provided a growing market for Cumberland coal, an increasing amount of local coal being converted into coke for the iron and steel works of the district; by 1912 more than a third of the coal was being used in this way.[117]

Year	*Coal Used in Coke Manufacture Tons*	*As % of Cumberland Coal Output*
1905	473,226	21·99
1906	488,135	21·94
1908	442,078	20·71
1909	539,083	23·34
1910	528,765	24·31
1911	607,306	26·44
1912	805,002	37·73
1913	936,071	40·34
1914	845,526	35·81

Despite the increased output of local coke the iron and steel industry in Cumberland was still largely dependent upon outside supplies. In 1907 five-sixths of the coke used in Cumberland ironworks came from the east coast.[118]

Colliery owners who constructed coke ovens often found it a profitable investment, particularly if they had an interest in a local ironworks. In 1909 the bye-product coke ovens at Watergate Colliery were said to be earning "a substantial profit" and but for the coke ovens erected at Clifton, to which small coal from Buckhill Pit was sent to be coked, it is doubtful if the latter colliery could have survived.[119] Bye-product coke ovens were a considerable improvement upon the old, wasteful beehive ovens. Larger outputs of coke were possible – in 1909 the hundred ovens at Harrington produced about 3,000 tons of coke per week for use in the nearby blast furnaces of the Workington Iron & Steel Co. Ltd. which owned both the colliery and ironworks at Harrington.[120] Coal was used more economically, larger amounts of coke being obtained per ton of coal consumed than with beehive ovens. On average 65 tons of coke were produced per 100 tons of coal by the Koppers ovens at Flimby Colliery between 1909 and 1912 and at St Helens Colliery between 1910 and 1912;[121] in Cumberland in 1905 473,226 tons of coal yielded 204,620 tons of coke, or 43·2 tons of coke per 100 tons of coal.[122] Improved techniques led to a reduction in production costs. At the Whitehaven coke ovens costs of production were:[123]

Year	*Coke Made (Tons)*	*Cost per Ton*
1885	18,834	2s. 10·05d.
1886	20,553	2s. 5·55d.
1887	19,660	2s. 6·90d.

At Flimby Colliery (1909-12) and St Helens Colliery (1910-12) the average cost of producing a ton of coke fell to 1s. 10. 69d.[124]

In addition to the manufacture of coke the extraction of bye-products such as benzol, tar and sulphate of ammonia proved highly profitable. It was calculated in 1912 that the 40 Koppers bye-products ovens at St Helens coking 74,751 tons of coal per annum would show a profit of £20,915. 12s. 10d. or 5s. 7·153d. per ton of coal used, and that the 40 similar ovens at Flimby Colliery coking 84,488 tons of coal per annum would show a profit of £13,747. 5s. 1d. or 3s. 3·051d. per ton of coal used.[125] These profit figures, of course, include the profits of both the coke and the bye-products. It is not surprising, therefore, that the number of beehive ovens in Cumberland fell from 664 in 1905 to 186 in 1913 and the number of Koppers ovens increased from 80 in 1908 to 323 in 1913.[126]

By 1914 there had been an important change in the control of mining. The landowners – the Senhouses, Curwens and Lowthers – had ceased to work their coal and had leased it to others. Falling profits and lack of capital made it impossible for the Senhouses and the Curwens to meet the growing costs of mining operations and even the Lowthers had to relinquish their grip on the industry. It seemed prudent to withdraw from a speculative industry and to accept colliery rents, a process which involved neither risk nor further capital commitment.

The landowners were replaced by iron-masters who wished to free themselves from dependence upon local coal owners by acquiring their own collieries. To supply coal to their furnaces at Cleator Moor the Whitehaven Hematite Iron Co. in 1847 sub-leased Cleator Moor Colliery, which was worked until 1871, and took over Bowthorn Colliery in 1869, raising coal there until 1894. In 1873 the West Cumberland Iron Co. took over Clifton Colliery and in the following year the Melgramfitz and Buckhill Collieries; these three collieries were controlled by the company until 1887. The Maryport Hematite Iron Co. leased Bolton Colliery 1872-86 and Ellenborough Colliery 1878-91, but both these mining ventures proved unprofitable. A Whitehaven firm, the Lonsdale Hematite Iron & Steel Co., controlled the Gilcrux and Crosby Collieries between 1883 and 1896. Harrington Colliery was taken over in 1865 by Messrs. Bain, Blair and Paterson of the Harrington Ironworks and in 1909 both the colliery and the ironworks became part of the Workington Iron & Steel Co. To complete the vertical integration of the industry the iron-masters also leased iron mines and limestone quarries to provide the necessary raw materials for their furnaces. The ownership of collieries ensured continuity of coal supplies at cost price and profits from the sale of surplus coal helped to offset the effects of a depression in the iron trade.

MINING METHODS

Much of the coal raised by the coastal collieries was won from workings under the sea and the distance from the shaft was long. At Whitehaven in 1869 undersea workings stretched two miles, the deepest being about 150 fathoms below the sea, although coal was being worked at all depths down to 60 or 70 yards. The bord and pillar system was then employed, the bords being six yards wide and the pillars twenty yards wide.[127] Some of these workings in 1872 were driven nearly four miles from the shaft and extended about two miles seaward from low water mark. Ventilation presented no problems although provided by a furnace of only moderate size.[128] It was said of these workings in 1908 that "they are well outside the danger zone, being about 100 fathoms deep, and no sign of any water coming through. The present workings at Whitehaven extend along the shore a distance of 4¾ miles, and at right angles to high water mark 3¾ miles."[129] As depths increased the size of the pillars increased. In 1905 the pillars of coal left in the first working under the sea were twenty-five yards square and the bords six yards wide. Thus 35 per cent of the coal was obtained and 65 per cent left in the pillars; but a system of panel working enabled the bulk of the coal left in the pillars to be removed.[130]

The longwall system of extracting coal was introduced at Ellenborough Colliery in 1868[131] and this example was followed elsewhere in the county; at No. 2 Pit St Helens in 1870, at Lowther Pit, Clifton in 1874 and at Allhallows in 1876. By this method all the coal was removed in a single operation; the workings were pushed forward in a long continuous line and as the colliers advanced the goaf (i.e. the space from which the coal had been taken) was packed with stone and slack on which the roof was allowed to settle. The coal was brought out through the goaf via gateways which were twelve yards apart and supported on each side by walls of stone. Only in the gateways was height made for the movement of coal by ponies or hand-trailing. This system was invariably employed in working thin seams. When the coal was undermined the pressure of the strata above helped to bring it down in large blocks and since no pillars were required nearly all the coal could be got away.[132]

In a labour intensive industry such as coal mining mechanization was needed to reduce wage costs and this need became more urgent with the marked decline in productivity between 1889 and 1914. But the colliery owners in Cumberland were reluctant to embark upon mechanized mining. The early coal-cutters were expensive, liable to break down, costly to maintain and their ability to lower production costs was doubtful. Moreover, the abundant supply of labour was sufficient in itself to delay the adoption of new methods of extracting coal. In an attempt to reduce labour costs a coal-cutting machine was introduced at Bertha Pit in 1899[133] and in the early years of the twentieth century experiments were made with others elsewhere in the county. They had little success and by 1913 only 1 per cent of Cumberland coal was cut by machine. Reports of the Inspectors of Mines show how little progress had been made in Cumberland towards mechanized mining before the outbreak of the first world war:

Year	*No. of Coal cutters*	*Coal Cut Tons*
1902	2	16,245
1903	2	6,487
1904	1	1,585
1905	2	7,695
1906	6	16,592
1907	4	5,200
1908	1	2,242
1909	1	220
1910	–	–
1911	3	606

Conveyors at the face had yet to make their first appearance in a Cumberland colliery.

Alterations were made in the underground movement of coal. It was stated in 1885 that stear-trailing had almost disappeared in west Cumberland and that tubs were taken directly to the face. Only in exceptional circumstances, e.g. a fault, was it necessary to resort to the stear.[134] R. F. Martin at Whitehaven introduced compressed air haulage on the main roads in William Pit although horses continued to be employed hauling coal near the face. At the four working pits there were 124 horses underground in 1880 and 136 in 1883 distributed in the following way:[135]

	1880	*1883*
William	63	56
Wellington	27	44
Croft	11	13
Henry	23	23

The compressed air haulage system was subsequently superseded by the "endless rope" system, first at Wellington Pit in 1889 and then at William Pit in 1890-1.[136] By the end of the century this method of hauling coal underground had been adopted in most mines in Cumberland. Sets, or trains of tubs, coupled to a clutch bogie, were attached to an endless steel wire rope. The clutch was screwed tightly to the rope and as it gripped the moving rope, which was driven by a steam engine at bank, it drew the tubs to the foot of the shaft. From there the tubs were drawn up in cages to the surface.[137] At Wellington Pit in 1910 the endless rope was 11,000 yards long and tubs were attached to it in sets of twenty, the rope travelling at about 3 m.p.h. when coal was being hauled.[138] All underground horses at Whitehaven Colliery were withdrawn in 1904 and men instead of horses had to haul coal near the face.

Cumberland colliery proprietors were slow in adopting improvements in winding coal to the surface and baskets continued to be used to a late date. As early as 1837 John Buddle had suggested that at Workington Colliery sheet iron tubs with slides in the shaft should be substituted for baskets in drawing coal.[139] The new method seems to have been first employed at Henry Pit, Broughton Moor in 1853 where the shaft was fitted "according to the modern and best approved system for drawing the Coal with Tubs and Cages." In the same year both shafts of the two working pits at Cleator Moor Colliery were fitted with tubs and cages. At Dean Moor as late as 1860 a horse gin was at work drawing the coal.[140] This was soon replaced by a steam engine and in 1864 a single decker cage came into use.[141] Coal winding with tubs and cages began at Whitehaven at the Wellington Pit in 1865, but the old-fashioned baskets were still being used at William Pit in 1875.[142] In 1871 a very powerful winding engine was erected at the new Henry Pit which raised four tubs on twin decked cages. Within six years tubs and cages were installed in all the Whitehaven pits and the use of baskets abolished; for this R. F. Martin was largely responsible.[143] When, in 1902, it was found that the drawing shaft of Wellington Pit, which was only ten feet in diameter, was inadequate to deal with the quantity of coal being raised it was decided as a temporary measure to meet the inadequacy to add a third deck to the cages for winding coal until a new shaft was sunk.[144]

In Cumberland the ventilation of mines by means of a furnace at the foot of the upcast shaft remained the usual practice until 1870. In that year T. E. Forster was responsible for the successful introduction of mechanical ventilation into the county when he installed a fan ventilator at Wellington Pit and erected a Guibal fan at Duke Pit.*[145] This work was continued by R. F. Martin who erected Guibal fans at the William and Kells Pits, and by the Whitehaven Colliery Company. A report on the colliery in 1889 says, "The James Pit underground furnaces which ventilated the Henry Pit workings and which were supplied with coal got in the neighbourhood of the Railway Tunnel, have been stopped by the company. The whole of the air from both Henry and William Pit workings now comes up William Pit upcast to the Fan. This change has made a great improvement in the quantity of air, and is a much safer and cheaper way."[146] By 1893 all the underground furnaces at the Whitehaven Colliery had been replaced by Guibal fans.[147] In the early years of the twentieth century ventilators of the Walker indestructible type, with fans 24 feet in diameter, were installed at the William and Wellington Pits.[148]

THE MINERS†

The expansion of the coal mining industry would have been impossible without a parallel increase in the supply of manpower in the mines; in Cumberland miners had yet to be

* J. H. Clapham, *Economic History of Modern Britain* (Cambridge, 1931-8), II, 102, ascribes the introduction of the Guibal fan to R. F. Martin, but he was not appointed viewer until 1874.

† All population statistics in this section, except where otherwise stated, are taken from, or are calculations based upon information provided by the Censuses of Population 1801-1931.

replaced by machines. Throughout the nineteenth century the population of the county increased, although at a lower rate than for England and Wales as a whole. Between 1801 and 1821 the population grew rapidly, from 117,230 to 156,124, and then more slowly until 1871 when it reached 220,253. The next decade showed a marked increase of 30,394, or 13·8 per cent. From then onwards further growth was slight; between 1891 and 1901 the population was almost stationary and then followed an actual decline in the total population.

Population of Cumberland

Year	*Population*	*% Intercensal Change*	*% Intercensal Increase in England & Wales*
1801	117,230		
1811	133,665	+14·0	14·00
1821	156,124	+16·8	18·06
1831	169,262	+ 8.4	15·80
1841	178,038	+ 5·2	14·27
1851	195,492	+ 9·8	12·65
1861	205,276	+ 5·0	11·90
1871	220,253	+ 7·3	13·21
1881	250,647	+13·8	14·36
1891	266,549	+ 6·3	11·65
1901	266,933	+ 0·1	12·17
1911	265,746	− 0·4	10.89

In the coastal towns of the coalfield there was a significant increase in the population; within a century the inhabitants of Maryport and Workington quadrupled and those of Whitehaven more than doubled:

Year	*Maryport*	*Workington*	*Whitehaven*
1801	2,932	5,716	8,742
1811	3,134	5,807	10,106
1821	3,514	6,439	12,438
1831	3,877	6,415	11,393
1841	5,311	6,045	11,854
1851	5,716	6,280	14,190
1861	6,150	6,467	14,064
1871	6,938	8,413	13,298
1881	8,126	14,361	13,374
1891	12,410	25,192	19,370
1901	11,897	26,139	19,324
1911	11,418	25,092	19,044

In its distribution and fluctuations the population reflected the economic history of the county. As the coal and iron industries developed the population tended to concentrate in the coastal area where minerals were to be found and from which they could be easily exported. For census purposes Cumberland was divided into nine Registration Districts, four of which – Wigton, Cockermouth, Whitehaven and Bootle – covered the industrial belt of West Cumberland. Between 1801 and 1881 the population of these four districts increased by 140 per cent compared with an increase of 83 per cent in the rest of Cumberland:[149]

Registration District	*1801*	*1881*
Wigton	14,907	23,440
Cockermouth	24,093	56,789
Whitehaven	19,960	59,292
Bootle	4,321	12,225
Total	63,281	151,746
Rest of County	53,949	98,901

Some of the villages which depended primarily upon the coal mining industry for their existence recorded a remarkable rise in the numbers of their inhabitants:

Year	*Aspatria & Brayton*	*Flimby*	*Dearham*
1801	327	273	403
1811	478	271	443
1821	632	376	515
1831	761	404	736
1841	988	546	1,037
1851	1,123	555	1,209
1861	1,210	1,178	1,509
1871	1,778	1,713	1,935
1881	2,408	2,123	2,246
1891	2,714	2,415	2,598
1901	2,885	2,482	2,147
1911	3,339	2,487	2,127

The varying fortunes of individual industries in Cumberland are amply illustrated by the occupational statistics contained in the decennial Censuses of Population. An ever-increasing number of coal miners recorded between 1841 and 1911 is testimony of the continued growth of the industry. The intercensal rate of increase for coal miners was greatest between 1881 and 1891, when it was 36·4 per cent, the average national rate being 36 per cent. After 1891 the number of Cumberland coal miners continued to rise, but at a lower rate than the rate of increase for the mining industry as a whole. Between 1891 and 1901 the Cumberland coal miners increased by 7·6 per cent, and 21 per cent in the next decade, whereas the number of coal miners in England and Wales together increased by 25 per cent and 36 per cent respectively.[150] Male coal miners enumerated in the county were:

Year	*No.*	*Year*	*No.*
1841	1,823	1881	5,278
1851	3,579	1891	7,199
1861	4,274	1901	7,748
1871	4,427	1911	9,374

Throughout the nineteenth century Cumberland was enriched by an influx of workmen from Scotland and Ireland and from nearly every county in England. Irishmen worked in the coal mines and in the 1840's found employment in railway construction and as dockers; there were 4,881 natives of Ireland in Cumberland in 1841 but they were outnumbered by Scots of whom there were 6,355. Ten years later the number of native Irishmen in the county had doubled to 9,866, these being a small part of the huge exodus which followed the great famine in Ireland. It was Ireland which provided the biggest supply of immigrants who reached a maximum of 14,093 in 1881. Scots furnished the next largest supply of labour which also reached its maximum in 1881 when 12,207 Scots were living in Cumberland. English immigrants, for the most part, came from the nearby counties of Westmorland, Yorkshire, Lancashire, Durham and Northumberland. As could be expected many Irishmen settled in the coastal towns where jobs were immediately available thus sparing them the hardship of going further afield in search of employment. How many Irishmen came to work in the collieries it is impossible to say but it is certain that a large proportion of these immigrants became coal miners. A witness stated to the Royal Commission on Mines (1862-4) that in the Cleator district, "The trade . . . has developed so rapidly lately that there is a great deal of imported labour and a great deal of new labour which goes into the collieries is Irish labour, and therefore you have generally speaking an inferior class of work people to deal with."[151] Cleator Moor became the home of the chief Irish community in Cumberland and even now it is referred to as "Little Ireland". About 1850 Cleator Moor was a collection of rows of squalid cottages hurriedly built to house a rush of immigrants who brought with them the debased living standards of their native country and quickly created an Irish ghetto. According to Caine, "Not only was there no water, except what was brought in carts . . . There was no sewerage. The streets were ill-laid, with scarcely any pavements. The houses were not drained. Sanitary conveniences were almost unknown. Pig-styes abounded on all sides, "the pig, the crathur, as clane as any Christian", was not infrequently an inmate of the houses, along with the people themselves . . . Cleator Moor was, at that time, significant of a terribly rough and semi-barbarous community."[152]

After 1841* the proportion of immigrants to native Cumbrians increased and by 1871 immigrants accounted for one-fifth of the population of the county.

Population of Cumberland

Year	*Native Cumbrians*	*Immigrants*	*% of Immigrants in Cumberland*
1841	153,770	24,268	13·63
1851	162,115	33,377	17·07
1861	166,786	38,490	18·75
1871	174,799	45,454	20·63
1881	191,428	59,219	23·62
1891	207,913	58,636	21·99
1901	213,100	53,833	20·16
1911	213,627	52,119	19·61

The origin of these immigrants and the fluctuations in numbers are shown in the following table:

* In 1841 2,209 of Whitehaven's 11,854 inhabitants were immigrants.

Immigrants in Cumberland

Birthplace	*1851*	*1861*	*1871*	*1881*	*1891*	*1901*	*1911*
Scotland	7,443	9,025	10,348	12,207	11,826	10,713	9,931
Ireland	9,886	10,529	11,870	14,093	9,698	6,371	4,403
Wales	230	412	932	1,252	1,217	966	813
Westmorland	3,910	3,924	2,701	4,453	4,949	4,752	4,735
Lancs.	3,906	4,888	5,652	6,977	8,197	8,094	4,341
Yorks.	1,221	1,438	1,732	2,535	3,487	3,511	1,905
Cheshire	232	285	521	617	691	762	511
Durham	909	1,109	1,497	2,265	2,489	2,648	2,250
Northumberland	2,427	2,621	2,851	3,403	3,322	3,358	2,352
Derby	98	92	168	285	1,196	920	594
Staffs.	167	200	535	755	964	871	459
Notts.	64	71	86	171	301	296	147
Warwicks.	104	111	201	310	349	361	157
Devon	107	166	508	1,030	864	699	406
Cornwall	88	125	1,141	1,967	1,580	1,146	766

This admixture of non-Cumbrians was reflected in the variety of religious denominations in Cumberland – Irish Catholics, Scottish Presbyterians, Methodists from Yorkshire and Cornwall, Baptists and Congregationalists from other areas added their chapels to the existing Anglican churches.

In the latter half of the nineteenth century the size of the labour force in some collieries increased notably. An analysis of the distribution of manpower at the Whitehaven Collieries in 1859 gives the following figures:[153]

Pit	*Hewers*	*Others Underground*	*Above Ground*	*Total*
Wellington	240	245	130	615
Croft	86	133	50	269
William	154	155	60	369
Countess	53	67	20	140
Wreah	18	12	10	40
Coke Ovens	—	—	20	20
Granary Yard	—	—	63	63
Total	551	612	353	1,516

In 1880 474 hewers were employed, 495 in 1884 and 538 in 1885.[154] By 1900 the number of employees at the three working pits – William, Wellington and Croft pits – had risen to a total of 2,181.[155]

The Mines Act of 1842 forbade the employment underground of women and of boys under ten years of age. Young persons were further protected by the Education Act of 1870 which led to the provision of state elementary schools and the Education Act of 1880 which made elementary education compulsory. As a result of these measures young boys ceased to work in the pits. The number of boys aged 12-13 employed in Cumberland collieries underground fell from fifty-eight in 1874 to seven in 1884; in the same period the number of boys 13-16 similarly employed fell from 465 to 385.[156]

After 1842 women continued to work at the collieries as surface hands until recent times. The Censuses of Population enumerated the female workers at Cumberland collieries as follows:

Year	No.	Year	No.
1851	72	1891	164
1861	104	1901	225
1871	169	1911	308
1881	115	1921	327

In the 1840's colliery owners seeking workmen had to compete with railway contractors. Thirty years later the competition came from the iron industry. As the mining of iron ore and the manufacture of iron increased in the county after 1870 a shortage of manpower made itself felt. Then, as now, the employer who could provide houses for his workers was likely to get them, and, equally important, keep them. To maintain production the colliery owners were obliged to embark upon housing programmes. The West Cumberland Iron Co., acting upon the advice of G. B. Forster, built forty miners' cottages in 1873-4 to attract employees to their Clifton collieries.[157] Seventy houses were built by the Moresby Coal Co. in 1880 for the men at Oatlands Pit, Pica and twenty more in the early part of this century.[158] The Lowthers had built for their miners since the eighteenth century in order to facilitate the recruitment of labour and when they leased the Whitehaven collieries in 1888 to Messrs. Bain & Co. they transferred 578 miners' cottages to them at the same time. Under the Lowthers these cottages were rent-free. The Whitehaven Colliery Co. Ltd., which leased the Whitehaven pits in 1913, built over 300 houses for their employees. Other colliery proprietors charged a small rent. At Broughton Moor Colliery, for example, there were twenty-two workmen's cottages let at 1s. 3d. – 2s. per week in 1871.[159] Not all miners lived in tied cottages for some coal owners did not make any attempt to house their employees. The advantages to the employer of having his men living in his houses were obvious and well understood; they are clearly stated in a letter from a pit manager, 10 April 1913, "By having men in Colliery houses the Manager is able to get a better class of men and to keep them. They will sacrifice a great deal rather than move, they are more dependent on their work, one can enforce strict discipline in the Colliery, they are content with a rather less rate of wages, no inducement need be given."[160]

Between 1850 and 1914 there were wide fluctuations in the wages of coal miners. In 1851 the Cumberland miner was earning 2s. 6d. to 3s. per day, a shilling a day less than in the other northern counties, an influx of Irish labourers being responsible for the depression of wages.[161] This was less than the Cumbrian miners had received in the 1840's. Later wages rose, the average daily rate between 1865 and 1870 being 4s. reaching 7s. between 1870 and 1875,[162] the increase in the latter period resulting from the shortage of labour which followed the expansion of the iron industry. At Whitehaven piece rates for hewers in 1871 and 1878 were:[163]

Pit and District	*Hewing prices per ton*				*Average Wage per day*	
	Nov. 1871		*March 1878*		*Nov. 1871*	
	s.	*d.*	*s.*	*d.*	*s.*	*d.*
William Pit						
Nova Scotia	1	5½			6	4½
Delaval	1	5½	1	6¾	5	2¾
Forster	1	6¼	1	6¼	5	3
Average	1	5¾	1	6¾	5	7¼

Pit and District	*Hewing prices per ton* *Nov. 1871*		*March 1878*		*Average Wage per day* *Nov. 1871*	
	s.	*d.*	*s.*	*d.*	*s.*	*d.*
Wellington Pit						
Victoria	1	8¼			4	7¼
Old Brake	1	8¼	1	7	4	7¼
Harton	1	8¼	1	8½	4	7¼
Martins	1	8¼	1	8	4	7¼
Bannock Bd.	1	8¼	1	11	4	7¼
Average	1	8¼	1	8½	4	7¼

There was a marked fall in wages after 1875, the average daily wage between 1875 and 1885 being only 4s. In 1885 the daily wages in the Cumberland coal mines were:[164]

Men (piecework)	2s. – 5s.
Men (day wages)	2s. 9d. – 4s.
Boys	1s. – 2s. 9d.
Women	1s. – 1s. 6d.

According to a Board of Trade report coal hewers in Cumberland in 1886 earned an average of 24s. 8d. per week and worked 48 hours per week; in 1891 their average net wage per shift was 6s. 9d. working 8½ hours from bank to bank.[165]

To strengthen their bargaining position the miners combined to form the Cumberland Miners' Association in 1872. It was not easy to organize all the Cumberland miners despite the efforts of Andrew Sharp, its able and courageous secretary. Some were reluctant to pay union dues and more feared to join a movement that would lead to strikes and the loss of a livelihood. Too many miners, especially in the Whitehaven area, lived in tied cottages and the resistance of the men could be broken by summary eviction. Further, union members who were involved in strikes were blacklisted by the colliery owners and found it difficult to obtain work. Seeking a docile labour force employers naturally preferred to hire non-unionists. Among other items discussed at a meeting of the Miner's Association 2 April 1903 was "unfair treatment of Union Men when seeking work at other Collieries to that which they have been employed, and why preference is given to those outside the Association."[166]

The Association was formed at a difficult time for the coal trade was soon to enter a period of depression, Cumberland coal having to compete in Ireland with coal from Scotland, the Scottish coal being sold at much lower prices. Hence the union in its early years had to face repeated demands for wage reductions. Nevertheless it gave maximum support to lodges and individuals involved in disputes with employers – the minutes of nearly every meeting of the Association show this under the heading "Support Claims". Every effort was made by the union to recruit more members and Andrew Sharp showed himself a skilful negotiator and powerful advocate on their behalf. In 1880 the West Cumberland Miners' Association had 1,000 members, the number rising to 3,259 in 1890 and 6,326 in 1910.[167] Like other unions it was mainly concerned with improving the working conditions of its members, securing compensation for men injured at work and for the dependants of those killed in the mines and combating non-unionism.

A fall in the price of coal precipitated a clash between masters and men in 1874; in

March of that year the coal owners in West Cumberland cut miners' wages by 10 per cent. At Brayton Domain No. 3 pit, where a 4-5 ft. seam was being worked, the men had an additional grievance for they were ordered by the manager to leave ten inches of floor. This they regarded as equivalent to a reduction in earnings of 25 per cent as well as making working conditions more difficult. They came out on strike in April. Within days all strikers who inhabited colliery houses were given notice to quit and county court eviction orders were quickly obtained in response to the owners' plea that the cottages were needed for other workmen. After six weeks the strike ended. It was agreed that the bottom ten inches would be worked and that the miners would ensure that all metal and stone would be separated from the coal.

Whilst this strike was in progress the employers announced a further reduction in wages of 15 per cent. Thus the miners in West Cumberland were faced with a twenty-five per cent wage cut in less than two months. At various collieries the men went on strike until the dispute was referred to an arbitrator, Rupert Kettle. His award shocked the colliers – hewers' wages were reduced $17\frac{1}{2}$ per cent, other underground workers 15 per cent and engine men 6d. per day or just over 9 per cent.[168]

Another wage cut of 10 per cent to take effect on 9 December was announced by the Cumberland Coal Owners' Association. This reduction was said to be absolutely necessary and could not be a matter for arbitration. By the middle of the month about 2,000 miners were on strike. These were members of the Miners' Association; in the Whitehaven area, where there were few unionists, the colliers remained at work. The strike lasted fifteen weeks although the miners, at a meeting held 11 January 1875, had declared their willingness to accept a 10 per cent reduction in wages if it came to them as a result of arbitration. The owners insisted on the ten per cent reduction and refused to consider arbitration. The miners considered that they were fighting for an important principle, the settlement of disputes with the employers by arbitration instead of having always to accept without question whatever the employers decreed. When the miners voted by 1,046 to 79 to remain on strike the owners replied by giving notice of a further reduction of 5 per cent in the miners' wages. Some violence accompanied this strike; blacklegs were assaulted and some had their house windows smashed. A particularly ugly incident took place at Westray Pit where the ropes attached to the cage were partially cut. Had the cage carried a load of men they would all have been killed but fortunately the cuts were seen by maintenance workers before any more harm was done. This led to two miners being charged with attempted murder. If the plan had succeeded a brother of one of the accused would have died. The two defendants were acquitted of the charge of attempted murder but sentenced to twelve months' hard labour on a charge of cutting the cage ropes with intent to damage and render useless the machinery of a coal pit.

The strike dragged on, inflicting much hardship on the miners and their families. It became clear that no man would be employed unless he accepted the 10 per cent reduction in wages. All the masters would say was, "Accept the 10 per cent reduction and leave the other 5 to arbitration." Many colliers left Cumberland to seek employment elsewhere, others, in desperate need, drifted back to work. By 17 March 1875 the strike was over, economic pressure compelling the strikers to resume work on the employers' terms.[169]

Although strikes were frequent at individual pits in the last quarter of the century the Cumberland miners as a whole did not plunge blindly into strikes to maintain or improve their position. On many occasions they resorted to arbitration. Thus in 1875 when the coal owners were seeking another reduction in wages both sides agreed (10 November) to refer the matter to two arbitrators and when these failed to reach a decision an umpire, Serjeant Wheeler, made the award. Wheeler stated that in 1871 hewers received on average 1s. 6·29d. per ton and in 1875 this had risen to 2s. 1·7d., a nominal advance of 38 per cent since 1871, but since the miners in 1875 were working thinner seams the real advance was 29½ per cent. At the same time the price of coal had risen from 6s. 6·26d. per ton in 1871 to 7s. 6·38d. per ton in 1875, an increase of 15½ per cent. Hence the hewers had gained 14 per cent over the coal owners. In 1875 their average earnings were 6s. 3d. for an eight hour day, from which was deducted 2.56d. per day for pit gear and 1·8d. per day was paid by the miners to a sick benefit club. Wheeler awarded a 10 per cent reduction in the wages of hewers, deputies, drivers and trappers, and a 5 per cent reduction in the wages of furnacemen, onsetters and horse keepers.[170]

The decline in the coal trade led to further wage cuts in 1876 when the Cumberland Coal Owners Association proposed a reduction of 15 per cent in underground wages and 10 per cent in surface wages. This was referred by both masters and men to the arbitration of A. MacDonald and G. B. Forster who awarded an 8 per cent reduction in the wages of hewers, trappers and drivers and a 10 per cent reduction in the wages of all other underground workers.[171] Three years later a brief strike took place to resist another 10 per cent reduction in wages. About 3,000 miners joined the strike, among them the hewers at Whitehaven who had not been on strike for thirty-five years. They objected not only to the fall in wages but also to what they claimed were excessive deductions for "dirt", i.e. unclean coal found in the tubs they sent to the surface. The strike lasted only a few days. The miners' union was not strong and since the coal trade was depressed not even a powerful union could have kept up wages. Nor did the miners have any funds of their own to fall back on and since many of them lived in cottages provided by their employers continued refusal to work would have resulted in eviction.[172] Andrew Sharp strongly denounced what he considered to be the "injustice" of the coal owners. He advocated restriction of output as one means of putting pressure on the employers and advised miners to emigrate to the colonies where, as former agricultural workers, many of them would quickly find employment.[173]

A change came in the system of wage payments in Cumberland 31 October 1879 when the sliding scale was first adopted by masters and men. Wages were now to be regulated every quarter by a sliding scale based upon the average selling price of coal, the average selling price of July-September 1879, which was 4s. 6·19d. per ton, being taken as the standard price. Rates of wages for the various classes of mineworkers in the same period were taken as the standard rates of wages. For every rise or fall of 4d. in the selling price wages for the following quarter were to be increased or reduced according to this scale:

	Increase
1st advance of 4d.	2½ per cent
2nd advance of 4d.	2½ per cent
3rd advance of 4d.	5 per cent
4th advance of 4d.	2½ per cent
5th advance of 4d.	2½ per cent
6th advance of 4d.	5 per cent

For the seventh and every subsequent advance of 4d. wages were to be increased by 2½ per cent.[174] Wages were regulated by a succession of sliding scales until the end of 1889 and until the end of 1888 miners' wages were above the standard wage.[175]

During the prolonged depression in the coal trade after 1873 the miners had found that arbitration to settle wage disputes resulted in a series of wage cuts. They therefore welcomed the sliding scale system in the hope that when coal prices increased their wages would rise automatically.

Higher wages, it seemed, could be obtained without resorting to strikes. For the employers the new system afforded an opportunity to lower wages during a depression period without being involved in labour disputes.[176] Unfortunately the trade depression continued and the miners came to see that sliding scale agreements did little to improve their lot. Under this system there were large fluctuations in wages as shown in the percentage additions to the standard wages of the Cumberland hewers at the end of the following years:[177]

1888	10	1903	40
1889	30	1904-6	37½
1890-93	40	1907	55
1894-97	30	1908	52½
1898	32½	1909-11	47½
1899	40	1912	52½
1900-1	60	1913-14	65
1902	50		

What the miners needed was a new system capable of ensuring more stable wage rates. Accordingly they terminated the sliding scale agreements in 1889 and chose to have their wages determined by a conciliation board. In April 1890 a Board of Arbitration and Conciliation was formed, consisting of an equal number of representatives of employers and employees, with an official referee, to regulate percentage variations of wages. By the end of the year wages were 40 per cent above standard. When prices began to recede in 1891 the employers claimed a 10 per cent reduction in wages, a claim rejected by the men.[178] The Conciliation Board was terminated in April 1891 but the 40 per cent continued to be paid until the strike of 1892, a strike settled by a most exhaustive arbitration.[179] The Cumberland Coal Commission was set up to adjust "the standard rates of wages for piece and daywork, having regard to the rates being paid for similar work during 1887, and to the rates now being paid in competitive districts, taking into consideration any differences in the hours and conditions of employment, and any changes that have taken place in the nature of the work since the above date." The owners in putting their case for a reduction emphasized the natural difficulties of mining in Cumberland – the limited extent of the coalfield, its many faults and its thin and dirty seams – competition from Scotland and Lancashire in the Irish market, the decline in output per man in Cumberland and the shorter hours worked there.[180] The award was a 7½ per cent reduction of the 1887 standard rate for piece workers and day workers had their 40 per cent advance on the 1887 standard rate reduced to 28 per cent.[181] As has been pointed out, the 7½ per cent reduction of standard rates for piece workers was roughly equivalent to a 10 per cent reduction in earnings.[182] Although wage rates could be fixed earnings could not be fixed. What mattered was continuous employment. In 1888 the Whitehaven colliers were said to "have only had employment for about three days a week for the last few weeks, and this continued want of work following the long depression of trade has affected them very severely."[183]

Trade unionism at Whitehaven had always been strongly opposed by the employers, an attitude which provoked an extraordinary diatribe in *The Northern Counties Gazette* of 21 June 1884 against the management of the colliery following an attempt to form a trade union there. In an article headed "Colliery Office Terrorism" the "tyranny" of R. A. Robinson, the principal colliery agent was savagely attacked,

> "The Whitehaven collier has to submit to vexatious reductions of wages without any explanation as to the reason thereof. He knows that he is cheated by his employers through a system of unfair deductions in the quantity of coal sent up the pit shaft . . . He is not permitted to think for himself . . . He is expected to take his cue from the colliery office in the recording of his vote, whether it be for a guardian of the poor, trustee of the town and harbour, or member of Parliament . . . and the greatest crime he can be guilty of is to join a union . . . a meeting of the colliers was held in Whitehaven on Saturday last . . . and the result of this meeting was that nearly 150 colliers decided to form a Union . . . on Thursday last the Colliery Office policeman was actually despatched to serve every man who had joined, or was suspected to have joined the Union, with a fortnight's notice to quit the tenement he holds under the Trustees of the Lowther Estates."

Robinson brought an action for libel against the proprietor of *The Northern Counties Gazette* and was awarded £150 damages. It was stated at the trial that the miners were given notice to quit their houses not for joining a union but for restricting output.[184]

Union officials found it difficult to recruit members at Whitehaven, but the transfer of the colliery to Messrs. Bain & Co. gave a new impetus to the movement. R. W. Moore reported in 1889, "The Miners' Union which was almost defunct at the time of the transfer of the collieries has now become very strong. Nearly all the hands, both above and below ground are members of it. The arbitrary treatment the men have received from the manager is undoubtedly the cause of this and the numerous strikes."[185]

Strikes by coal miners were usually caused by colliery owners attempting to impose wage cuts or refusing to grant wage increases. In 1885 miners at the Watergate, Moresby, Robin Hood, Broughton Moor and Ellenborough collieries went on strike in opposition to the employers' demand that there should be a fourteen days' notice system and that all wages due to men dismissed for misconduct should be forfeited. At Ellenborough the men remained on strike for six weeks until the colliery owner agreed that a man dismissed for misconduct should receive his wages up to the date of dismissal.[186] A strike to resist wage reductions affected the entire coalfield for three months in 1892 and, as stated above, the wages dispute was referred to the Cumberland Coal Commission for settlement. The Miners' Federation paid out £17,929 to support the strike in Cumberland.[187] In the following year the Cumberland miners again went on strike, this time to obtain an increase in wages. Many of the colliery owners showed themselves willing to grant pay increases and their employees soon returned to work. At Whitehaven, however, the strike lasted longer and was accompanied by violence, troops and additional police having to be called in to maintain order after attacks on the houses of blacklegs. The strike ended when the Whitehaven men were promised that their rate of pay would be brought up to 40 per cent above the standard of 1888.[188]

The withdrawal of horses and the institution of the manhandling of tubs at Croft Pit, Whitehaven in 1903 led to a prolonged dispute, eventually settled by the Conciliation Board which decided in favour of man-trailing and fixed a price for it. The Whitehaven Colliery Co. decided in 1904 to adopt the same system of trailing at the Wellington and William Pits, a decision which provoked a strike which lasted over fourteen weeks. The employers were adamant in their refusal to allow the dispute to go to the Conciliation Board; in their view the question whether men or horses should haul the trucks of coal underground involved the principle of management and was outside the jurisdiction of the Conciliation Board. If the men accepted the removal of horses the Conciliation Board could fix the payment for man-trailing. The strikers insisted that they should not be expected to do the work of horses and that this was a matter for the Conciliation Board. They were supported by their union which provided 10s. per week per man and 1s. per week for each child. As happened so often, near-starvation ended the strike; in its latter stages strike pay fell to 2s. 6d. per week and hunger compelled the strikers to return to work on the employers' terms.[189]

Early in 1912 the miners in Cumberland and throughout the country were voting for or against strike action to obtain a minimum wage for all underground workers. By 4,918 votes to 813 the Cumberland miners voted in favour of a strike.[190] The Miners' Federation drew up a schedule of minimum rates for coal-getters in each mining district; for Cumberland the minimum rate was fixed at 6s. 6d. per day. A national coal strike lasting five weeks, fully supported by the Cumbrian miners, followed the failure of the Miners' Federation to persuade the coal owners to accept the minimum wage principle. The government intervened and quickly passed the Minimum Wage Act by which minimum wages for underground workers were to be fixed in each mining district by a Joint District Board consisting of representatives of both employers and workmen. This Act did not satisfy all miners since the minimum rates were not included in it and had yet to be determined by the District Boards. In Cumberland the miners voted by 2,980 votes to 1,897 against resuming work. The executive committee of the Miners' Federation recommended a return to work and the national strike soon ended, the Cumberland miners going back to the pits 9 April.[191]

The minimum daily wage fixed for mine workers in Cumberland in 1912 may be compared with the minimum daily wage in June 1914:[192]

	1912		*1914*	
Adults	*s.*	*d.*	*s.*	*d.*
Hewers	6	0	6	8
Deputies	6	0	6	8·58
Shotfirers, firemen	5	10	6	6·35
Engine drivers, pumpmen, horsekeepers	4	2	4	7·97
Onsetters, shift labourers	4	7	5	1·56
Rolleywaymen	4	9	5	3·79
Oilers	4	0	4	5·72
Bogie hands	3	9	4	2·37
Boys				
Drivers, couplers, lamp carriers, engine boys, clippers, pump boys	2	0	2	2·86
Trailers, brakers	2	9	3	0·94
Apprentice hewers	4s. – 5s. 6d.		4s. 5·33d. – 6s. 1·34d.	

A comparison of the average wage per shift shows that in 1914 all categories of mining labour in Cumberland, except women and girls, earned less than the average for the United Kingdom:[193]

	Average wage per man per shift			
	Cumberland		United Kingdom	
Underground adults	*s.*	*d.*	*s.*	*d.*
Piecework coal getters	8	1·72	8	9·97
Coal getters on day wage	6	9·18	6	11·91
Putters, fillers, hauliers, trammers	5	11·48	6	5·44
Timbermen, stonemen, brushers, rippers	7	0·01	7	6·01
Deputies, firemen, examiners	7	2·98	7	8·34
Other underground labour	5	7·90	5	9·40
Surface adults				
Winding enginemen	6	8·05	7	1·87
Other enginemen	5	4·33	5	7·13
Stokers, boilermen	4	9·73	4	11·44
Pitheadmen	5	0·49	5	1·85
Persons on and about screens	4	7·84	4	9·25
Mechanics, joiners, blacksmiths	5	4·45	5	7·46
Other surface labour	4	3·68	4	9·84
Youths and boys				
Underground	3	5·54	3	6·91
Surface	2	0·96	2	5·42
Women and girls	2	1·54	1	11·41

It may be noted that in Cumberland mining disputes rarely concerned hours of work. Since 1877 an eight hour day had been customary and it was reported that, "Practically the only difference that a legal eight hour day would make in this district would be to substitute a rigid time limit for the more elastic system now in operation which allows men to remain rather longer at work if they desire to do so, and which enables the older men to take their own time in going to and returning from their working places."[194] Hence the 1908 Act, which fixed a maximum working day of eight hours for underground workers, made little difference to the length of the working day for miners in Cumberland.

It was not only the miners who organized themselves into a body to protect their interests, the employers did precisely the same thing. Since 1842 the Cumberland colliery owners, in an attempt to avoid competition between themselves, had met regularly to fix the price of coal. As trade unionism developed in the county the Cumberland Coal Owners' Association became not only an instrument to regulate coal prices but also a weapon to resist the miners' claims for better working conditions. The miners, crippled by their poverty and lack of education, found their employers to be formidable opponents, men of wealth and education, fortified by a judiciary predisposed in their favour. For the miners it was a struggle for some form of economic justice, for the colliery owner it was a struggle for maximum profits.

An examination of the rules of the Coal Owners' Association*[195] shows that its members were obliged to adopt a common policy in respect of wages and hours of work and to support each other should strikes take place. Rule 5 states,

"The objects of the Association shall be to discuss, consider and regulate all matters relating to the coal trade, with the view of obviating unnecessary competition between the members, and in regard to wages and hours of working to make such regulations between the members, and between the Association and the workpeople, as may from time to time be agreed upon by not less than two thirds of the whole voting power of Assocation.

Rules 10-12 defined what action should be taken to deal with strikes:-

"10. So soon as a dispute arises at any Colliery, causing, or likely to cause a claim on the Association, the owner of such Colliery, or his agent, must at once communicate with the Secretary of the Association, whose duty it shall be forthwith to summon a meeting of the Association to examine and investigate the circumstances, and determine whether such Colliery shall be taken under the protection of the Association or not. So long as the owner of the Colliery on strike conforms strictly to the instructions of the Association, he shall continue to be under the protection of the Association.

11. The owners of any Colliery which shall be required by the Association to resist unjustifiable demands of the workmen, and thereby be put upon strike, shall, so long as the owners remain under the protection of the Association, as provided by clause 10, be indemnified by the Association for all expenses and losses of profit or otherwise arising out of or incurred during the period of such strike, the same to be ascertained and awarded by two impartial Viewers, ... In the event of any other strike at any Colliery, this Association may, if two-thirds of the voting power at any meeting duly convened think fit, and upon a pledge having been given by the owners of the Colliery to make no settlement without the consent of the Association, agree to pay a subsidy to such owners during the continuance of the strike, so long as the owners conform strictly to the instruction of the Association.

12. No general district strike or lock-out shall be agreed to without the consent of two-thirds of the whole voting power of the Association.

Rule 13 obliged members to enforce the Association's regulations concerning wage rates and hours of work:

"No member of this Association shall make any alteration in the general rate of wages paid to his workmen, or in the hours of working at his Colliery, without the sanction of the Association, or refuse to concur in any alteration of the general rate of wages or hours of labour resolved by the Association in General Meeting."

With few members it was easy to organize the Association and to reach decisions. Once locked in a struggle with the impoverished miners the wealthy colliery owners could usually hold out for a successful outcome.

Mining accidents continued to take their toll of human life. Reports of the divisional inspectors show that 950 employees were killed in Cumberland coal mines between 1851 and 1914:

* The rules referred to here are those adopted by the Coal Owners' Association in 1890.

Cause	*No. killed*
Falls of roof, stone, coal	343
Explosion	238
Crushed by trams, tubs, waggons	140
Fell down shaft	46
Other shaft accidents	13
Cage	36
Machinery	19
Shot firing	9
Gas	6
Kicked by horse	5
Inclined plane	3
Miscellaneous	92

As can be seen, the biggest single cause of mining fatalities was falls of roof, or stone or coal. During this period, except for the year 1905, men were killed in this way every year. This would suggest that not enough attention was paid to the provision of props and girders. Carelessness, as always, was an important factor in many accidents, e.g. shaft accidents. The repetitive nature of many mining accidents shows a perverse disregard of regulations and an unwillingness to learn from past errors. Explosions of firedamp killed over 200 men and too often it was neglect of regulations by employers and workmen which accounted for these deaths – the men using candles or removing the tops of their lamps in gassy areas and the employers failing to provide adequate ventilation in the pits.

The high mortality rate led to the appointment of government inspectors to inspect the mines, their main function being to raise safety standards. Unfortunately there were too few of them and each had too many collieries to inspect. Matthias Dunn*, one of these inspectors, found that he and his colleagues, when attempting to enforce safety regulations, had to contend not only with the obstructiveness of colliery owners and the carelessness of miners, but also with the class bias of the magistracy, men of property reluctant to convict other men of property and with little regard for the lives of obscure colliers. When colliery owners or managers were summoned for breaches of the regulations magistrates often dismissed the charges or imposed nominal fines. An explosion at Hope Pit, Cleator Moor (18 July 1862) killed a miner and Dunn charged the owners with not providing adequate ventilation in the pit, a charge which was dismissed. Four men were killed by an explosion in the same pit in May of the following year. In Dunn's words, "Four lives were sacrificed through the utter neglect of my caution and remonstrances, both after the previous explosion in the Hope Pit in July 1862, and at the hearing of the summons in 1863 (February)."[196] After a gas explosion at Allhallows Colliery (10 July 1877) which killed a trailer the colliery manager was charged with failing to ensure sufficient ventilation and was fined £5.[197]

In view of its long and alarming record of deaths caused by the explosion of fire-damp the managers of Whitehaven Colliery in the nineteenth century should have done their utmost to provide the most efficient system of ventilation possible. Regrettably, this did

* Dunn was appointed in November 1850. He had to inspect collieries in Northumberland, Durham, Cumberland and Scotland.

not happen. An explosion of fire-damp at Croft Pit caused the deaths of five men in 1855.[198] A similar explosion in the Countess district of William Pit in 1882 killed four men and this was followed by the prosecution of the colliery agent and the pit manager for failing to provide adequate ventilation. The magistrates dismissed the charge, a decision which displeased the divisional inspector of mines, J. Willis, who stated in his annual report, "I must say that I felt much surprise at the dismissal of the case . . . . I am of opinion that all mining cases should be tried before a stipendary magistrate. In small towns, political, professional, and commercial relationships are sometimes very powerful."[199] Conditions at William Pit did not improve. A report by R. W. Moore[200] (28 November 1900) says of the district in which the 1882 explosion took place, "I have to report that the ventilation of the Countess district Main Band workings, in William Pit, is in a very unsatisfactory state." Seven years later five men were killed by an explosion of fire-damp at William Pit. At the inquest upon these victims J. B. Atkinson, divisional inspector of mines, stated that there had been breaches of the Coal Mines Regulation Act and emphasized that the ventilation of the pit was insufficient.[201]

In 1888 thirty miners were killed by an explosion of fire-damp at St Helens No. 3 Pit. This followed the firing of a shot which ignited a blower of gas and set fire to the coal.[202] The worst accident in the history of Cumberland mining occurred 11 May 1910 when 136 men were killed by an explosion of fire-damp and the resulting fires at Wellington Pit, Whitehaven. To prevent the fire from spreading the pit was sealed off and the dead men entombed behind the sealing wall. The official report on this accident[203] made harsh comments on both men and management. Discipline in the mine was not strict enough, the scheme of ventilation was "not well chosen" and the ventilation of the pit inadequate. Both officials and workmen showed great laxity in dealing with gas, men preferring to continue working when gas was present rather than leave their places and there was no uniform system for reporting and dealing with gas wherever it appeared. Further, not enough had been done to prevent the accumulation of coal dust which helped to propagate the explosion. Professor Galloway asserted bluntly,

> "So far as I can see, nothing can be said in palliation of the state of affairs shown to have existed before the explosion; the safety of the men, which must be the paramount consideration in all mining operations, imperatively demands the absence of explosive gas from the neighbourhood in which they are at work; and a return to conditions the same as or similar to, those which formerly obtained in this mine must, by some means or other, be made absolutely impossible in the future."

Magistrates continued to show undue leniency towards employers when dealing with cases arising out of mining accidents. After an explosion of fire-damp in the Lickbank seam of Clifton colliery (4 August 1914) which killed three men, the colliery agent, manager and under-manager were charged with failing to ensure that working places had been examined for gas and failing to keep full and accurate records relating to the presence of gas in the colliery. The defendants pleaded guilty. Although three men had been killed and the explosion had travelled around the whole of the Lickbank seam the Workington magistrates considered that only a technical breach of the Act had been committed and that the payment of costs would meet the case.*[204]

* In other cases nominal fines were imposed when neglect of safety regulations had led to fatal accidents, e.g. the owner and manager of Harrington Colliery were fined a total of £5 after two miners had been killed by an explosion of fire-damp 31 March 1883.

Despite improvements in ventilation, safety lamps and shot-firing, the adoption of tubs and cages instead of baskets and the appointment of inspectors to enforce safety regulations, mining in the latter half of the nineteenth century remained the most dangerous of industrial occupations. Yet little was done to provide for the dependants of miners killed in the pits. Too often they had to rely upon appeals to the charity of the public. Similarly, there was little help for the many miners who suffered from respiratory diseases and other ailments caused by the nature of their work.

COSTS AND PROFITS

The detailed reports of T. E. Forster and G. B. Forster provide useful information concerning the working costs of a number of collieries in Cumberland and demonstrate the cost variations from one pit to another, and sometimes the differences in working costs of different seams in the same pit. Their reports give these figures of the costs of raising a ton of coal at the following collieries:[205]

Year	Melgramfitz		Clifton		Dearham Colliery Lonsdale Pit		Dearham Colliery Lowther Pit	
	s.	d.	s.	d.	s.	d.	s.	d.
1868	3	8·7	4	4	3	9½	4	7¼
1869	3	7·8	4	0·5	3	7¼	4	5¼
1870	4	8·3	4	10·7	4	1½	5	3¼
1871	4	7·5	4	4·9	5	2¾	6	7½
1872	5	1·5	5	9·5				

At Seaton Moor colliery in 1868 coal was raised out of the Ten Quarters seam at a cost of 4s. 9½d. per ton, but in the Metal Band seam of the same pit the working cost amounted to 7s. 3d. per ton.[206] An examination of the working costs at Whitehaven Colliery for eight half years, June 1880 – December 1883, shows that costs fluctuated from year to year:[207]

Half year ending:	Coal Sold Tons	Cost per Ton Sold s.	d.
June 1880	139,051	7	2·9
Dec. 1880	159,122	6	6·9
June 1881	153,576	6	7·0
Dec. 1881	161,346	6	5·3
June 1882	141,484	7	5·5
Dec. 1882	142,479	7	7·3
June 1883	133,090	8	4·6
Dec. 1883	142,090	8	1·6

The greatest annual output in the history of Whitehaven Colliery when it was directly controlled by the Lowther family was in 1886 when it produced 417,039 tons of coal.[208] Yet two years later the colliery passed into other hands. The decision to relinquish the colliery is explained by the fact that this was a time of falling prices and shrinking profit margins. In 1884 the working cost per ton of coal amounted to 7s. 6d. and the average price was 7s. 10$\frac{1}{8}$d., leaving a profit of 4$\frac{1}{8}$d. per ton. In the following year the working cost per ton amounted to 6s. 10$\frac{3}{8}$d. and the average price was 7s. 0$\frac{3}{16}$d., leaving a profit of only 1$\frac{13}{16}$d. per ton.[209]

Geological features made the Cumberland coalfield a high cost area. Thin seams and frequent faults gave rise to haulage difficulties and delayed mechanization, particularly the use of machine cutters. In 1913 only 1 per cent of Cumberland coal was cut by machinery compared with 8 per cent for Great Britain as a whole.[210] As the industry grew older, thicker seams tended to become exhausted and coal had to be won at greater depths and at longer distances from the shaft. At Whitehaven in the early twentieth century coal was worked four miles from the shaft. Because the steepness of the gradient made riding impossible the men had to walk to and from the face, the time taken by miners from the surface to the face and back varying from twenty minutes to three-and-a-half hours. Since an eight hour day was customary little time was left for productive work. This colliery had to haul its coal further than any colliery in England which added appreciably to working costs.[211] Excessive gas, especially in the Whitehaven pits, also helped to increase the cost of production.

Cumberland colliery owners frequently complained of the heavy burden imposed upon them by mining royalties. Normally lessees had to pay a fixed annual rent (sometimes called a certain or dead rent), often whether the mine was wrought or not. A tonnage rent or royalty was also payable, either on total output or sales, and this varied occasionally with the type of coal raised or with the seam from which it came, or was fixed as a proportion of the selling price. Additional payments had to be made when the movement or working of coal necessitated crossing territory not comprised in the lease – underground wayleaves if the lessees had to move coal along the roads of adjacent mines belonging to someone else; shaft rent if the coal was brought up another proprietor's shaft, and surface wayleaves if the coal was carried across land belonging to another. A worker of minerals might have to pay a rent to the proprietor of the land upon which his pit was situated if the minerals (usually called "foreign" minerals) were not the property of the landowner and were brought up the pit and over the surface of his land. Further payments – airleaves and waterleaves for the purpose of ventilation and drainage – had sometimes to be made when adjoining pits were in separate ownership.

In 1889, on an output of 1,740,737 tons of coal in Cumberland and Westmorland, £41,705 was paid in royalties and £2,000 in wayleaves.[212] At that time the average royalty per ton of Cumberland coal was 6d., an average that was exceeded only in west Scotland where it was 7d. per ton.[213] On the Lonsdale estates the average royalty was 7d. per ton, a charge justified by the earl's viewer, R. W. Moore, on the ground that it included interest on capital since Lord Lonsdale had worked his own collieries for some time and the lessees therefore had the benefit of the capital invested by him in the development and equipment of the mines.[214] The percentage of royalty to selling price was higher in Cumberland than in any other mining district in England and Wales – in 1888 it was 10·2 per cent,[215] and although this was less than the percentage of selling price taken by iron ore royalties it undoubtedly had an adverse effect upon the sale of Cumberland coal in a competitive market. At Whitehaven Colliery for the seven years 1906-12 there was an average yearly output of 559,263 tons on which was paid an average annual royalty of £24,023 or 10.309d. per ton on all coal raised. The average selling price of all Whitehaven coal during these years amounted to 8s. 5·274d. per ton.[216]

The older inland collieries e.g. Bolton, with their thin seams and diminishing output, found it difficult to work at a profit and had to close down. Other collieries with thicker seams such as Workington were abandoned because they were wrought out. Lord Leconfield's royalty at Cleator Moor was exhausted by 1871. Messrs. Barker & Harrison had spent over £20,000 in opening it out and hardly recovered their outlay, whilst the Whitehaven Hematite Iron Co. lost many thousands of pounds in working the colliery.[217] A letter, 4 September 1888, from the secretary of the St Helens Colliery & Brick Works Co. to Lord Lonsdale's agent complained that "since the Company was formed in 1877 no dividend had been paid whilst we paid in Royalty Rents the sum of £20,539. 12s. 7d."[218] In the early 1890's many colliery owners were seeking a reduction in their rents. Natural difficulties, low prices, high wages and Scottish competition were the usual reasons put forward for rent reductions, but these requests were rarely heeded. One of these colliery owners was the Maryport Hematite Iron & Steel Co. which had taken over Ellenborough Colliery in 1878 and worked it until 1891 when it had to go into liquidation after spending over £30,000 in developing the colliery. The receiver of the colliery, in an effort to keep the colliery going, tried to obtain a reduction in tonnage rents. He was unsuccessful and the colliery had to close.[219]

Profits varied with output, working costs and the selling price of coal. At Melgramfitz Colliery output and profits between 1865 and 1868 were:[220]

Half year ending	*Output Tons*	*Profit*			*Profit per Ton*	
		£	*s.*	*d.*	*s.*	*d.*
30 Sept. 1865	26,586	568	19	1		5·1
31 March 1866	31,473	1,517	4	0		11·5
29 Sept. 1866	38,430	2,246	3	1	1	2
30 March 1867	31,082	2,058	19	10	1	3·8
28 Sept. 1867	28,420	1,188	16	0		10
28 March 1868	22,996	516	7	5		5·4

Cumberland colliery owners believed they were handicapped by the shorter working hours which obtained in the county. According to J. D. Bain, Cumberland hewers worked 12½ per cent shorter hours per shift than Scottish hewers and 22 per cent less than Lancashire hewers. Off-hand men in Cumberland worked 8 per cent less time per shift than those in Scotland and 13 per cent less time than those in Lancashire.[221] The eight hour day, which had been adopted in the Cumberland coalfield much earlier than in many other districts, was not regarded as a blessing by the employers.

Strikes added to the difficulties of the colliery owners and, if prolonged, could convert profit balances into losses. A circular letter addressed to its members by the Cumberland Coal Owners' Association notes that at a meeting of the Association held at Workington, 30 May 1892, it was unanimously resolved, "That a concurrent application be made by all the members of the Association to their respective Lessors for a substantial deduction from the half year's Rents next payable, on account of the serious losses resulting to the

Lessees from the present strike of the Miners, which at the end of the week will have lasted for three months." This letter adds, "It was estimated that the cost of keeping the Pits unwatered and in working order will, up to that date, have been equal on the average to at least one half year's Royalty Rent, whilst the disastrous effects of the stoppage by the disorganization of local industries and the loss of the Irish Markets will seriously hamper the Lessees for a long time to come after work has been resumed upon any terms.[222]

The effects of the 1912 strike can be seen in the accounts of the St. Helens Colliery Company:[223]

Year	*Profit**			*Loss*		
	£	*s.*	*d.*	£	*s.*	*d.*
1904	1,621	6	1			
1905	93	13	7			
1906	5,346	13	7			
1907	5,394	7	3			
1908	3,785	17	1			
1909	735	4	3			
1910				112	2	5
1911				1,236	9	2
1912				4,206	9	8

* These figures do not take into account the profits made from coke ovens.

Although few statistics are available, reports and correspondence show that the industry in Cumberland suffered from a high rate of absenteeism. A vistor to Allhallows Colliery in 1899 found forty-seven out of 110 hewers absent or 42·7 per cent and this, in his view, was "no doubt due largely to the very high wages now prevailing." Absenteeism in some collieries in 1899 was so bad that on one or more days in each week the output of coal was reduced by nearly one half, and, in the same year, the Cumberland Miners' Association was induced to join the Coal Owners' Association in issuing a circular to the miners deploring the prevalence of absenteeism and pointing out that it meant a diminished output, prevented orders being completed punctually and caused off-hand men to be paid in full on a reduced output.[224] The colliery owners complained that, exclusive of general holidays, absenteeism of hewers in the Cumberland district amounted to 18·35 per cent or 9·54 weeks per man per year.[225]

Until 1881 colliery owners in Cumberland had benefited from the increase in productivity characteristic of the industry generally since 1860. In Cumberland output per person employed rose from 182 tons per annum in 1874 to 282 tons in 1880, an increase of 55 per cent. In the following year this figure rose to 286 tons, the highest annual output per employee obtained in the years before 1914. Productivity declined in the next decade, from 284 tons in 1882 to 214 tons in 1891. A prolonged strike in 1892 was responsible for the low output of 186 tons per employee in that year. This was followed by an improvement on productivity, from 242 tons in 1893 to 265 tons in 1899. Thereafter there was a decline in productivity; in 1913 output per employee was only 210 tons.[226]

Labour disputes, which became more frequent after 1880, led to the loss of much production and pit explosions, like strikes, often made temporary pit closures necessary. More coal was lost with the growth of absenteeism and the absence of hewers was particularly damaging. Hewers, whose earnings depended on their output of coal, were tempted when wage rates were high to absent themselves from work or to exert themselves less strenuously to obtain a wage packet large enough to meet their needs.

As pits increased in depth and extent and working faces became more distant from the shaft more men not directly engaged in coal getting had to be employed e.g. to maintain the miles of underground roads and in safety work. The growing proportion of non-hewers helps to account for the decline in the annual output per employee. In Cumberland, as in other mining districts, there were too many old, ill-equipped and badly planned pits with dwindling reserves of coal and poor production results. Yet their owners were reluctant to close them and their survival also contributed to the deterioration in productivity.

Transport costs continued to hamper the inland collieries, even after the introduction of railways. Coal at Broughton Moor Colliery in 1871 was raised to the surface at a cost of 4s. 10¾d. per ton, but by the time the coal had reached Maryport harbour the cost had risen to 6s. 11d. per ton. Rail transport to Maryport cost 1s. per ton and wayleaves, royalty rents, taxes and salaries together accounted for the additional 1s. 0¼d. At that time Broughton Moor coal was sold at Maryport for 8s. 6d. per ton, leaving a profit of 1s. 7d.[227] At Dearham Colliery in 1892 working costs amounted to 8s. 11·78d. per ton; since the coal realized 7s. 6·06d. per ton there was a loss of 1s. 5·72d. on every ton of coal drawn to bank. The waggon road from the colliery to Dearham Bridge cost about 3d. per ton and the M. & C. Railway Co. charged 9d. per ton for moving the coal from Dearham Bridge to Maryport, a distance of 2½ miles.[228]

High railway charges gave rise to the building of the Cleator & Workington Railway to compete with existing railways. In 1873 the Whitehaven, Cleator & Egremont Railway decided to increase the rates on iron ore by 7 per cent and on coal and pig iron by 11 per cent. The infuriated ironmasters, colliery owners and traders, led by the Earl of Lonsdale, Lord Leconfield and H. F. Curwen whose estates contained large mineral deposits, decided to promote a new railway from Cleator Moor to Workington (via Moresby Parks and Distington). Another branch was to strike north and link up with the M. & C. Railway on their Derwent branch from Bullgill to Cockermouth. By July 1878 the line was open for mineral traffic. Whilst trade was brisk it carried a heavy volume of traffic – it brought Durham coke to the furnaces at Cleator Moor and Distington from Siddick or Workington and distributed their pig iron; it carried all the output of Moresby Colliery and the coal from Broughton Moor on the northern extension line.[229]

Both the landowners who worked their own coal and the colliery owners who leased coal royalties could not continue to extract coal indefinitely. It was a diminishing asset subject to increasing costs – as the coal was won at greater depths and at longer distances from the shaft the costs of haulage, winding, ventilation and the maintenance of underground roads were raised. Once the coal had been removed new reserves had to be found elsewhere. Landowners could not rely upon a permanent income from their coal royalties; this part of their income could change very rapidly as the Curwen family, for example, discovered. Nevertheless, despite the many difficulties of the industry, it was possible for some landowners to profit from it considerably.

Between 1869 and 1885 profits at Whitehaven Colliery varied widely from year to year, falling sharply in 1884 and 1885; the profit of over £75,000 in 1873 was the highest in the history of the colliery. These profits can be compared with the rents derived from the Lonsdale coal royalties leased to other colliery owners:[230]

Year*	Whitehaven Colliery Profits £	s.	d.	Coal Royalty Rents £	s.	d.
1869	19,925	10	3	11,733	14	7½
1870	15,873	15	8¼	13,505	13	8
1871	11,727	9	9½	13,392	17	9½
1872	26,723	13	6½	12,910	15	1
1873	75,285	10	10	15,398	16	5
1874	48,991	9	6	14,941	11	0
1875	26,342	9	10	10,877	2	1½
1876	11,054	17	8½	10,684	1	0
1877	6,168	10	1	11,730	10	0
1878	7,667	4	0½	10,767	19	3
1879	12,643	14	6½	9,935	13	5
1880	13,785	8	3	9,987	3	2
1881	23,101	18	5½	13,390	3	4
1884	3,477	9	10			
1885	2,359	5	7½			

* Year ending 30 June

After 1888 the Lowther family ceased to work any of their own coal. Nevertheless they obtained considerable revenue from their coal royalties. In 1896 colliery rents amounted to £29,282·3s. and from 1900 to 1913 the annual income derived from coal royalties exceeded this figure:[231]

Year	Colliery Rents £	Year	Colliery Rents £
1900	46,707	1907	39,378
1901	53,237	1908	42,776
1902	43,699	1909	42,791
1903	44,285	1910	38,111
1904	41,353	1911	33,266
1905	42,142	1912	40,949
1906	39,017	1913	48,737

The Leconfield coal royalties, although smaller and less valuable than those belonging to Lord Lonsdale, yielded a substantial income to the lessor:[232]

Year	Income £	Year	Income £
1853	3,766	1874	6,940
1854	4,335	1876	5,925
1855	3,840	1877	4,785
1856	3,444	1878	5,024
1857	3,430	1879	3,552
1860	3,610	1885	6,531
1861	4,328	1886	5,375
1862	4,346	1887	5,928
1863	4,623	1888	5,818
1864	4,553	1889	5,262
1865	5,916	1890	6,067
1869	5,724	1891	6,818
1870	4,800	1892	6,243
1871	5,739	1893	5,631
1872	7,409	1894	4,968
1873	9,428		

THE COAL TRADE

In the latter half of the nineteenth century Cumberland coal had to meet increasing competition in the Irish market. Local colliery owners found that they had to lower the shipping price of their coal when traders were struggling against the price cutting of their competitors in Ireland. Lower prices affected adversely not only the income of the colliery owner but also, as the reports of estate agents show, the mineral rents of those landowners whose leases of their coal royalties linked tonnage rents to coal prices. At Maryport a waggon of coal of 59 cwts. cost 23s. 3d. in 1854, rising to 24s. 6d. early in 1855. A reduction in the price of Scottish coal sold in Ireland made it necessary to reduce the price of coal at Maryport to 22s. 6d. in April 1855 and this was followed by a further reduction in 1856 to 20s. 6d. per waggon.[233]

In January 1857 the Cumberland colliery proprietors agreed to raise the price of their coal to shippers in the Irish trade. Many traders refused to pay the increased price, preferring to lay up their vessels and for several weeks very little coal was shipped. Some colliery owners in the Maryport and Workington districts therefore decided to charter vessels on their own account and instructed the captains to sell the coal for what it would fetch in Dublin and other markets. It proved to be an unwise decision for the captains, sure of their freight, showed little regard for the best interests of the coal owners and were content to dispose of their cargoes as speedily as possible at any price. This intrusion by the colliery owners into the coal trade was brief and unsuccessful. At a meeting of their associaton in March 1857 it was resolved that the practice of freighting vessels should be discontinued.[234] However, by the end of the year the price of a waggon of coal at Maryport had been raised to 21s. 6d.[235]

Railway construction stimulated the development of the inland collieries and increasing output led in turn to a demand for improved harbour facilities. Coal shipments from Maryport rose from 191,883 tons in 1850 to 307,381 tons in 1854 and this, together with the increasing size of ships, gave rise in the early 1850s to an agitation for the building of another dock. The harbour trustees decided to construct a "floating dock" i.e. a wet dock the gates of which would hold the water in the dock from one high tide to the next, thus allowing loading to go on without stoppage.[236] Begun in 1854, the dock was officially opened 20 October 1857 and was the first wet dock in Cumberland.[237] It soon became evident that it was filling a long-felt want for there was an immediate increase in coal shipments which in 1867 amounted to 480,247 tons, the largest quantity shipped in one year from any port in Cumberland. Reports of the Maryport customs officers show that it was the insufficiency of harbour accommodation that had hampered the coal trade. While the dock was being built the harbour and channel were deepened and the railway company re-arranged the whole of their system of lines running to the dock, building two bridges which effected a considerable saving of distance, and erecting hurries to accelerate the loading of vessels.[238] It was this dock which enabled Maryport to replace Whitehaven as the leading port in Cumberland.

Within six years of the opening of the Elizabeth Dock the demand for additional accommodation was renewed, the new dock being inadequate to deal with the increasing trade of the port. A public meeting held in Maryport 13 February 1863 passed resolutions calling for increased accommodation for shipping and for the imposition of supplementary rates on imports and exports to provide the necessary funds. A deputation presented a memorial on behalf of the merchants and shipowners of Belfast to the harbour trustees asking that they should take steps to construct another dock on the north side of the harbour. The trustees, whilst agreeing that a new dock was desirable, opposed the suggestion that extra rates should be levied since this would deter ships from using the port, a point of view supported by the coal proprietors of the Maryport district.[239]

In 1865 the trustees were preparing a new Dock Bill and Thomas Page reported on a proposed new dock. His report dealt at length with the inadequacy of existing arrangements and the growing volume of trade; vessels, depending on the wind, arrived in fleets and crowded the harbour, often damaging the harbour and each other; as ships jammed the docks loading became more difficult and lengthy and expensive delays were frequent.[240] The application to Parliament (1866) failed however, owing largely to inadequate estimates of costs, but two years later a second application was successful. Unfortunately a period of depression set in and efforts to obtain the necessary capital proved abortive, a contract entered into having to be abandoned. In the end the powers lapsed and it was ten years before new powers were sought.[241]

At Workington attempts to provide more facilities for ships were successful and a new dock, the Lonsdale Dock, was opened in 1865. A similar need arose at Whitehaven in the next decade as the ore and iron traffic increased and in 1876 the Queen's Dock was opened. But both these docks were small and unable to deal with large vessels.

When a thriving iron industry was added to the coal mining industry in the 1870's further improvements were needed at the Cumberland ports. Until 1868 Maryport had been regarded primarily as a port for the shipment of coal, but this changed as blast furnaces appeared in the district. From this time coal exports from Maryport diminished, a decline in trade that was compensated by the export of iron and later by the import of ore and the export of steel rails. This traffic required larger ships and better accommodation for them. Increasing prosperity and the pressure of local traders induced the Maryport harbour trustees to renew their application to Parliament for powers to build a new dock and in 1879 their application was successful. Work on the Senhouse Dock began in 1880 and after four years' work it was officially opened 27 May 1884.[242] By 1887 coal had ceased to be the chief element in the trade of Maryport, the import of iron ore and the export of pig iron and steel competing in importance with the traditional export of coal. Its import of hematite exceeded that of the other Cumberland ports combined and its export of coal, although reduced, was not often surpassed by that from the neighbouring ports.

Until 1865 the bulk of Cumberland coal was exported; after that date the percentage of coal shipped declined rapidly:

Year	*Percentage of coal shipped*	*Year*	*Percentage of coal shipped*
1855	75·01	1895	29·39
1865	53·19	1905	29·03
1875	32·11	1913	26·23
1885	28·22		

With the development of the iron industry in Cumberland a larger proportion of the coal was retained in the county for use in the furnaces. Although Cumberland coal, for the most part, produced an inferior quality of coke an increasing amount of it was converted into coke for the local iron and steel works; by 1912 more than a third of the coal was being used in this way. Railway construction also contributed to the reduction in coal shipments for coal was moved out of the area by rail as well as by sea. Railways transported growing quantities of coal from West Cumberland collieries for land sale – 371,795 tons in 1875, rising to 808,934 tons in 1913.[243] Between 1880 and 1914 the Maryport & Carlisle Railway sent as much coal for land sale as it did to Maryport for shipment.

In the latter half of the nineteenth century coal shipments from Cumberland fluctuated widely, from a maximum of 788,314 tons in 1867 to a minimum of 333,713 tons in 1872. After 1870 until the turn of the century the general trend was downwards; between 1901 and 1913 coal shipments remained steady at over 600,000 tons per annum. Individual ports showed large variations in coal shipments. At Whitehaven only 89,033 tons were shipped coastwise in 1873, the lowest quantity recorded since 1744, but in 1897 299,980 tons were shipped coastwise, the greatest amount in the history of the port. Statistics of coastwise shipments from the Cumberland ports show that in this period Whitehaven was losing its position as the leading port:-[244]

Year	*Whitehaven* *Tons*	*Workington* *Tons*	*Maryport* *Tons*	*Carlisle* *Tons*	*Total* *Tons*
1851	226,266	95,703	201,676		523,645
1861	183,366	128,959	380,304		692,629
1871	130,962	64,684	404,205		599,851
1881	214,642	37,235	229,960	30,428	512,265
1891	167,907	89,501	435,412	15,625	708,445
1901	253,271	114,053	236,098	8,307	611,729
1911	223,109	110,565	294,538	11,837	640,049

By the end of the nineteenth century Cumberland coal had ceased to dominate the Dublin market. G. H. Liddell, the colliery agent to Whitehaven, says in a letter dated 16 January 1886,

> "The great drawback to the Colliery during the past year has been the falling off in the shipments of coal to Dublin. During the year 1879 95,000 tons of Whitehaven coal were sent to Dublin being 11% of the total imports for that year. Since then the quantity of Whitehaven imports has steadily decreased until last year it only amounted to 37,300 Tons, which is only 4·3 of the total imports.
>
> I think it is not putting it too strongly to say that we have now practically lost the Dublin trade . . . . . Unless some radical change is made in the state of our relations with the Dublin market I fear it is hopeless to expect the Collieries to do any good in the present state of Trade."[245]

Keen competition from colliery owners in Scotland and Lancashire, who were cutting prices to the disadvantage of the Cumbrians, accounted for the decline of coal shipments from Cumberland to Ireland. In 1890 and 1891 the total exports of Cumberland coal were less than a third of the amount of Scottish and Lancashire coal sold in Dublin and Belfast only. A flourishing export trade was necessary to the well-being of the Cumberland coalfield for the small population of the county provided a limited market for household coal. But even in Cumberland Scottish and Lancashire coal found buyers, over 100,000 tons of it being imported into the county in 1891.[246] Thus the Cumberland colliery owner had not only to compete with his neighbours for the local and Irish markets, but he had also to face competition in both from Scotland and Lancashire.

For Cumberland coal there was really only one export market, viz. Ireland. In 1898 the competing districts sent the following proportions of their output to Ireland:[247]

Cumberland	*Lancashire*	*Scotland*	*North Wales*
14·47%	3·47%	3·63%	0·54%

The Irish market, therefore, was of much greater importance to Cumberland than to its competitors who could use Ireland as a dumping ground for their surplus coal.* Lancashire and Scotland now dominated the Irish market; Cumberland's share was relatively small as can be seen from the proportions supplied by the different districts:[248]

Year	*Cumberland* %	*Lancashire* %	*Scotland* %	*North Wales* %
1891	13·82	40·40	45·40	0·38
1898	15·83	44·86	38·39	0·92

Exports of Cumberland coal outside the United Kingdom were always small. After 1868 they never reached 10,000 tons in any year and in some years only trifling quantities were exported.

Notes and References

1 *VCH*, 364.

2 R. W. Moore, "Whitehaven Collieries", *Trans. of the Federated Insitute of Mining Engineers*, vii (1893-4), 637.

3 *Ibid.*, 637. *VCH*, 365.

4 D/Lons/W, Lease of St. Bees School royalty.

5 D/Lons/W, Forster report on Lonsdale Collieries 1880-1.

6 *Ibid.*

7 D/Lons/W, Lease of Whitehaven Collieries.

8 D/Lec, Mineral Report 1887.

9 *Whitehaven News*, 16 Feb. 1888.

10 D/Lons/W, Forster report on Lonsdale collieries 1888, R. W. Moore to G. B. Forster, 28 Dec. 1888.

11 T. Eastwood, *The Geology of the Whitehaven and Workington District* (1931), 132.

12 *VCH*, 365.

13 D/Lons/W, Forster Reports on Lonsdale Coal Royalties, Minutes 1904-25.

14 D/Lons/W, Colliery lease.

15 D/Lons/W, Whitehaven Colliery records.

16 D/Lons/W, Coal Royalty Rent Account Books.

17 D/Lec, Dixon reports 1853-73. *VCH*, 366.

18 Forster, Shelf 53, vol. 13, Copy of colliery lease.

19 D/Lons/W, Coal Royalty Account Books, D/Lec, Dixon report 1869.

20 Forster, Shelf 53, vol. 13.

* A letter from the Cumberland Coal Owners' Association in 1892 states that Scottish coal was then being exported to Ireland at 3s. per ton f.o.b. less than Cumberland coal.

[21] D/Lons/W, Coal Royalty Rent Account Books. C. Caine, *Cleator and Cleator Moor: Past and Present* (Kendal, 1916), 208.
[22] *VCH*, 372.
[23] D/Lons/W, Coal Royalty Rent Account Books; R. W. Moore's report on Asby Colliery, 7 Oct. 1980.
[24] D/Lons/W, Misc. Reports 1870-1917.
[25] *VCH*, 372.
[26] *VCH*, 367.
[27] Whitehaven Library MS 48, Workington Colliery Book 1799-1853 – Statement of Alvan Penrice.
[28] *VCH*, 367.
[29] D/Cu, Workington Colliery Coal Shipped Books.
[30] D/Lons/W, Forster report on Curwen royalty 1869.
[31] D/Lons/W, Forster reports on Curwen royalty 1873-6. *VCH*, 367.
[32] D/Cu, Harrington Colliery Coal Shipped Books.
[33] D/Cu, 4/235 Harrington Colliery lease.
[34] *VCH*, 368.
[35] D/Lons/W, Forster report on Curwen royalty 1874.
[36] *VCH*, 368.
[37] D/Cu, Harrington Colliery Royalty Accounts 1867-95.
[38] *Supplement to Stock Exchange Gazette*, 19 Aug. 1909.
[39] D/Lons/W, Curwen Mineral Leases.
[40] *VCH*, 369.
[41] Whellan, *op. cit.*, 481.
[42] Forster, "West Cumberland 1872-5", report Jan. 1873.
[43] *Ibid.*, Sept. 1874.
[44] *Ibid.*, 31 Oct. 1873.
[45] *Ibid.*, Sept. 1874.
[46] D/Lons/W, Clifton Colliery lease.
[47] D/Lec, Dixon report 1853.
[48] D/Lec, Dixon reports 1856-74; Mineral Report 1886. *VCH*, 370.
[49] Carlisle Record Office DX/425/11.
[50] *VCH*, 370.
[51] D/Lec, 92/25 Dixon report on Wythemoor Colliery, 1 Dec. 1874.
[52] D/Lons/W, Forster reports on Curwen royalties 1878-80. Royal Commission on Mining Royalties 1889-93, *Final Report*, App. 91.
[53] D/Cu, 4/239 Wythemoor Colliery lease 1904.
[54] D/Le,c Mineral Report 1908.
[55] D/Le,c Mineral Reports 1909-14.
[56] D/Le,c 92/18 Report on Haycastle Colliery 1872. *VCH*, 371.
[57] D/Lons/W, Unlisted papers.
[58] D/Lec, Dean Moor colliery lease; Dixon reports.
[59] D/Lec, Dixon reports 1853-79.
[60] D/Lec, Dixon reports 1889-93.
[61] D/Lons/W, Coal Royalty Rent Accounts.
[62] *VCH*, 373.
[63] D/Lons/W, Coal Royalty Rent Accounts.
[64] *Ibid.*, *VCH*, 373.
[65] Forster, Shelf 53, vol. 11, Report on Seaton Moor Colliery 1868. *VCH*, 374.
[66] *VCH*, 374.
[67] D/Lons/W, Coal Royalties Minutes 1904-25.
[68] D/Sen, Netherall 19/21, 19/434.
[69] *Ibid.*, W. B. Peat to H. P. Senhouse, 4 May 1891.
[70] *Whitehaven Advertiser*, 15 Sept. 1894.
[71] D/Lec, Dixon reports 1853-6, 1860, 1862, 1869.
[72] D/Lec, Dixon reports 1873-4, 9. Adam's report 1875. Forster, "West Cumberland 1872-5", Sept. 1874.
[73] D/Lec, Mineral Reports 1885-1910.

[74] *Ibid.*
[75] D/Lec, Mineral Report 1894.
[76] D/Lec, Mineral Reports 1898-1910.
[77] D/Lec, Mineral Report 1913.
[78] D/Lec, Mineral Reports 1907-14.
[79] Forster, Shelf 51, vol. 5, Report on Dearham Colliery.
[80] D/Lons/W, Forster report on Lonsdale Collieries 1882.
[81] D/Lec, 298/16.
[82] D/Lec, Mineral Report 1886.
[83] D/Lons/W, R. W. Moore's report on Dearham Colliery, 10 Sept. 1892.
[84] D/Lec, Mineral Reports 1894-1914.
[85] D/Lons/W, Coal Royalty Rent Accounts. *Cumberland Pacquet*, 1 Oct. 1903. *VCH*, 376-7.
[86] D/Lons/W, Coal Royalty Rent Accounts.
[87] *VCH*, 376.
[88] D/Lec, 298/9, MB 4/12, Mineral Reports 1885-93. *VCH*, 378.
[89] *VCH*, 378.
[90] D/Lec, Mineral Reports 1891-4. *VCH*, 377.
[91] Forster, Shelf 53, vol. 13, Report on Flimby Colliery.
[92] *VCH*, 374.
[93] *VCH*, 374. D/Lons/W, Coal Royalty Rent Accounts.
[94] D/Lec, Dixon reports 1853-60.
[95] *VCH*, 378-9.
[96] D/Lec, Mineral Reports 1870-1910.
[97] D/Lons/W, Coal Royalty Rent Accounts. D/Lec, Mineral Reports 1907-1913.
[98] Forster, "Whitehall 1877".
[99] D/Lec, Mineral Report 1886.
[100] D/Lec, Mineral Reports 1886-1910.
[101] D/Lec, Mineral Reports 1913-17. *VCH*, 379.
[102] D/Lec, Dixon reports 1853-70.
[103] *VCH*, 379.
[104] D/Lec, Bolton Colliery lease. Forster "Whitehall 1882-3".
[105] D/Lec, Mineral Report 1886.
[106] D/Lec, Mineral Reports 1887-97.
[107] *Mineral Statistics of the United Kingdom*, 1854-96. *Mines and Quarries General Report with Statistics*, 1897-1914. *Reports of Inspectors of Mines*, 1873-1914.
[108] D/Lons/W, Output books.
[109] D/Lons/W, Output books. *VCH*, 363-4.
[110] *Mineral Statistics* 1854-96. *Mines and Quarries General Report with Statistics*, 1897-1914.
[111] *Report of the Royal Commission on the Coal Industry*, 1925, i, 117-18.
[112] *Mineral Statistics* 1854-78.
[113] Forster, "West Cumberland 1872-5".
[114] *VCH*, 365.
[115] D/Lec, Mineral Reports 1889-1913.
[116] *Mines and Quarries Reports*, 1905-14.
[117] *Ibid.*
[118] *Report of Departmental Committee on the Probable Economic Effects of a Limit of Eight Hours to the Working Day of Coal Miners* 1907, q. 18394.
[119] D/Lec, Mineral Reports 1909-15.
[120] *The Workington Iron and Steel Co. Ltd.* (1914), 10.
[121] D/Lons/W, Coal Royalties Minutes 1904-25.
[122] *Mines and Quarries Report*, 1905.
[123] D/Lons/W, Forster report on Lonsdale Collieries 1888.
[124] D/Lons/W, Coal Royalties Minutes 1904-25.
[125] *Ibid.*
[126] *Reports of Inspectors of Mines*, 1905-1913.

[127] *Report of the Royal Commission on Coal in the United Kingdom*, 1871, ii, 366-7.
[128] Forster, Shelf 51, vol. 16, G. B. Forster to J. Prestwich, 9 Nov. 1872.
[129] *Report of the Royal Commission on Coal Mines*, 1907-11, iv, q. 54604 – Evidence of T. P. Martin.
[130] *VCH*, 352.
[131] D/Sen, Netherhall 19/357.
[132] *VCH*, 352. T. S. Ashton and J. Sykes, *The Coal Industry in the Eighteenth Century* (Manchester, 1929), 26-7.
[133] D/Lec, Mineral Report 1899.
[134] Forster, Report on Allhallows Colliery, 23 April 1885.
[135] D/Lons/W, Forster report on Lonsdale Collieries 1884.
[136] D/Lons/W, Forster reports on Lonsdale Collieries 1889-91.
[137] *VCH*, 353.
[138] *Report on Wellington Pit explosion*, 1910. Cd. 5524 (1911).
[139] Buddle, vol. i, Report on Workington Colliery 3 Oct 1837.
[140] D/Lec, Dixon report 1853, 1860.
[141] D/Lec, 92/1.
[142] *Whitehaven Guardian*, 18 Nov. 1875.
[143] *VCH*, 365.
[144] D/Lons/W, Misc. Reports 1870-1917, Whitehaven 22 May 1902.
[145] *Carlisle Patriot*, 23 Dec. 1870. *VCH*, 356-7.
[146] D/Lons/W, Forster report on Lonsdale Collieries 1889.
[147] R. W. Moore, "Whitehaven Collieries", *Trans. of the Federated Institute of Mining Engineers*, vii (1893-4), 642.
[148] *Report on Wellington Pit explosion*, 1910. Cd. 5524 (1911). *Report on William Pit explosion*, 1947. Cmd. 7410 (1948).
[149] J. Jewkes and A. Winterbottom, *An Industrial Survey of Cumberland and Furness* (Manchester, 1933), 58.
[150] *Ibid.*, 62.
[151] Evidence of Thomas Ainsworth, q. 13163-4.
[152] Caine, *op. cit.*, 429.
[153] R. W. Moore, MS "History of Whitehaven Collieries".
[154] W/Lons/W, Forster reports on Lonsdale Collieries 1884-6.
[155] *VCH*, 365.
[156] Royal Commission on Accidents in Coal Mines 1886, *Final Report*, App. ii, 139.
[157] Forster, Reports on Clifton Collieries 1873-4.
[158] T. H. Bainbridge, "Population Changes over the West Cumberland Coalfield", *Economic Geography*, vol. 25 (1949), 130.
[159] Forster, Report on Broughton Colliery 1871.
[160] D/Lons/W, Misc. Reports 1870-1917, I. Brown to E. H. Wrigglesworth, 10 April 1913.
[161] *Report of Inspector of Mines*, 1851.
[162] Royal Commission on the Depression of Trade and Industry 1886, *Second Report*, App. Pt. ii 93.
[163] D/Lons/W Misc Colliery Papers.
[164] Royal Commission on the Depression of Trade and Industry 1886, *Second Report*, App. Pt. ii, 94.
[165] *Report of the Royal Commission on Mining Royalties* 1889-93, Report by Oswald Walmesley.
[166] Records of the Cumberland Miners' Association.
[167] Board of Trade, *Statistical Tables and Report on Trade Unions*.
[168] *West Cumberland Times*, 4 April–13 June 1874.
[169] *West Cumberland Times*, 28 Nov. 1874–23 Jan. 1875. *Whitehaven News*, Jan.-March 1875.
[170] *Carlisle Journal*, 29 Feb. 1876.
[171] *Whitehaven News*, 3 Aug. 1876.
[172] *Ibid.*, 8 May 1879. *Carlisle Patriot*, 9 May 1879.
[173] *Whitehaven Free Press*, 20 Sept. 1879.
[174] *Ibid.*, 25 Oct. 1879. *West Cumberland Times*, 1 Nov. 1879.
[175] Royal Commission on Mining Royalties 1889-93. *First Report*, Appendix 332.
[176] M. W. Kirby, *The British Coalmining Industry 1870-1946: A Political and Economic History* (1977), 13-14.

[177] Mining Association of Great Britain. *Statistical Review of the Coal Industry 1925*. Board of Trade, *Abstract of Labour Statistics of the United Kingdom.*
[178] Cumberland Coal Commission, *Minutes of Proceedings*, 1892, 6.
[179] J. W. F. Rowe, *Wages in the Coal Industry* (1923), 156.
[180] Cumberland Coal Commission, *Minutes of Proceedings*, 1892, 4-9.
[181] *Ibid.*, 131.
[182] Rowe, *op. cit.*, 156.
[183] *Whitehaven News*, 2 Feb. 1888.
[184] *Cumberland Pacquet*, 20 Nov. 1884.
[185] D/Lons/W, Forster report on Lonsdale Collieries 1889.
[186] *Whitehaven News*, 19 Mar.–30 Apr. 1885.
[187] R. P. Arnot, *The Miners: A History of the Miners' Federation of Great Britain, 1889-1910* (1949), 217.
[188] *Whitehaven News*, 27 July–21 Sept. 1893.
[189] *Ibid.*, 30 June–29 Sept. 1904.
[190] R. P. Arnot, *The Miners: Years of Struggle* (1953), 91.
[191] Records of the Cumberland Miners' Association. *Whitehaven News*, 29 Feb.–11 April 1912.
[192] Records of the Cumberland Miners' Association. *Report of the Royal Commission on the Coal Industry 1919*, iii, 114.
[193] *Report of the Royal Commission on the Coal Industry 1919*, iii, 104.
[194] Departmental Committee on the Probable Economic Effects of a Limit of Eight Hours to the Working Day of Coal Miners, *Final Report*, Cd 3505 (1907), Pt. I, 38-9.
[195] D/Lons/W, Misc. Colliery Papers.
[196] Dunn's, *Divisional Report*, 1863, 25.
[197] *Report of Divisional Inspector of Mines*, 1877.
[198] *Cumberland Pacquet*, 3 July 1855.
[199] *Report of Divisional Inspector of Mines*, 1882, 322-3. *Cumberland Pacquet*, 29 April, 19 Aug. 1882.
[200] D/Lons/W, Moore's Report on Whitehaven Collieries, 28 Nov. 1900.
[201] *Whitehaven News*, 28 Nov. 1907, 9 Jan. 1908.
[202] *Cumberland Pacquet*, 26 April 1888.
[203] *Report on the Explosion at Wellington Pit, Whitehaven, 11 May 1910*. Cd. 5524 (1911).
[204] *Report of Divisional Inspector of Mines*, 1914.
[205] Forster, Shelf 51, vols. 5, 17 – Colliery reports.
[206] *Ibid.*, Shelf 53, vol. 11.
[207] D/Lons/W, Forster report on Lonsdale Collieries 1884.
[208] D/Lons/W, Forster report on Lonsdale Collieries 1888.
[209] D/Lons/W, Forster report on Lonsdale Collieries 1886.
[210] *Reports of Inspectors of Mines.*
[211] Report of Committee on Eight Hour Day 1907, *Final Report*, ii, 268-73, evidence of T. P. Martin.
[212] Royal Commission on Mining Royalties 1889-93, *Final Report*, 10.
[213] *Ibid.*, *Final Report*, 5.
[214] *Ibid.*, *Final Report*, Appendix 84.
[215] *Ibid.*, *Final Report*, Appendix 96.
[216] D/Lons/W, Coal Minutes 1913-25.
[217] D/Lec, Dixon reports 1869-71.
[218] D/Lons/W, Forster report on Lonsdale Collieries 1888.
[219] D/Sen, Ellenborough Colliery letters and reports, W. B. Peat to H. P. Senhouse, 4 May 1891.
[220] Forster, Shelf 53, vol. 13.
[221] Cumberland Coal Commission, *Minutes of Proceedings*, 1892, 7-8.
[222] D/Cu, Misc. Colliery Papers.
[223] D/Lons, Coal Royalties Minutes 1904-25.
[224] D/Lec, Mineral Report 1899.
[225] Cumberland Coal Commission, *Minutes of Preceedings*, 1899, x.
[226] *Report of the Royal Commission on the Coal Industry 1919*, iii 122.
[227] Forster, Shelf 51, vol. 5.

[228] D/Lons/W, Forster report on Lonsdale Collieries, report by R. W. Moore on Dearham Colliery, 10 Sept. 1892.

[229] W. Mc. G. Gradon, *Furness Railway: Its Rise and Development 1846-1923* (Manchester, 1946), 33-100.

[230] D/Lons/W, Forster reports on Lonsdale Collieries; Misc. Reports 1870-1917; Statements of income from Lonsdale Collieries.

[231] D/Lons/W, Statements of income from Lonsdale royalties.

[232] D/Lec, Dixon reports; Mineral Reports.

[233] D/Lec, Dixon reports 1854-6.

[234] *Cumberland Pacquet*, 24 Mar. 1857.

[235] D/Lec, Dixon report 1857.

[236] Maryport Harbour records.

[237] *Cumberland Pacquet*, 27 Oct. 1857.

[238] Whellan, *op. cit.*, 332.

[239] Maryport Harbour records.

[240] T. Page, *Report on the proposed Harbour at Maryport* (1865).

[241] Maryport Harbour records.

[242] *Maryport Advertiser*, 30 May 1884.

[243] *Mineral Statistics of the United Kingdom. Mines and Quarries General Reports with Statistics.*

[244] *Ibid.*, *Report of the Royal Commission on Coal in the United Kingdom* 1871, iii, Appendix, 71-86.

[245] D/Lons/W, Forster report on Lonsdale Collieries 1886.

[246] Cumberland Coal Commission, *Minutes of Proceedings*, 1892, 4-5.

[247] Cumberland Coal Commission, *Minutes of Proceedings*, 1899, xi.

[248] *Ibid.*

CHAPTER VI

1915-1946

WHITEHAVEN DISTRICT

Developments at Whitehaven collieries quickly followed their acquisition by the Whitehaven Colliery Co. Ltd. A new pit, Haig Pit, was sunk about half a mile south of the harbour. This consisted of two shafts, about forty yards apart, – No. 5 (sunk August 1914-May 1916) and No. 4 (sunk October 1916-March 1918) – both sunk about 400 yards to the Main Band.[1] At Ladysmith Pit a battery of sixty-nine bye-product coke ovens was erected and a power station built to provide electricity for the pits.

During its twenty years the company had a troubled existence. Strikes in 1921, 1923, 1926 and 1931 diminished output severely and the imposition in July 1932 of an import duty of five shillings per ton on British coal by the Irish Free State was particularly injurious to Whitehaven. Frequent pit explosions exacerbated the situation; production was lost whilst repairs, often costly, were being made. Because of its liability to spontaneous combustion, which gave rise to heatings and fires, the Main Band seam in William Pit was sealed off in 1928 and from that time until 1941 most of the pit's output came from the Bannock Band seam.[2] By 1930 the company's financial position was poor and soon it had to cease production at two of its four pits – Ladysmith in December 1931 and Wellington in July 1932. A record output of 881,239 tons in 1927 was followed by annually decreasing amounts until 1933 when only 386,569 tons of coal were raised.[3] It would appear that by 1930 the company had lost its early vigour, only readily accessible coal was being won and little development was taking place. According to one engineer's report (24 June 1933) the Whitehaven coalfield for the past fourteen years

> "has never been tackled in anything like a proper way in view of the tonnage available, the high grade quality and the unique transport facilities which the Lowther Estates can put at the disposal of their Lessees.
>
> The present Company have adopted a lamentably weak general policy, and, whilst part of their area is at the present time yielding results which would astonish owners in most districts, this is largely because the area named has worked itself and in spite of bad management."[4]

In 1933 the company gave up the colliery after being served with a writ by the landlord to re-possess the leased property.

The colliery was taken over 11 August 1933 by the Priestman Whitehaven Collieries Ltd. for a term of ninety-eight years at an annual rent of £10,500 and a further rent of £3,581 per annum for the farms and cottages. Tonnage rents were 9d. for Main Band coal, 7d. for Bannock Band and 5d. for other seams.[5] Much work had to be done to make up for the neglect of earlier years and it was done quickly. The ventilation system at Haig Pit was re-organized. This involved making roads 4,900 yards long, the re-arrangement of the existing upcast shafts, the provision of an additional upcast shaft and the installation of a supplementary fan with a capacity of 300,000 cu. ft. per minute. No. 5 shaft, formerly used for ventilation purposes only, was equipped for coal winding. At William Pit the airways were improved and the underground equipment overhauled and modernized. At the same time the large Countess district in the Bannock Band seam was opened out and drifts driven over 470 yards towards the Six Quarters seam in the Lowca area. A completely new screening plant was set up at William Pit and additional plant at Haig Pit. Other surface improvements included a thorough overhaul of the coke ovens, locomotives and waggons and the provision of new workshops at Haig Pit and Ladysmith for repairing rolling stock. By 30 June 1935 £71,843 had been spent in modernization and improvements.[6]

Prior to 1933 it had become increasingly difficult to sell Whitehaven coal because of its decline in quality which resulted from inefficient screening and grading plant and a reduction in the number of workers employed in cleaning the coal. It was the policy of the new company to put on the market coal that was consistently reliable as to size and quality. This was achieved by the erection of new screening plant, the employment of sufficient screen hands to ensure the proper cleaning of the coal and samplers and chemists were engaged to check quality and grading from the pit to the point of delivery. These measures, backed by an extensive selling campaign, led to the recovery of lost markets and the acquisition of new ones. New business was obtained in east Cumberland, parts of Westmorland and south Scotland where Whitehaven coal had not previously been sold. The removal early in 1935 of the duties on British coal imported into the Irish Free State gave a further stimulus to the colliery. By the spring of that year all of Whitehaven's coal output could be sold and no stocks were carried. Similarly all the coke and bye-products made at Whitehaven were disposed of without difficulty.[7]

Despite its strenuous efforts to resuscitate an ailing colliery, all the more valuable because they helped to alleviate the chronic unemployment then prevalent in the area, the company had to close in October 1935. This was a severe blow to the Cumberland coalfield and more particularly to the town of Whitehaven, many of whose inhabitants depended for their livelihood upon the continuance of mining. It may be that the capital expended in colliery improvements in so short a time was more than the increase in coal output justified. For over a year the pits were idle whilst attempts were made to form another company which would re-open the colliery. In 1937 the colliery was taken over by the Cumberland Coal Company (Whitehaven) Ltd. in which the controlling interest was held by the Coltness Iron Co. Coal production was resumed in March 1937 and improvements in mining practice introduced. A plan to improve ventilation involved the construction of seven miles of airways and took several years to complete. Changes also took place on the surface. Pithead baths were provided and a centralized power-house was built at Haig Pit at a cost of £250,000 to supply steam, electricity and compressed air to the whole undertaking.[8] Output and profits quickly increased but the outbreak of war in 1939 prevented further progress:[9]

Year	*Output*	*Profit*		
	Tons	£	*s.*	*d.*
1937	285,151			
1938	598,819	61,718	15	9
1939	641,241	20,896	16	6
1940	607,829	809	14	0
1941	427,761	19,017	2	11

Apart from the strike year 1923 and the year 1936 when it was closed, Whitehaven Colliery between the two world wars remained the most important colliery in Cumberland, producing between twenty and forty per cent of the county's total output of coal. In the years 1925-28 and 1938-41 it raised more than a third of Cumberland's coal and this proportion reached 40·27 per cent in 1939.

At Walkmill Colliery, Moresby the average annual output of coal during the ten years 1911-20 was 124,469 tons. A Lonsdale royalty at Moresby was leased in 1917 by the Workington Iron & Steel Co. (which became a branch of the United Steel Cos. Ltd. in 1919) and worked to Harrington No. 11 Pit.[10] The United Steel Cos. Ltd., seeking to ensure its coal supplies, embarked upon a policy of acquiring collieries in Cumberland and in 1924 bought the ordinary shares of the Moresby Coal Co. and thus obtained control of the Walkmill and Oatlands pits although the name of the company was retained.[11] Output at Walkmill increased thereafter, exceeding 250,000 tons in 1927 and 1928 and until 1940 annual output was well above the average of 1911-20.

WORKINGTON DISTRICT

In 1937 the United Steel Companies began the Solway Pit, Workington to win coal, estimated at 200 million tons, from an undersea tract some ten square miles in extent. Two shafts, each twenty-one feet in diameter, were sunk through very soft ground by a Belgian firm using a freezing process to cope with the mud and water until hard rock was reached about 150 feet below the surface. Sinking was continued until a depth of about 1,000 feet had been reached.[12] To a firm of this kind the pit seemed ideally situated – it was close to blast furnaces, steel works, coke ovens and docks and a few miles away were hematite ore mines and limestone quarries. Yet many local miners were convinced that the pit had been sunk in the wrong place and they proved to be right. It was believed that there were eight workable seams and to exploit them the most modern methods were adopted. Between 1937 and 1946 £368,029 was invested in the pit.[13] Much was expected of the Solway Pit; all it yielded was disappointment. The coal was badly faulted, some of the seams were too thin and dirty to be commercially successful, development and working costs were too high and the output too small. This was probably the worst mining investment in the history of the Cumberland coalfield and but for nationalization the pit would have closed much earlier than it did.

To increase their supplies of coal the United Steel Companies leased two Lonsdale royalties at Parton, one in 1919 and the other in 1933.* The former royalty contained land coal and the latter undersea coal and both were worked to Harrington no. 10 Pit which had been sunk 650 feet to the Six Quarters seam. Two new pits were sunk – No. 11 in 1916 and No. 12 in 1921-22. No. 12 Pit was abandoned in 1932. The No. 5 or Micklam pit continued working. This was essentially a fireclay pit although it did produce some coal e.g. the average annual output 1934-8 amounted to 10,132 tons of

* 1919 lease: 50 years from 1 January 1919. Dead rent £200 per annum. Tonnage rents: Main Band 10d., Little Main and Six Quarters 6d., other seams 4d. 1933 lease: 60 years from 1 July 1933. Dead rent £1,000 per annum. Tonnage rents 4d.-10d.

fireclay and 1,448 tons of coal. Almost all the coal produced was used at the brickworks which provided fireclay blocks for the steel works at Workington.[14]

Apart from the strike years 1921 and 1926 the Allerdale Coal Co.'s William Pit at Clifton produced 120-160,000 tons of coal per annum between 1919 and 1927.[15] Like all other mining companies the Allerdale Company suffered badly from these strikes and in 1927 it had serious financial problems. In that year, according to a statement made by its secretary, the company had lost £110,000 on William Pit since its acquisition and the pit was then working at a loss of £3,000 per month. To reduce losses the company decided to lower wages both at William Pit and Buckhill Pit. For a month the men at both pits refused to accept the new terms, but, receiving no support from the Cumberland Miners' Association whose funds had been exhausted by strike payments in recent years, they had to agree to a nine per cent reduction in base rates for hewing, brushing and contracting.[16] This was followed by an improvement in production at William Pit, the annual output of coal increasing steadily, exceeding 200,000 tons between 1933 and 1939.[17]

Oatlands Pit, Distington, after some years of declining output, ceased production in 1932 and in 1934 the pit was abandoned. Dean Moor colliery continued to produce coal for local sale until its closure in 1951. Desipte attempts at intimidation this non-union colliery, which employed only about a dozen men, was kept going throughout the national strike in 1926. As a result of the quota system introduced by the Coal Mines Act (1930) the colliery, whose owners did not belong to the Cumberland Coal Owners' Association, had to reduce its ouput by ten per cent. It also lost a market for its small coal when a power station at Cleator Moor closed down in 1930. It was therefore decided to erect a plant to convert the small coal into ovoid briquettes. This plant, estimated to cost £2,000, was completed in 1931 and its briquettes sold readily.[18]

After 1911 coal and fireclay at Camerton were wrought by means of drifts. In 1918 the Millom & Askam Hematite Iron Co. acquired a controlling interest in the Camerton Coal & Firebrick Co. which was then working two drifts, River Drift and Wood Drift. Output in that year amounted to 14,002 tons of coal and 12,129 tons of fireclay and in the following year 8,538 tons of coal and 9,152 tons of fireclay. A third drift, Hall Drift, was begun in 1920 and driven from the Wood Drift towards the lower seams in the River Drift area. As output from the Lonsdale royalty declined an adjoining royalty belonging to R. Falcon-Cooke was opened out; between 1923 and 1938 all production was from this royalty. In 1939 the colliery was taken over by Messrs Thomas McKay & Sons whose principal aim was to mine the fireclay needed for their adjacent brickworks, any coal produced being entirely for their own consumption.[19]

Coal mining at Moorehouse Guards, Seaton, which had ceased in 1914, was resumed in 1921 when Thomas Lamb leased the Lonsdale royalty there. He and his four sons worked the White Metal band seam by means of a drift until 1928 when a pit was sunk. This small family business lasted until 1945 when the colliery was abandoned.[20]

During the ten years 1911-1920 the St Helens Colliery, Siddick had a average annual output of 161,151 tons of land coal and 58,871 tons of undersea coal. Although No. 2 Pit ceased production in 1921, when the colliery raised 203,000 tons of coal, output quickly increased to 300,000 tons, reaching a maximum of 382,000 tons in 1927. Except for the year 1932 more than half of the colliery's output between 1923 and 1935 came from undersea workings. Thereafter came a rapid decline in the yield of undersea coal, from 128,506 tons in 1936 to 12,468 tons in 1940.[21] In 1943 the colliery was bought by the United Steel Cos., a purchase which gave them control of all the coastal collieries from Risehow to Lowca.[22] Two years later the colliery was closed for extensive repairs, including the complete relining of both shafts, one of which was deepened to reach a six foot seam of Main Band coal. Not until April 1948 was the colliery re-opened.[23]

MARYPORT DISTRICT

Despite the stimulus to production provided by the first World War the Allerdale Coal Co. had to face increasing difficulties, particularly at Buckhill Colliery. In 1915 it was said that this colliery could not have continued on its own account without the help of the coke ovens at Clifton which were also owned by the company.[24] After the war there was a rapid fall in output and between 1925 and 1929 the company suffered heavy losses. The willingness of Lord Leconfield, the lessor, to accept a much lower rent than he was entitled to receive helped, but did not save the colliery. By 1931 the company had a bank overdraft of £85,000 and liabilities estimated at £100,000. To cut its losses the company closed down Buckhill Colliery in 1932.[25]

Much higher prices for coal enabled the Flimby & Broughton Moor Coal Co. to carry on its Bertha Pit at Broughton Moor a little longer, but in 1917 its properties were sold to the Risehow Colliery, Coking & By-Product Co. Ltd. who also took over the leases of their predecessor. Bertha Pit proved increasingly unprofitable and in 1923 the company surrendered the lease. In the following year the colliery plant was dismantled and removed.[26]

Among the assets acquired by the Risehow Colliery Co. in 1917 was the Alice Pit, Outfields and they had no more succes than at Bertha Pit. The Leconfield royalty was becoming exhausted, output falling from 20,106 tons in 1917 to 1,762 tons in 1919 and production for the most part depended upon adjacent royalties owned by Mr Dykes and Mr Walker. In 1920 the lessees complained that they were losing 10s. per ton, although this loss was made up by the government. Mining operations were hampered by a growing volume of water in the colliery which caused the local Inspector of Mines to issue a peremptory warning that unless the company were prepared to increase their pumping plant he would call upon them to close the mine as unsafe. Having lost money for some time the company were unwilling to invest more capital in pumping machinery and in 1921 the colliery was abandoned.[27]

A Lonsdale royalty at Dearham Bridge was leased in 1917 by the Birkby Coal Co. Ltd. to which was added, in 1922, an adjoining royalty belonging to the Dean and Chapter of Carlisle. From their Birkby Mine coal was raised from the Three Quarters seam and output was increased when the New Allbright Drift began to produce coal in 1929.[28] The company put down boreholes in a nearby Leconfield royalty and located the White Metal and Ten Quarters seams but did not take up the option of leasing the area.[29] Production continued until 1950 when the colliery was abandoned.

Between 1926 and 1929 Crow coal at Dearham was worked on a small scale, and not very successfully, by the Rowbeck Coal Co. Ltd. The firm did not employ more than ten persons and had to abandon Mill Drift in 1928 and Mill Pit in the following year.[30]

In 1917 the Flimby and Seaton Moor collieries were taken over by the Risehow

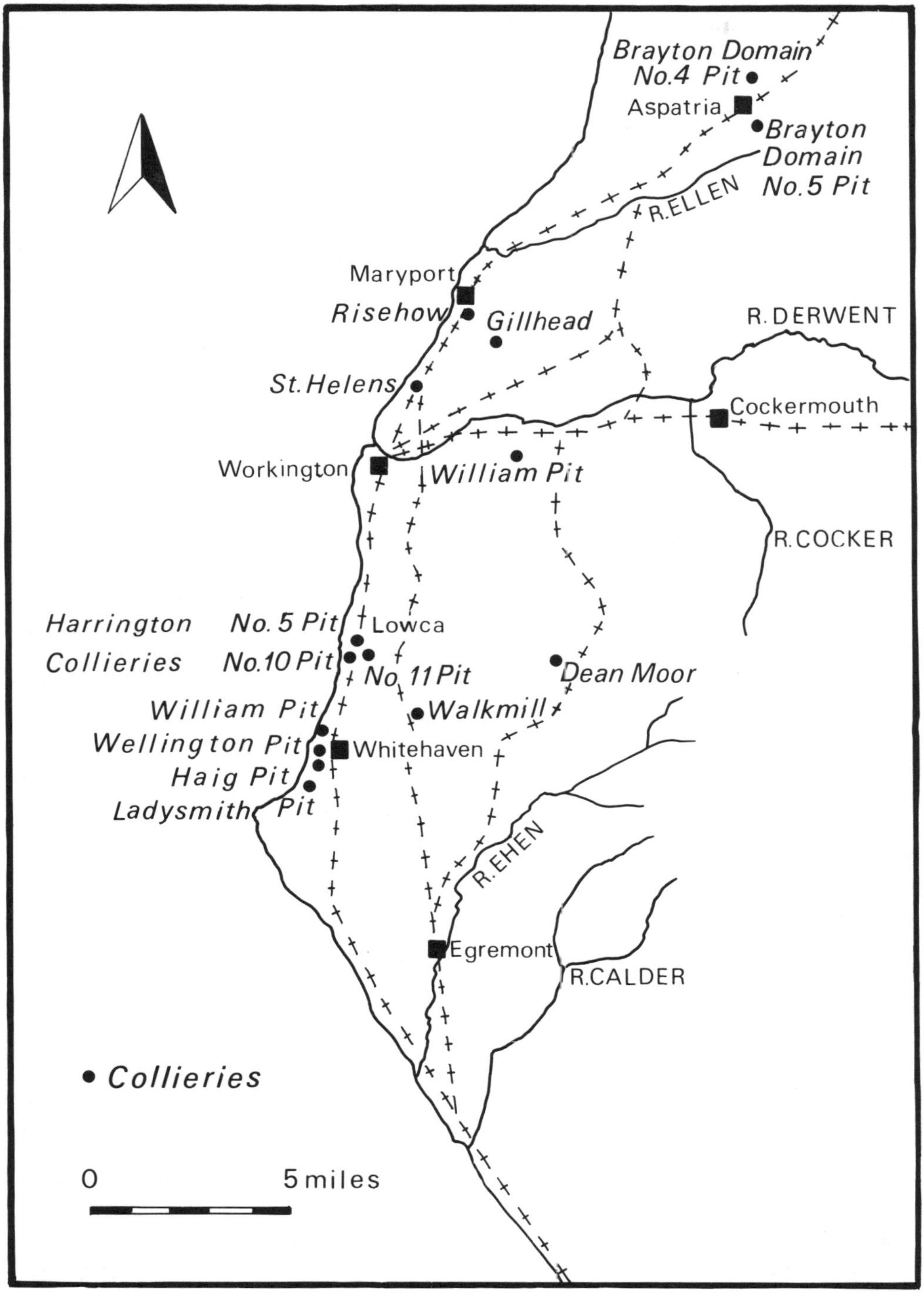

FIG. 5 – Coal Mines in Cumberland 1920.

Colliery, Coking & By-Product Co. Ltd. A new pit was sunk at Risehow 580 feet to the Three Quarters seam where coal production began in 1918.[31] To extend their control of the Cumberland coalfield the United Steel Cos. Ltd. bought a large share in the Risehow Co. in 1924.[32] Because of falling output mining ceased at Seaton Moor Pit in 1926. At Risehow the coal came from royalties belonging to the Lowther and Curwen families and the area worked by the colliery was enlarged by the addition of a section of Senhouse land leased in 1938. Nearby, the Gillhead Pit, which belonged to the United Steel Cos., was worked in conjunction with Risehow Colliery to which coal was sent by aerial ropeway.

A new venture began at Birkby in 1945 when Robert Moffatt & Sons opened out a drift mine, the Ellenbank Mine. Between twenty-five and thirty men were employed to work coal in the Six Quarters seam. The colliery had a life of only seven years; the exhaustion of the coal reserves and rapidly deteriorating working conditions compelled its closure in 1952.[33]

At Oughterside, between 1920 and 1938, a succession of companies found it impossible to mine coal at a profit. The lease of Oughterside Colliery was assigned to the Cumberland Coal, Power & Chemicals Co. Ltd. in 1920, but soon the firm found it necessary to go into voluntary liquidation and the lease was resumed in 1923 by the Oughterside Coal Co. Ltd. Mining proved unprofitable and from 1925 the company, in arrears with the rent, made strenuous efforts to sell the colliery, finally succeeding in 1928 when the colliery and coking plant were bought by the Oughterside Coal Co. (1928) Ltd. for about £10,000. This company had a brief and unhappy existence. Differences among the directors probably accounted for the fact that there was no attempt made to develop the colliery and mining was restricted to robbing the pillars. In 1930 this firm also went into voluntary liquidation and the colliery was carried on by the liquidator, with no better results, until 1933 when it was abandoned. Three years later mining began again at Oughterside when the Westmoor Colliery Co. Ltd. sank a drift mine at Westmoor End. Coal was worked from the Ten Quarters, Thirty Inch and Rattler Band seams until 1938, an influx of water from the old colliery making further production impossible. Another company, the Oughterside Coal Co. Ltd. formed in 1939, made the last attempt to win coal in this area. In sinking Westmoor Nos. 1 and 2 Mines three different seams of coal were found but most work was done in a seam of Yard Band coal three feet thick. Output was small and this, together with the difficulty of obtaining workmen, led to the abandonment of the enterprise in 1943.[34]

In the Aspatria district Joseph Harris continued to work his Brayton Domain Colliery at a profit. Between 1913 and 1918 the annual output of Nos. 4 and 5 Pits exceeded 200,000 tons, subsequently, as the coal was worked out, the yield was considerably reduced. No. 4 Pit closed in 1933 and No. 5 Pit in 1942, the Yard Band in both pits having been exhausted.[35]

During the First World War mining at Allhallows Colliery consisted mainly of robbing the large pillars left in the Yard Band. When this had been completed coal was next wrought in the Thirty Inch seam. The colliery was bought in 1920 by the United Steel Cos. who obtained a new lease which added 287 acres to the area held by their predecessors. Its annual output increased between 1921 and 1924 but declined rapidly thereafter, from 60,653 tons in 1924 to 6,909 tons in 1927. When it became clear that the colliery could not be run at a profit it was abandoned in 1928 and its equipment dismantled.[36]

In the 1920's attempts were made to mine coal in the Bolton area. An adit driven 100 yards by Tom Dand in 1924 reached the Ten Quarters seam at Pow Ghyll, Bolton where the coal was five feet thick and of good quality. After working this seam he was succeeded in 1926 by the Wigton Coal Co. Ltd. which sank a pit sixty-six feet to the Master Band. However, little development took place and coal getting amounted to no more than robbing back pillars of coal which had been left by earlier miners in old workings. Only twenty to fifty tons of coal were raised per week and in 1930 the colliery was closed.[37]

Various small drift mines, each employing a handful of men, were at work briefly during the 1930's. Isa Drift, Crosby, which belonged to Messrs. Williamson, was in production in 1932. White Metal Band coal was worked by this firm at two drifts in Great Clifton – at the Range Drift 1933-34 and at the Jubilee Drift 1935-37. Between 1931 and 1932 Four Feet coal was mined at Annie Drift, Winscales by the Winscales Colliery Co. Ltd. and Rattler Band coal at Bridgefoot Drift by John Turnbull. At Maurice Stobart's Ellen Hall Mine, Aspatria a little coal was won from the Metal Band between 1936 and 1937.[38]

WEST CUMBERLAND

Two world wars and a long period of severe economic depression subjected the coal mining industry in West Cumberland to considerable fluctuations. A patriotic rush to the colours on the outbreak of the First World War was a serious drain on manpower resources. In 1914 691 out of the 1,296 men employed at Whitehaven Colliery enlisted and at the small Wythemoor Colliery forty-four out of the eighty-two employed joined the army. During the war 3,682 men left the Cumberland coalfield to join the forces, a large proportion since only about 8,000 men were employed. These men, young and fit, were mainly hewers and trailers and their departure meant a lower output and increasing production costs per ton. To meet the shortage of labour underground workings were reorganized wherever possible; seams that were difficult to work were left and attention was concentrated upon those which yielded a larger output. In addition, coal cutting machines and conveyors were introduced in a few mines and their use extended in others. Development work such as drifting and making new roads was almost entirely abandoned and often the maintenance of the main roads underground was neglected to such an extent that it was only just possible to move the coal. A shortage of timber, scarcity of tubs, voluntary absenteeism, colliery disputes, a steep rise in the cost of materials and a lack of railway and shipping transport added to the difficulties of the industry.[39]

State control of a vital fuel in time of war was unavoidable and in February 1917 the coal mining industry was taken over by the government when a Coal Controller was appointed to regulate output and distribution. It was now the business of the state, not private enterprise, to maintain coal supplies, determine the level of capital investment, allocate scarce resources and "zone" coal distribution in order to reduce transport costs. The Coal Controller was also responsible for guaranteeing the payment of dividends to high cost companies out of the excess profits of the industry. The collieries, however, were to continue under their existing management subject to the right of the Controller at any time to take over the management of any undertaking.[40]

The output of Cumberland coal fell from 2,360,510 tons in 1914 to 1,904,324 tons in 1918 and continued to decline until 1921 when only 1,335,120 tons of coal were produced. Government control of the industry had failed to maintain output. Between the wars output was seriously diminished by strikes in 1921, 1926 and 1931 and by the closure of the Whitehaven pits in 1936. In the years 1922-24 and 1927-30 annual production exceeded two million tons; in 1927 Cumberland collieries produced 2,416,049 tons, the highest figure in the history of the coalfield. After 1930 output never again reached two million tons and the output for 1936, except for the strike year of 1926, was the lowest since 1875.

In 1919 a war-shattered Europe, rebuilding its industries, needed large quantities of coal and the almost insatiable demand for coal provided a powerful stimulus to British coal mining. By the end of 1920 the export boom had collapsed for the abnormally high price of export coal had been halved. Two fortuitous events, the American coal strike of 1922 and the French occupation of the Ruhr in 1923, made some recovery possible by leaving large markets open to British coal. These windfall gains could not last and after 1924 the industry had to face the full blast of competition for which it was not prepared. The age of easy sales was over and Cumberland collieries suffered along with the others. Statistics of coal output reflect the vicissitudes experienced in Cumberland.[41]

Year	*Output Tons*	*Year*	*Ouput Tons*
1914	2,360,510	1927	2,416,049
1915	2,248,810	1928	2,138,823
1916	2,279,295	1929	2,021,405
1917	2,161,550	1930	2,017,477
1918	1,904,324	1931	1,512,878
1919	1,842,153	1932	1,512,681
1920	1,847,699	1933	1,420,697
1921	1,335,120	1934	1,564,204
1922	2,190,499	1935	1,537,432
1923	2,114,037	1936	1,189,948
1924	2,188,795	1937	1,343,093
1925	1,985,442	1938	1,568,175
1926	976,354	1939	1,592,300

Between 1889 and 1914 there had been a marked decline in the productivity of the British coalfields as measured by the annual output per person employed. During the First World War and for the following three years the position grew worse, demonstrating the inefficiency of the industry; the annual output per person employed fell each year from 265·5 tons in 1915 to 144·2 tons in 1921. In the Cumberland coalfield the picture was the same, a continuing fall in productivity, from 206·6 tons in 1914 to 157·6 tons in 1920 and to 115·4 tons in the strike year 1921. In Cumberland, except for the years 1926 and 1931 when major strikes took place, there was a steady improvement in productivity between 1922 and 1938, but the annual output per head was always less than the national average.

Annual Output per Person Employed[42]

Year	*Cumberland* *Tons*	*United Kingdom* *Tons*
1914	206·6	234·3
1920	157·6	183·8
1925	179·9	220·5
1930	206·8	261·8
1935	239·9	288·8
1938	235·5	287·0

Productivity is measured more accurately in terms of output per manshift. Such statistics of output per manshift as are available show a decline in productivity from 1914 to 1921, particularly during the period 1917-21 when the government controlled the industry, which may justify one writer's assertion that the results of this control were "little short of disastrous".[43] Various factors combined to bring this about. During the war years the industry had been greatly neglected and was in a backward state. Little had been done to improve the layout of pits, to improve roads above and below ground or to increase mechanization. Higher prices for coal made it possible for employers to meet demands for higher wages without investing in expensive machinery to reduce labour costs. The enlistment of the young and fit men in the armed forces meant that there was an ageing labour force unable to maintain the output of earlier years. Further, it may be that enhanced wages were not conducive to maximum effort on the part of the mineworkers.

After 1921 there was a little improvement in the output per manshift. This resulted from the closure of the least productive pits, the return to the eight hour day and the growth of mechanized mining as more machine cutters and conveyors came into use. The Cumberland coalfield shared in this improved productivity but its output per manshift was always below that of the United Kingdom generally. Apart from the geological difficulties which hampered production a major cause of low productivity was the fact than in many Cumberland collieries there were unusually long distances from the shaft to the working face, particularly in the Whitehaven pits. Much time was wasted walking to and from the face leaving the miners less time per shift for coal getting. For the same reason less coal could be hauled in each shift from the face to the shaft.

Output per Manshift[44]

Year	*Cumberland* *Cwts.*	*United Kingdom* *Cwts.*
1914*	18·92	20·32
1920	11·89	14·54
1925	15·41	18·02
1930	16·38	21·62
1935	17·49	23·35
1938	17·23	22·4

* Month of June only.

For Cumberland and for every other coalfield the 1930's were an age of depression, of closed collieries and unemployed miners. Basically the cause was a fall in the demand for coal both at home and abroad. In foreign countries British coal had to complete with coal from Germany, Poland and the Netherlands and it could not do so effectively. Technically the British industry lagged behind its competitors; mechanized mining here was much less highly developed than in other countries where production was concentrated in the larger and more efficient mines and where working costs were much lower than ours. In Great Britain there were too many collieries, many of which were too small to operate profitably and others where the accessible coal had been worked out and thinner and deeper seams had to be worked at higher production costs. Economies in the industrial consumption of coal further lessened demand. More gas and coke per ton of coal used, more efficient use of waste gases and economies in coke consumption in the manufacture of iron and steel all injured coal mining. In addition, coal had now to complete with other fuels, particularly oil.

The Cumberland coal mining industry was not affected by the growing competition for continental markets, but it suffered severely from the increasing competition between the British coalfields for the Irish market, traditionally its most important market. An import duty of 5s. per ton on British coal imposed by the Irish Free State in 1932 led to a sharp reduction in British coal exports to Ireland and, as competition intensified, to Cumberland coal obtaining a smaller share in the declining Irish market. Coal exports from Cumberland to Ireland fell from 442,200 tons in 1931 to 328,807 tons in 1935.[45] In Cumberland itself an appreciable tonnage of coal was in the 1930's being imported from other mining districts in the north. Between 1921 and 1931 the population of the county declined from 273,173 to 263,151, a fall of 3·7 per cent, which also helped to reduce the demand for local coal.

Colliery owners attempted to meet foreign competition by seeking to impose wage cuts and to lengthen the working day. This precipitated the prolonged national strike in 1926 and although it ended in a victory for the owners it did not solve the problems of the industry. With productive capacity greatly in excess of demand, unfettered competition would lead to lower prices and wages and to unemployment for many miners. Some policy of restriction therefore seemed desirable. The Coal Mines Act of 1930 was designed to reduce competition within the country by regulating output and by introducing a certain amount of price control which together would ensure profitable working. A Central Council, representing all the coal owners, was set up to control the output of the industry as a whole. It allocated production quotas, which were revised quarterly, to each of the seventeen districts into which the country was divided. In each district an executive board of local coal owners allocated quotas to individual collieries and fixed minimum prices for all classes of coal. What happened in Cumberland is shown below:[46]

Year	*Allocation Tons*	*Output Tons*	*% by which ouput fell short of Allocation*
1931	2,281,664	1,739,800	23·75
1932	2,102,702	1,714,521	18·46
1933	1,775,610	1,628,682	8·27
1934	1,837,250	1,798,834	2·09
1935	1,919,334	1,787,674	6·86
1936	1,846,818	1,398,731	24·26
1937	1,663,900	1,598,839	3·91
1938	1,899,076	1,883,087	0·84

The Coal Mines Act of 1930 did not have much effect in Cumberland since the quota system did not at any time limit production in the county, the owners always having a sufficient quota allocated to them to raise and sell all the coal they could produce. The Act did not establish a national price maintenance scheme and price competition between districts continued unabated. It did, however, lead to the control of district prices but in Cumberland coal prices had been controlled for many years by the local Coal Owners' Association and price-cutting had never been a major problem. It may be that the Act did enable less efficient pits to survive longer than they might otherwise have done.

In 1900 there were forty-three collieries working in Cumberland; by 1919 there were only thirty-five, this number being further reduced to thirty in 1925 and to twenty-five in 1938. Exhaustion of coal reserves and the abandonment of the least profitable collieries accounted for this diminution. The ownership of collieries became concentrated into fewer hands and a few large firms were responsible for most of the county's output and for the employment of the majority of the miners in the district. It has been said that "in 1930 98 per cent of the output was produced by 8 undertakings operating 21 mines."[47] Since 1920 an outstanding feature of the coalfield has been the expansion of ownership by the United Steel Cos. Ltd. To their collieries at Lowca and Gillhead were added Allhallows Colliery in 1920, Outlands, Moresby and Risehow Collieries in 1924, Solway Colliery in 1937 and St Helens Colliery in 1943. Apart from the year 1932 this firm's Cumberland collieries between 1930 and 1935 produced more than one third of the county's total output of coal and between 1936 and 1940 this proportion exceeded forty per cent. After the acquisition of St Helens Colliery the firm had an even larger share of the coal production of Cumberland.

Output of United Steel Cos. Cumberland Collieries[48]

Thousand Tons

Year	*Output*	*Year*	*Output*
1930	691	1936	682
1931	681	1937	658
1932	449	1938	664
1933	505	1939	643
1934	549	1940	634
1935	608		

Other iron and steel firms were concerned with coal mining, e.g. the Millom & Askam Hematite Iron Co. Ltd. took over Camerton Colliery in 1918 and the Barrow Hematite Steel Co. Ltd. had large interests in the Allerdale Coal Co. In Cumberland the connection of coal mining with the iron and steel industry was particularly close; in 1933 nearly half of the county's coal output was consumed by the United Steel Cos. Ltd.[49] In addition to the coal produced in its own mines this firm drew upon the output of other local mines. A report on the Cumberland coalfield (19 November 1931)[50] shows that about sixty per cent of the output of the Allerdale Company was carbonized and sold to the United Steel Cos. and twenty per cent of the output of the Whitehaven collieries was disposed of in the same way.

The Cumberland coalfield was now largely dependent upon the local iron and steel industry, its biggest single customer. Ireland, the traditional market, although less important than in the nineteenth century, continued in this period to absorb between one-quarter and one-third of the county's output of coal.

COKE

Coke Production in Cumberland[51]

Year	*Coal Carbonized* *Tons*	*Coke Output* *Tons*	*No. of Ovens* *in use*
1914	845,526	480,544	383
1915	841,388	468,582	385
1916	868,512	524,947	515
1917	952,480	570,232	390
1918	803,987	505,848	388
1919	722,221	450,735	386
1920	785,102	496,590	390
1921	Figures not available		
1922	527,007	362,672	350
1923	762,113	517,786	390
1924	716,631	489,286	390
1925	625,294	429,686	319
1926	225,169	154,967	115
1927	746,289	512,478	362
1928	658,845	454,712	333
1929	657,235	449,716	303
1930	656,121	448,158	302
1931	393,914	267,619	213
1932	349,192	236,975	198
1933	292,761	194,655	148
1934	523,768	353,088	262
1935	541,551	363,750	262
1936	501,082	337,054	224
1937	591,917	418,057	138
1938	567,400	393,444	162

These statistics reflect the economic history of West Cumberland at this time. The needs of the iron and steel industry during the First World War led to an increase in coke production, over 500,000 tons in each of the three years 1916-18; in 1917 the record total of 570,232 tons of coke was made. Only twice since the war did output exceed half a million tons, in 1923 and 1927. Strikes in 1926 and 1931 and the slump in the iron and steel industry 1931-33 account for the fall in coke production in these years. Coke production began to recover in 1934.

Coal mining in Cumberland depended to a large extent upon the fortunes of the iron and steel industry and its demand for coke. Between 1914 and 1938 about one-third of the county's coal output was converted into coke for use in local furnaces.

Percentage of Cumberland Coal Carbonized

Year	%	*Year*	%	*Year*	%
1914	35·8	1923	36·0	1931	26·0
1915	37·4	1924	32·7	1932	23·0
1916	38·1	1925	31·4	1933	20·6
1917	44·0	1926	23·0	1934	33·4
1918	42·2	1927	30·8	1935	35·2
1919	39·2	1928	30·8	1936	42·1
1920	42·4	1929	32·5	1937	44·0
1922	24·0	1930	32·5	1938	36·1

After 1916 beehive ovens were no longer used and from 1922 onwards only bye-product ovens, mostly of the Koppers type, were employed in the manufacture of coke. The Workington Iron & Steel Co. decided to close its old coke ovens at Harrington and Moresby and to concentrate its coking operations in a new, single unit at Workington, adjacent to the Derwent blast furnaces, which could be fed direct by belt conveyor from the proposed new Solway Pit nearby. This unit, consisting of a battery of fifty-three Woodhall-Becker regenerative ovens and the associated bye-product plant, was erected at a cost of about £450,000 and came into service in July 1936. It had a designed capacity of 6,000 tons of furnace coke per week compared with the 4,800 tons per week output of the 160 ovens at Harrington and Moresby which were closed down.[52] The company's output of coke amounted to:[53]

Year	*Thousand Tons*	*Year*	*Thousand Tons*
1930	234	1936	241
1931	193	1937	317
1932	90	1938	334
1933	131	1939	256
1934	169	1940	315
1935	232		

After the replacement of the beehive ovens by regenerative ovens not only were substantial profits made from the sale of bye-products but there were also economies in the use of coal. The following table shows increasing quantities of coke being made per 100 tons of coal carbonized in West Cumberland.

Year	*Tons*	*Year*	*Tons*	*Year*	*Tons*
1914	56·8	1923	67·9	1931	67·9
1915	55·6	1924	68·2	1932	67·8
1916	60·4	1925	68·7	1933	66·4
1917	59·8	1926	68·8	1934	67·4
1918	62·9	1927	68·6	1935	67·1
1919	62·4	1928	69·0	1936	67·2
1920	63·2	1929	68·4	1937	70·6
1922	68·8	1930	68·3	1938	69·3

The output of coke was absorbed almost entirely by the local iron and steel industry.

It did not, however, suffice to meet the needs of the industry. In 1938 the Workington Iron & Steel Co., the largest coke producing concern in the area, obtained about ninety per cent of its coke supplies from Cumberland and the remaining ten per cent from the North-East Coast. At the Millom & Askam Ironworks about seventy-four per cent of the coke needed there came from Durham and twenty-six per cent from Cumberland.[54]

1939-1946

In the Second, as in the First World War, state control of mining operations was late in coming and it took a major fuel crisis in 1942 to stir the government into action. A new department, the Ministry of Fuel and Power, was then created to supervise the industry and its minister, assisted by a Controller-General and a National Coal Board, became responsible for the operation of the mines. The country was divided into eight producing regions, each with a Regional Controller, assisted by a Regional Coal Board, who was empowered to control the collieries and to give directions to their managers. But the colliery owners continued to retain financial control of the industry. Again governmental authority failed to bring about an improvement in output or productivity in the coal mines when it was most needed.

For the Cumberland coalfield and for the mining industry as a whole the war years were a period of contraction in manpower, output and productivity. In Cumberland two collieries were closed, Brayton in 1942 and Oughterside in 1943, the closure of the latter being hastened by the lack of mineworkers. A little was done to find additional sources of coal. Development work was continued at the new Solway Colliery and a small drift mine, Ellenbank Mine, was opened at Birkby in 1945.

A shortage of miners was the major cause of production problems and this, in part, stemmed from pre-war experience. For over a decade before the war the numbers employed in Cumberland mines fell steadily, from 11,622 in 1927 to 6,658 in 1938, a fall of 42·71 per cent. Low wages, frequent unemployment and the monotony and bad housing of the mining villages made the industry unattractive to young men who sought jobs outside mining. For a war economy there were not enough miners in 1939 and until May 1941, when they were forbidden to leave the pits, their numbers were further diminished by enlistment in the armed forces or by migration to other industries offering higher wages. In 1945 there were almost a thousand fewer miners working in Cumberland than in 1939, a drop of 14·68 per cent. Not only were there fewer miners but they also worked less effectively. A sharp rise in absenteeism and a fall in the productivity of face workers helped to bring about a continuous decrease in annual output; in 1945 less than two-thirds of the 1939 tonnage as produced.[55]

Year	*Output* Tons	*No. of employees*	*Annual ouput per person employed* Tons
1939	1,592,300	6,716	237·0
1940	1,524,600	6,471	235·6
1941	1,209,600	5,584	216·6
1942	1,106,600	5,521	200·4
1943	1,074,000	5,510	194·9
1944	1,049,500	5,622	186·6
1945	999,900	5,730	174·5
1946	1,016,300	5,772	176·0

Both nationally and locally there was a serious decline in the ouput per manshift:[56]

Year	Cumberland		Great Britain	
	At the face	*Overall*	*At the face*	*Overall*
	Tons	*Tons*	*Tons*	*Tons*
1939	2·23	0·84	3·00	1·14
1940	2·17	0·81	2·97	1·10
1941	2·18	0·75	2·99	1·07
1942	2·16	0·74	2·91	1·05
1943	2·06	0·71	2·86	1·03
1944	2·02	0·68	2·70	1·00
1945	2·09	0·67	2·70	1·00
1946	2·10	0·69	2·76	1·03

It will be seen that the output per manshift in Cumberland was always much below the national average. In the national table of productivity the Cumberland coalfield was near the bottom, its performance being slightly better than that of the Bristol and Somerset coalfields.

Various factors contributed to the fall in productivity. Mining called for tremendous physical effort, especially from underground workers, and as the years went by an ageing labour force found it increasingly difficult to meet the demands made upon it. Younger workers recruited from other industries were not accustomed to the strain imposed by employment in the mines. The dissatisfaction of the miners with their lot had much to do with the fall in productivity. They strongly resented the industrial bondage forced upon them by the Essential Work Order of May 1941 which to them not only implied some kind of social inferiority but also denied them the opportunity to better themselves by moving to higher paid jobs in other war industries. Until the summer of 1942 miners' wages lagged behind those paid to many other industrial workers and inspired a great deal of bitterness in mining areas. A discontented body of miners meant less productive miners.

During the war there was a continuing growth of absenteesim; for Great Britain the percentage of absenteeism rose from 6·9 in 1939 to 16·3 in 1945; in Cumberland absenteeism rose from 10·9 per cent in 1943 to 14·6 in 1945 and to 15·01 in 1946.[57] Undoubtedly the discontent of the miners was responsible for much of this absenteeism. There were, however, other causes, difficult to identify with absolute certainty and whose effects it is impossible to measure. During the depression period miners had become accustomed to short time working and had lost the habit of working a full week. The resumption of full time working between 1939 and 1945 was, in terms of physical effort required, too much for some miners who felt they had to miss shifts. When wages increased in the latter years of the war there were those who preferred to take time off to higher wages, an adequate wage packet did not necessarily demand that a full week should be worked. It may be that some miners absented themselves because the old industrial discipline had disappeared and it now seemed that they were indispensable and almost impossible to dismiss. What is certain is that absenteeism was a major problem that was never satisfactorily dealt with and, despite higher wages and increased mechanization, there was a decrease rather than an increase in productivity.

MINING METHODS

Coal owners in Cumberland were late in adopting machinery in their collieries; for too long coal mining was essentially a "pick and shovel" industry. This may have been due to conservatism, or to an unwillingness or inability to invest capital in the new machines. Between the wars when some collieries had to close and others were run at a loss there was a marked reluctance to invest in an ailing industry. The existence of many faults, frequent changes in gradient, long haulage roads, often old and badly planned, all militated against the early and extensive use of machines to cut and convey coal.

The few coal-cutters in use before 1914 had not been very successful. However, the percentage of Cumberland and Westmorland coal cut by machine rose from 1 per cent in 1913 to 20 per cent in 1931, increasing annually thereafter until 1938 when it reached 72 per cent. Until 1936 the percentage of machine-cut coal was below the national average but from that year until 1945 it was above the national average.*[58]

Year	*No. of machines*	*Coal Cut by Machines Tons*	*% of Total Output Cut by Machines*	
			Cumberland and Westmorland	*Great Britain*
1913	11	29,963	1	8
1922	24	166,802	7	15
1931	51	304,622	20	35
1938	80	1,128,765	72	59
1942	80	861,344	78	66
1945	87	901,375	76	72

Not all miners welcomed mechanization. To the old fear of being displaced by machines were added objections to the "speeding-up" of work and disputes sometimes arose about reduced tonnage rates paid for machine-cut coal.

Conveyors arrived late in Cumberland. In 1919 the lessees of Allhallows Colliery were experimenting with a fifty yards length of conveyor in the 30 Inch seam.[59] In 1921 there were only two conveyors in the county and none in use between 1925 and 1928. After 1928 conveyors were installed in the pits in increasing numbers, reaching 100 in 1945, the percentage of coal mechanically conveyed in Cumberland and Westmorland rising from three per cent in 1930 to fifty-five per cent in 1945, a percentage much smaller than that for the industry as a whole.

Cumberland and Westmorland[60]

Year	*No. of Conveyors*	*No. of Gate-end Loaders*	*No. of Power Loaders*	*Tonnage Conveyed Mechanically*	*% of Output Conveyed Mechanically*	
					C. & W.	*Gt. Brit.*
1930	8	1		67,987	3	17·4
1935	22	5		262,874	17	43·0
1940	38	1		624,082	41	61·1
1943	58	–	5	529,944	49	66·3
1945	100	2	14	659,765	55	71·0

* Separate statistics for Cumberland of coal cutting machines, electrical equipment, conveyors and washing plants are not available.

During the Second World War power loaders* were introduced into a number of collieries. These new machines both cut coal and loaded it on to the conveyor which ran alongside the face and were in time to revolutionize the industry; by cutting down the manhandling of coal they greatly reduced the physical effort demanded of miners, accelerated the advance of the face and made possible large savings in the use of manpower. Power loaders first came into use in Cumberland at the Solway Colliery and by the end of the war there were 14 power loaders at work in the county. In 1945 11·03 per cent of the total ouput in Cumberland was power loaded compared with the national percentage of 1·3 per cent.[61]

In this period useful progress was made in the adoption of electrical equipment. In 1913 the total horse power of electric motors in the collieries of Cumberland and Westmorland amounted to 4,213; by 1945 this figure was increased to 26,645 h.p.[62] Electric cap lamps were introduced in 1929 and within a few years they were being used throughout the coalfield. This new practice meant that more miners were without means of testing their working places for gas. Attempts were made to invent a convenient and reliable gas detector which could be carried by some of the workmen, but the flame safety lamp continued to be accepted by the miners as the best means of testing for gas.

Many of the fatal accidents in Cumberland collieries were caused by falls of roof and to reduce this danger steel props, which were more efficient than timber supports, were introduced in the 1920's. Local miners regarded them with suspicion believing that the wooden props were safer because they made creaking noises which gave warning of roof movements. The conversion from timber to steel supports was a slow process. In 1938 of the underground roadways in the pits of Cumberland and Westmorland twenty-two miles were supported by steel arches, thirteen miles by straight girders, seventy-eight miles by timber and other means and thirty-one miles were unsupported. At the coal face there were only forty-six steel props and 704 steel bars.[63] Subsequently the conversion programme was accelerated and by 30 June 1945 of the 176 miles of underground roads in the Cumberland collieries forty-six miles were supported by steel arches, thirty-three miles by girders and seventy miles by timber and other means; twenty-seven miles were unsupported. At the working face there were 3,654 steel props and 3,097 steel bars or straps.[64]

In Cumberland mines an excessive amount of time and energy was consumed by the men walking to and from their working places. In 1945 this loss of time, which might have been used more productively, was estimated to the equivalent to at least 250 manshifts per day. In the later years of the Second World War something was done to remedy this situation. At Haig Pit a man-riding set came into operation capable of carrying the miners two-and-three-quarter miles inbye in nine minutes. In 1945 a new man-riding shaft was put to work at Clifton Colliery and a new man-riding drift was about to come into use. At the Solway and Risehow collieries preliminary work was going on to prepare for the introduction of locomotives underground to move both men and coal.[65]

Throughout this period Cumberland led other coalfields in the provision of clean coal, the proportion of washed coal in the county being well above the national average. In 1927 38·8 per cent of Cumberland coal was cleaned compared with 20·5 per cent for

* Particulars in respect of power loaders were not collected separately prior to the year 1943.

Great Britain as a whole; in 1945 the percentages amounted to 71·6 and 47·4 respectively.[66] Much less enterprise was shown in providing facilities for washing the miners. The first pithead bath, at Harrington No. 10 Pit, did not appear until 1929 and the second at Siddick in 1933. It may be noted that in 1933 there were thirty mines at work in Cumberland. Baths were installed later at other collieries and proved a great boon to the miners, most of whom lived in cottages without bathrooms. Until pithead baths were provided miners at the end of the shift had to travel home in dirty clothes and wash themselves as best they could in a tin bath in front of the kitchen fire.

THE MINERS

Persons Employed in Cumberland Collieries[67]

Year	*No.*	*Year*	*No.*
1914	11,422	1929	9,823
1919	11,355	1930	9,752
1920	11,722	1931	8,633
1921	11,561	1932	7,070
1922	11,496	1933	6,459
1923	11,805	1934	6,922
1924	12,044	1935	6,406
1925	11,036	1936	4,962
1927	11,622	1937	5,990
1928	10,489	1938	6,658

It will be seen from the table above that the Cumberland coalfield between 1914 and 1927 provided employment for more than 11,000 persons, the number of mineworkers reaching a record total of 12,044 in 1924. Until 1928 there were more miners at work than in the years immediately preceding the war. After 1927 there was an almost continuous decline in employment which reached its nadir in 1936 when only 4,962 persons were working, the lowest figures since accurate manpower statistics first became available in 1874 and less than half the number employed during the years 1910-1928. The complete cessation of production at Whitehaven Colliery throughout 1936 was responsible for the large reduction in the active labour force in that year. Since 1928 the number of mineworkers in the coalfield has always been below 10,000. The re-opening of Whitehaven Colliery and the sinking of Solway Pit in 1937 quickly led to a substantial improvement in employment figures.

The "Occupations" volume of the 1921 Census provides a table of the number of miners per thousand males aged 12 and over in different areas of the county and the degree to which these districts then depended for their livelihood upon local collieries is clearly shown:

Urban Districts		*Rural Districts*	
Arlecdon and Frizington	110	Alston	83
Aspatria	509	Brampton	102
Cleator Moor	20	Cockermouth	353
Cockermouth	30	Whitehaven	205
Harrington	198	Wigton	109
Maryport	249		
Whitehaven M.B.	328		
Wigton	20		
Workington M.B.	86		

The same source analyses the different kinds of work done per thousand male miners:

Superintending staff	52	Makers and repairers of roads	99
Hewers and getters	549	Other underground workers	79
Conveyors of material to the shaft	103	Surface workers	118

The proportion of hewers and getters, 549, can be compared with the corresponding figures for Northumberland 415, Durham 397 and the West Riding of Yorkshire 531.

For Cumberland the decade 1928-38 was one of extreme suffering, high unemployment and widespread human misery and this was particularly true of the coastal strip where industry was concentrated. This was an age of pusillanimous politicians and egregious economists who helped to convert once prosperous districts into economic deserts which were then designated "Special Areas", a pointless euphemism to replace the former term "Depressed Areas" which was too near the truth for official comfort. West Cumberland was one of these Special Areas and there all industries suffered during the depression period, but none more than the mining industry. Excessive dependence upon one industry proved calamitous in the towns and villages scattered throughout the coalfield. In 1932 44·3 per cent of insured coal miners in Cumberland and Westmorland were unemployed, 33 per cent in December 1934 and 21·2 per cent in 1938. For the industry generally the percentage of unemployed, insured coal miners in these years was 34·0, 28·1 and 15·5. In 1938 the percentage of unemployment in the Cumberland coalfield was the highest in all the mining districts in the country.*[68]

The hardships endured at this time were felt most in Maryport whose plight was truly desperate. Here iron manufacture and shipbuilding had ceased, neighbouring collieries had closed down and the opening of a large dock at Workington in 1927 had killed the trade of Maryport, once the busiest port on the Cumberland coast. It was said of Maryport in 1933 that, "The town is, for the most part, living on public funds."[69] In the following year 57·3 per cent of its insured population were out of work and 1,684 men, mainly colliers, were wholly unemployed.[70] By 1937 the worst of the depression was over, the coal mining industry was beginning to recover and miners were slowly returning to work, but the level of unemployment of mineworkers was still cruelly high.

Miners who lost their jobs tended to remain unemployed for long periods. Mining was largely an hereditary occupation and many miners never thought of a career outside the pits – if a pit closed a miner and his sons would seek work in other collieries or remain unemployed. Further, collieries were often situated in isolated villages where alternative employment did not exist and in a period of depression the miners were almost unemployable outside the mines. An almost total lack of facilities to re-train miners for other work helped to prolong unemployment. Attempts were made after 1928 to transfer unemployed Cumberland miners to Kent and other coalfields. These attempts failed, most of the Cumbrians quickly returning home because they disliked the working conditions in other coalfields and, above all, because they were homesick.[71]

The decrease in the demand for coal meant that there was a surplus of manpower in

* The number of unemployed mineworkers in West Cumberland fell from 3,351 in November 1934 to 1,703 in September 1938. (*Report of the Commissioner for the Special Areas in England and Wales for the year ended 30th September 1938*, Cmd. 5896 (1938) 104).

the coalfield destined to be unemployed for many years. Intermittent production in some pits, particularly at Whitehaven where over 2,000 miners were employed, aggravated the problem of unemployment caused by the permanent closure of other pits. Both Haig and Wellington Pits, which were connected to each other, were closed for two months following an explosion at Haig Pit in 1927. Precisely the same thing happened in 1931 and on each occasion about 1,400 men were thrown out of work. At William Pit there was no production between September 1928 and December 1929 and, even worse, all work at Whitehaven Colliery stopped in October 1935, the colliery remaining idle until March 1937. Under-employment was also a feature of the depression period. When the coal trade was slack the miner's working week was often restricted to four shifts or less and his meagre wage packet became even more slender.

Those who survived this era of appalling suffering and despair still remember, with an almost corrosive bitterness, the poverty they were compelled to endure, often for years at a time, and the blank monotony of prolonged unemployment. In the grey streets of the mining villages grey faced ex-miners and their families went hungry and aged prematurely; if their bellies were empty the pawnbrokers' shops were full. Enormous physical and psychological damage was inflicted on the unemployed. Poverty, malnutrition and bad housing conditions lowered resistance to disease. Many miners lived in old and damp terraced cottages, often back to back, without baths, an indoor water supply or indoor sanitation. In these slum conditions disease flourished and in Cumberland the rates for tuberculosis and infant mortality were both considerably above the national average.

It is not surprising that the county population fell from 273,173 to 263,151 in 1931 and to an estimated 255,280 in mid-1939.[72] The chief cause of this decline was migration to other areas, particularly by younger men seeking to escape from what appeared to be permanent unemployment. Some found work in other British coalfields and some in the mines of Canada and South Africa. A few enlisted in the armed forces where there was at least a certainty of food, shelter, clothing and a little money.

One act of compassion by a great landowner should be recorded. In 1932 Lord Leconfield allowed outcrop coal at Broughton Moor to be worked by unemployed miners of that area in order that they might obtain fuel for their own use. For this coal the men paid a nominal royalty of 1d. per ton and in addition they had to promise not to sell the coal or work it deeper than six feet from the surface. This concession ensured that those who took advantage of it at least had warmth in their homes.[73]

Between 1914 and 1939 there were wide variations in the average earnings of miners. During the First World War and the early post-war years when coal was urgently needed wages rose sharply. Because the general cost of living by 1 August 1917 had risen 80 per cent since the outbreak of war and miners' wages by only about 40 per cent the Miners' Federation sought an increase in wages for its members. The government, then in charge of the industry, awarded an increase from 17 September 1917 of 1s. 6d. per day to all workmen aged sixteen and over and 9d. per day to those under sixteen. An exactly similar advance was granted from 30 June 1918. These were national, flat rate increases given by the government, not district increases conceded by colliery owners. Flat rate increases lessened the differential between the wages of the lower paid and those of the more highly skilled workers. Some colliery managers disliked these additional payments since they did not accord with the principle of payments by results.[74] A comparison of the average earnings per shift of mineworkers in the United Kingdom for the month of November in 1913 and in 1918 shows that earnings had almost doubled in the intervening period:[75]

	November 1913		November 1918	
	s.	*d.*	*s.*	*d.*
All underground workers	7	1	13	6
All surface workers	4	8	9	5
All workpeople	6	7	12	6

In the Cumberland coalfield the average wage per shift in November 1918 was as follows:[76]

	s.	*d.*
Underground Adults		
Piece-work coal getters	17	6·25
Coal getters on day wage	12	8·41
Putters, fillers, hauliers and trammers	13	4·36
Timbermen, stonemen, brushers and rippers	14	2·23
Deputies, firemen and examiners	14	4·85
Other underground labour	12	9·80
Total Adult labour	15	10·45
Surface Adults		
Winding enginemen	12	4·64
Other enginemen	10	11·37
Stokers and boiler men	10	0·09
Pitheadmen	10	11·78
Persons on or about screens	10	1·37
Tradesmen i.e. mechanics, joiners, blacksmiths, etc.	11	8·27
All other surface labour	10	1·34
Total Adult Labour	10	8·46
Youths and Boys		
Underground	7	10·57
Surface	5	4·65
Women and Girls	6	8·29

Early in 1919, while the industry was enjoying boom conditions, the Miners' Federation put forward sweeping demands – the nationalization of the mines, a six instead of an eight hour day and a 30 per cent increase in gross wages, exclusive of the 3s. per day war wage. Terms offered by the government were rejected and the miners voted almost six to one in favour of a national strike; in Cumberland 10,579 miners voted for the strike and 913 against.[77] To avert a strike the Sankey Commission was set up to investigate the miners' claims. An interim report, immediately accepted by the government, gave the miners a seven hour day and a flat rate increase in wages of 2s. per day for those aged sixteen and over and 1s. per day for those under sixteen. From 16 July 1919 pieceworkers received an advance of 14·2 per cent in their wages to compensate for the loss of earnings resulting from the reduction of working hours from eight to seven per day. Hours of surface workers were reduced to 46½ per week, exclusive of meal times.[78] Colliery owners resented having to pay not only the general increase in wages but also the same wages for a seven hour day as they did formerly for an eight hour day. Much to the disgust of the miners, to whom by this time nationalization had become almost an article of faith, nothing was done to nationalize the mines.

As the cost of living continued to rise sharply further wage concessions became necessary. A 20 per cent advance on gross wages, exclusive of the "War Wage" and the "Sankey Wage", was awarded to all colliery workers from 12 March 1920, subject to a minimum increase of 2s. per day for those aged eighteen and over, 1s. for those aged sixteen and under eighteen and 9d. for those under sixteen.[79]

In May the government raised substantially the prices of domestic and industrial coals and in so doing added to the still growing cost of living. To resist this the miners demanded higher wages and a reduction in coal prices. When these demands were rejected the miners voted in favour of a national strike; in the Cumberland district they voted 6,677 for and 1,802 against strike action.[80] The strike, postponed for a time to allow further negotiations between the miners and the government, began 16 October and lasted until 3 November. Negotiations continued whilst the strike was in progress and terms were agreed. From 4 November 1920 the daily wages of mineworkers aged eighteen and over were to be increased by 2s., those aged sixteen and under eighteen by 1s. and those under sixteen by 9d. This increase was conditional upon output reaching a certain level, the "datum line". Further increases, payable after 3 January 1921, would depend upon further increases of output. This was a temporary arrangement and a permanent scheme for the settlement of wages was to be worked out by 31 March 1921.[81] The miners returned to work after voting upon these terms; the Cumberland miners voted 3,547 for and 2,853 against resuming work.[82]

Negotiations soon began between the miners and the colliery owners about the new wages scheme. The miners wanted a National Wages Board and a National Pool of Profits which would ensure that the level of wages would depend upon the national financial position of the industry and not upon that of individual districts; in this way efficient and prosperous collieries would subsidize the less efficient and poorer collieries. The mine owners wished to abandon national wage agreements and to return to district wage settlements. With the collapse of coal export prices in the winter of 1920-1 smooth and orderly negotiations were no longer possible for the government found that they were running the industry at a loss of about £60 million per annum and therefore announced their intention to hand back the collieries to their owners at the end of March, instead of at the end of August as previously arranged. The owners had now to negotiate with the men a wages settlement to operate after 31 March and in view of the heavy losses incurred by the mining industry substantial wage reductions were certain to take place.[83] The government's announcement came as a tremendous shock to the miners who now stood to lose the wage advances gained during the period of control. In each district the local association of colliery owners gave the men notice to quit and put forward new terms of employment. Although the miners realized that wages would be reduced they were not prepared for the drastic wage cuts proposed by the employers. A national stoppage of the industry was inevitable. It was calculated that a Cumberland miner earning £4. 15s. 4d. for working 5½ shifts per week in March 1921 would, under the owners' new proposals, earn only £2. 11s. 9½d., little more than half his previous wages.*[84] The lock-out of the miners began 31 March and lasted until 1 July. In Cumberland the local Miners' Association provided financial assistance for its members – 15s. per week for each miner and 2s. per child. Its funds dwindled rapidly and within six weeks the allowances had to be reduced to 2s. 6d. per man and 1s. per child.[85] A government offer of a £10 million subsidy to prevent large reductions in wages was rejected by the miners, the Cumberland miners voting 5,168 to 1,548 against the offer.[86] However, the miners, after exhausting their funds had to admit defeat and return to work on the employers' terms. Not only did they have to accept wage cuts but they had also to abandon a national wages system in favour of district wage agreements.

Following the 1921 stoppage a new system was devised and to apply it the coalfields were divided into thirteen wage districts of which the Cumberland coalfield was one. The old sliding scale system, based upon selling prices, did not allow for changes in production costs or in the volume of trade and therefore did not allow for changes in profits. Under the new system the miners' wages consisted of two elements – the basis rate and the percentage addition to it. The basis rate was fixed for each miner and depended upon his grade and place of work; accordingly it could vary from pit to pit or from seam to seam. The percentage addition was settled in each district after periodical ascertainments of the "net proceeds", i.e. the sum remaining from the gross proceeds after deducting the cost of standard wages,† the cost of standard profits (equal to 17 per cent of standard wages) and the costs of production other than wages. The net proceeds were then divided between profits and wages in a fixed proportion, the proportion going to wages being the percentage addition to the basis wage. The agreement of 1 July 1921 also embodied a minimum wage formula; in no district, whatever the result of the ascertainment, were wages to be paid at rates less than 20 per cent above the "standard", i.e. above a level approximately equal to the rates in force in 1914. Further, where the rate of wages determined in any district did not provide a subsistence wage to low-paid day wage workers additions to it would be made in the form of an additional allowance per shift, such allowances to be determined by district boards, or, on failure to agree, by an independent chairman.[87]

Disappointed though the miners were by the terms of the agreement they did derive some benefit from it. They were now to share in the profits of the industry for their wages would no longer depend upon the selling price of coal alone but would be related to profitability. Miners working in high cost and less efficient collieries would share in the proceeds of the most efficient and prosperous collieries within their districts. They had once hoped for a national pool of profits to determine the level of wages, now they had at least a district pool of profits.

An examination of the average earnings per shift and per annum of the mineworkers shows wide differences between earnings before and after 1921:[88]

* In 1921 average earnings per shift in Cumberland fell from 18s. 10¾d. in the quarter Jan. - March to 10s. 0¾d. in the quarter Oct.-Dec. (*Report of Chief Inspector of Mines, 1921.*)

† Standard wages were the district basis rates existing 31 March 1921, plus the district percentages payable in July 1914, plus, in the case of pieceworkers, the percentage additions which were made consequent upon the reduction of hours from eight to seven.

	Per shift				Per annum					
	Cumberland		Great Britain		Cumberland			Great Britain		
	s.	d.	s.	d.	£	s.	d.	£	s.	d.
1919	14	4			212	0	0	191	7	5
1920	17	1	16	10¼	235	14	1	223	6	8
1922	9	10½	9	11¾	124	3	8	125	2	0
1923	9	4½	10	1	122	13	5	134	6	1

After 1920 average earnings fell sharply and did not again reach the level of 1920 until 1942.

In September 1923 the average gross earnings per shift of mineworkers in Cumberland amounted to:[89]

	s.	d.
Underground Adults		
Coal getters	12	4·89
Putters, fillers, hauliers and trammers	7	5·24
Timbermen, stonemen, brushers and rippers	11	1·00
Deputies, firemen and examiners	11	7·84
Other underground labour	8	2·78
Total adult underground labour	11	2·81
Surface Adults		
Winding enginemen	10	2·39
Other enginemen	8	3·22
Stokers and boiler men	7	7·59
Pitheadmen	7	11·96
Persons on and about screens	7	1·73
Tradesmen i.e. mechanics, joiners, blacksmiths and other skilled workmen	8	2·59
All other surface labour	7	4·10
Total Surface Adult Labour	7	9·86
Total Adult Labour	10	5·88
Boys		
Underground	4	10·70
Surface	2	11·78
Total Boys	4	4·37
Women and Girls	3	6·81
Aged and Infirm Persons	6	1·08
Persons on Light Work in Receipt of Compensation	6	6·55
Grand Total (All Workers)	9	2·40

Strikes at individual collieries were not uncommon and usually of short duration, but at Whitehaven in 1923 the 2,500 miners employed there went on strike for seventeen weeks. This followed an announcement from the owners, 1 May, that new methods of working were to be adopted at the Whitehaven pits. Working places were to be driven four yards wide instead of six yards as was the practice hitherto and new hewing prices were to be drawn up. H. Watson Smith, the colliery manager, believed narrower working places would lead to quicker development and help to preserve the roof. He refused to consider the men's view that they should have one-third in money value for the loss of one-third of their working places and asserted that it was for the owners alone to decide how the colliery should be worked. His autocratic and overbearing attitude did not help matters, nor did his unwise assumption that his subordinates were necessarily his inferiors. The strike began 15 May and eight days later Smith informed a miners' deputation that:

1. The Company reserved the right to operate any method they thought best for the working of the pits.
2. All customs and systems prejudicial to the working of the colliery were to be abolished.
3. Cavilling was to be abolished.
4. There was to be a uniform scale of confiscation of coal for dirt sent up the pit, the scale being that in use in the Main Band of Wellington Pit.
5. The Company would have the right to choose any tub or as many tubs as they liked for picking purposes.
6. Any man absent for two days in any week without a reasonable cause would be dismissed.

If these conditions were accepted the question of hewing prices could be referred to the Cumberland Conciliation Board. The real issue between the men and their employers was the abolition of cavilling. This was an old custom by which the men drew lots for working places at the end of every quarter; it was a matter of chance who got the best working places but the system did ensure some sort of equality and justice. To the employers cavilling cost time and lost production at the beginning of every quarter when the miners had to accustom themselves to new working places. To the miners cavilling was an old and cherished custom and they preferred a haphazard selection of working places to direction by officials who might be influenced by their likes and dislikes. In his dealings with the miners Smith proved adamant, his inflexibility prolonging and embittering the strike. His policy seemed to be one of starving the men into surrender.

After the breakdown of negotiations riots occurred at Whitehaven (2 and 3 July) beginning with the smashing of the colliery office windows, followed by attacks on the fire station and police station and the looting of shops. Police were stoned by the mob; for nearly five hours there was continuous conflict between police and public. Additional police contingents arrived from Yorkshire and Lancashire to quell the rioters and about sixty policemen and 200 rioters and spectators were injured. Order was restored and sixty-three persons, including seven women, were charged with the offence of riot and committed for trial at the next Carlisle Assizes. Of these, twenty-four were discharged and the rest sentenced to terms of imprisonment varying from one to twelve months.

Negotiations were resumed between the management and the men but they were fruitless. Eventually (24 August) the miners accepted the Company's suggestion of arbitration and agreed in the meantime to resume work on the old terms and conditions. Sir A. L. Dickinson was appointed arbitrator. He determined:

1. The Company have the right to work the pits by any system or method they decide.
2. Cavilling to end at the ballot of Christmas 1924.
3. Management to have the right to test any tub at their option and to impose a fine of 1s. for every 18 lbs. of dirt in a tub with a further 1s. fine for each additional 18 lbs. of dirt.
4. A workman absent for two days in one week without satisfactory reason would be deemed to have broken his contract of service.
5. Tonnage payments for cutting coal were fixed for the different seams and the hewers were to receive additional payments for setting timber and for each yard the working place advanced.

The men returned to work in the second week in September after suffering much hardship, many of them having been compelled to apply for poor relief.[90]

Differences about wages between the miners and the coal owners continued on a national level in 1924. After lengthy negotiations the 1921 national wages agreement was modified in June 1924. Standard profits were reduced from 17 to 15 per cent of standard wages and the miners' proportion of surplus proceeds was raised from 83 to 88 per cent. The minimum addition to standard wages was raised from 20 to $33\frac{1}{3}$ per cent and subsistence wages were raised by one-eighth.[91]

As coal prices fell in 1925 much of the mining industry was being carried on at a loss and at the end of June the owners announced their intention to end the 1924 wages agreement. What they wanted was a reduction in wages and a return to the eight hour day. The Miners' Federation refused to meet the owners to consider the latter's new proposals or to participate in a Court of Inquiry set up by the government to investigate the dispute. A threat by the Trades Union Congress to support the miners by calling a general strike compelled the government to offer a temporary subsidy to enable the colliery owners to continue to pay existing wage rates and to set up a Royal Commission to inquire into the industry. The government subsidy, which helped to postpone the clash over wages and hours, cost £23 million, equivalent to 2s. 9d. per ton on all coal sold during the period of subvention; for Cumberland the subsidy amounted to 5s. 6d. per ton.[92]

The report of the Samuel Commission (published 10 March 1926) did nothing to ease the situation. Its recommendations that coal royalties but not the mines should be nationalized, that existing working hours should be maintained and that some wage cuts were desirable satisfied neither side. Meetings between the Miners' Federation and the Mining Association in March and April showed that agreement was impossible; the owners demanded wage reductions and a longer working day, the miners refused to make any concessions. The owners announced that the existing terms of employment would cease 30 April, the day on which the government subsidy would end. The miners were now locked out and the General Council of the T.U.C. called for a general strike in their support from midnight 3 May. Nine days later the general strike was called off unconditionally by the T.U.C., leaving the miners to fight their own battles.

For nearly eight months the miners obstinately remained on strike, tightening their belts as the months dragged on. They and their families suffered enormously, many becoming dependent upon poor relief. At the beginning of the strike the Cumberland Miners' Association paid 10s. per week to adults on strike together with a weekly allowance of 2s. per child. But by 18 May strike pay was reduced to 5s. per man and children's allowances were abolished. Only one Cumberland colliery, Dean Moor, a small, non-union pit, continued working throughout the strike. Despite their penury the rest of the miners were obdurate in their determination to hold out against the employers. However, even they had to succumb; prolonged poverty sapped their resistance and it was hunger that compelled their surrender. In October men started trickling back to work; by the middle of November about 250 miners had resumed work. In November representatives of the Cumberland miners and coal owners met twice to discuss terms for a return to work. The owners' final terms, and it was the owners who were able to dictate terms, stipulated:

1. Hours of work underground to be eight per shift, five days per week and $6\frac{1}{2}$ hours on Saturday.

2. Surface workers to work forty-nine hours per week.
3. Minimum percentage on basis rates to be 35 per cent until the end of March 1927.
4. Subsistence allowances up to a maximum of 6s. 10d. to be paid until the end of March 1927.
5. This agreement to last three years from 1 January 1927 and thereafter terminable by either of the parties giving three months' notice.

A ballot of the miners resulted in a vote of 2,124 for and 2,242 against accepting these terms, a majority of 118 preferring to continue the strike. This majority, according to the rules of the Cumberland Miners' Association, was insufficient and the strike was therefore ended 2 December, the miners returning to work on the following day.[93]

For the miners the strike ended in total failure; they had to accept the employers' terms – wage reductions, longer hours and district instead of national wage agreements. Union funds, national and local, were almost exhausted. It was reported in 1927 that the Cumberland Miners' Association during the previous seven years had paid £150,343 to support members in dispute with their employers and as a result was unable to provide strike pay for the miners at Clifton and Buckhill pits who were then resisting wage cuts.[94]

A little of the ground lost by the miners was regained in 1930 when the Labour government passed a Coal Mines Act which reduced working hours to 7½ per day. This was continued by the Mines Act of 1931, a disappointment to the miners who had assumed that following the expiry of the 1926 Eight Hour Act 8 July 1931 a seven hour day would automatically be restored.

Despite the wage reductions and longer working day imposed since the 1926 strike Cumberland coal owners were operating at a loss and at a meeting of the District Wages Board in June 1930 they brought forward a claim for a reduction in the minimum percentage payable on the basis rate from 30 to 15 and for reductions in the subsistence allowances. These proposals the miners' representatives refused to consider. The owners gave notice that the wages agreement would be terminated 30 November 1930 and for some months the collieries were worked under temporary agreements by which the hours of work were varied but which did not involve any wage changes.[95]

Continuing losses compelled the owners to raise the question of wage reductions again in 1931. In April 1931 the minimum wage rates paid to the various classes of workmen in the Cumberland coalfield were:[96]

	s.	*d.*
Pieceworkers:		
Hewers	8	0
Daywagemen:		
Deputies	8	1·47
Firemen, Shotfirers	7	10·77
Shiftmen	7	8·07
Rolleywaymen	6	5·15†
Shift labourers, Onsetters	6	2·46†
Horsekeepers, Pumpmen, Engine Drivers	5	7·70†
Oilers and Greasers	5	4·99†
Bogie Hands	5	0·92†
Boys		
Couplers, Drivers, Lamp Carriers, Pumpers, Clippers Engine Boys	2	8·49
Dilliers, Trailers, Brakers	3	8·68

† Subsistence Wages:

Male daywage workers aged 21 and upwards, 6s. 10d. per day.
Male daywage workers aged 18 and under 21, 6s. 6d. per day.

After further negotiations it was agreed that the wages dispute should be submitted to the National Industrial Board set up under the Coal Mines Act of 1930, the miners' representatives undertaking to accept the Board's report, whatever it might be, and to recommend its acceptance to the men. The case was heard by the Board, 7 and 8 May, the owners again claiming a reduction in the minimum percentage from 30 to 15 and a reduction in the subsistence wage for adult daywage workers from 6s. 10d. to 6s. 6d. per day. The Board's report, issued 19 May, recommended a reduction of 7½ per cent in the minimum percentage rates but no change in the subsistence rates. At a formal meeting of the District Wages Board ten days later to receive and adopt the report the miners' representatives went back on their earlier promises and refused to accept the report or to recommend its adoption to the men.[97] Several ballots were then held by the Cumberland Miners' Association to decide whether the new terms should be accepted. The ballot of 3 June showed 3,734 against and 520 for the National Board's report. In a second ballot, 18 June, 3,404 miners voted for strike action and 1,590 against; this was seventy-four over a two-thirds majority necessary for a strike. An analysis of the voting figures showed that four collieries had voted against a strike, three others did not have a two-thirds majority and one or two large collieries did not have a very large margin over two-thirds. A third ballot was taken in which 670 voted for and 1,464 against the new terms. The executive council of the union decided not to publish the figures but to say that over 5,000 had voted and that there was above a two-thirds majority for a stoppage of work. The men at Whitehaven, Clifton and Moresby had not voted; nevertheless it was decided to add the votes of the lodges which had not voted to those of the men who had voted in favour of a strike, giving a figure of 5,031 for a strike and 1,463 against.[98]

That the miners' leaders had broken their commitment to accept the report of the National Board was bad enough, but their falsification of the results of the ballot was much worse. They were responsible for a strike which should never have begun and which had only lukewarm support from the miners themselves. The strike began 24 June and lasted until 13 August. It achieved nothing, the men having to accept the 7½ per cent reduction. The miners made the mistake of going on strike "with the sun on their backs" and they had very little sympathy from the public; it was generally recognized that the industry was in difficulties and that some wage cuts were necessary if the collieries were to survive. Although its funds were low the union provided its members with strike pay of 10s. per week together with an allowance of 1s. per child. Men began to return to work before the strike was officially ended; a reduced wage was preferable to strike pay and the humiliation of parish relief. When the strike ended not all the strikers were re-employed – 350 were paid off at Moresby, twenty at Buckhill and thirteen at Oughterside.[99]

Between 1931 and 1939 miners' wages improved slightly. From 11 June 1934 subsistence wages per shift in Cumberland were increased from 6s. 10d. to 7s. 1d. for men, from 6s. 6d. to 6s. 8·85d. for men aged 18 and under 21 and from 3s. 8d. to 3s. 9·61d. for women.[100] In the following year the Miners' Federation launched a campaign to secure a general increase of 2s. per day for men and 1s. for boys. A national ballot showed that an overwhelming majority of the miners were prepared to strike if necessary to enforce this claim; the Cumberland miners voted 2,175 to 211 in favour of a strike. The situation was eased when, at the request of the Mining Association, the major industries offered a voluntary increase of 1s. per ton on their contract prices for coal.

This enabled the coal owners to make concessions, not on a national basis as the miners wanted, but on a district basis and a strike was averted. In the different mining districts the men received flat rate increases of 6d. to 1s. per day; in Cumberland adults received 6d. and boys 3d. per day, the additions dating from 1 January 1936.[101]

The minimum percentage addition to the basis rates in Cumberland was raised from 22½ in 1936 to 27 in 1937 and to 30 from 17 May 1938. Subsistence wages were again increased in 1937; from 7s. 1d. to 7s. 3½d. for men, from 6s. 8·85d. to 6s. 10·25d. for youths and from 3s. 9·61d. to 3s. 10·61d. for women. From 16 October 1937 subsistence rates in Cumberland were merged into the basis rates.[102] Miners in the county obtained holidays with pay for the first time in 1938. Men got six days holiday per year and £3 pay, males under 21 and women got £1. 10s. 0d. holiday pay; in 1943 the payments were increased to £3. 15s. and £1. 17s. 6d. respectively.[103]

The average cash earnings per manshift of mineworkers in the Cumberland coalfield can be compared with those for the industry as a whole:[104]

Year	*Cumberland*		*Great Britain*	
	s.	*d.*	*s.*	*d.*
1918[a]	12	4½		
1919	14	4		
1920	17	1	16	10¼
1921[b]	14	5¾	15	11
1922	9	10½	10	0
1923	9	4½	10	1
1924	10	3½	10	8
1925	10	8	10	6
1926[c]	10	7½		
1927	9	0½	10	1
1928	8	8¾	9	4
1929	8	7¾	9	3
1930	8	9¼	9	4
1931	8	8¾	9	2
1932	8	9¼	9	2
1933	8	8	9	2
1934	8	8	9	2
1935	8	10¼	9	3
1936	9	7¼	10	0
1937	9	11¾	10	8
1938	10	9¼	11	3

[a] July-December 1918.
[b] Average of quarters Jan.-March and Oct.-Dec.
[c] August 1925-April 1926 inclusive.

As can be seen, earnings fluctuated between 1918 and 1927, remained stable between 1928 and 1934 and thereafter increased annually. The Coal Mines Act of 1930 contributed to the stability of earnings by introducing a system of price control which made it possible to maintain prices; in the latter years of the decade as prices increased so did earnings. Throughout this period, except for 1920 and 1925, the average earnings of Cumberland mineworkers were below the national average. Figures of earnings per manshift are, however, an incomplete guide to the amount of total earnings which depended upon the number of shifts worked; when the demand for coal declined full-time work was not available. During the depression period short time working was common in the Cumberland coalfield and imposed severe hardship on all mineworkers, especially those on minimum wages or subsistence wages. Those who worked four shifts or less per week could not earn a decent living and their wages were further diminished by "offtakes"* which were the same in amount whether the miners worked one or six shifts per week. When few shifts were worked these offtakes came to a large proportion of the small weekly earnings.

Statistics of average annual earnings provide a better guide to the financial position of the mineworkers than do figures of average earnings per shift.

Average Annual Earnings of Cumberland Mineworkers[105]

Year	*£*	*s.*	*d.*	*Year*	*£*	*s.*	*d.*
1919	212	0	0	1932	113	2	4
1920	235	14	1	1933	115	1	0
1922	124	3	8	1934	120	10	4
1923	122	13	5	1935	123	19	0
1924	131	2	10	1936	136	10	0
1925	126	16	6	1937	142	5	4
				1938	147	11	8

Earnings slumped after the peak year of 1920; in 1922-3 they were little more than half and in 1932-3 less than half the amount earned in 1920. After 1932 there was a continuous if small increase in annual earnings. With such low incomes the miners were bound to suffer although their sufferings were mitigated to some extent by the fall in the cost of living.

Since "average earnings" take into account the earnings of all mineworkers – adults and boys, skilled and unskilled, the highest and lowest paid – they shed little light on the earnings of each of the many categories of workers. It is unfortunate that statistics of the earnings of each grade of mineworkers were not compiled.

1939-1946

As the war progressed the government had to intervene to ensure an adequate supply of manpower for the mining industry. By the Essential Work Order, 15 May 1941, every colliery became a scheduled undertaking and no man could leave or be dismissed from his employment without the consent of the National Service Officer. The miners were thus bound to the pits, a restriction they disliked intensely. At the same time they were guaranteed a weekly wage whether short-time was being worked or not; if the miners could not leave their jobs they had, in fairness, to be sure of a weekly wage.[106] To check

* Offtakes were deducted weekly from the miners' wages to pay the checkweighman, family doctor, meet the costs of explosives, lamps, pick sharpening and to maintain the local Miners' Institute.

the continuing decline in manpower and output conscripts in 1943 were given a choice of entering the mines or serving in the armed forces. Since this added insufficient recruits to the industry a form of industrial conscription, devised by Ernest Bevin, was adopted. In a new kind of lottery ten per cent of the men called up were directed to work in the pits.

To meet increases in the cost of living flat rate additions were made to the wages of all mineworkers on a national basis. These war additions, agreed between the Miners' Federation and the colliery owners, amounted to 8d. per shift for adult workers from 1 November 1939, with a further 5d. per shift from 1 January 1940. An agreement between the workers and the employers, 20 March 1940, laid down how these additions should be calculated. For every increase of one point in the cost of living index wages were to be raised by 0.7d. per shift for adult workers and these wage increases were to be paid nationally. Later cost of living increases per shift were:[107]

1	April	1940	4d.	1	Jan.	1941	6d.
1	Oct.	1940	5d.	1	July	1941	4d.

In May 1941 the miners failed to obtain a national £4 per week minimum wage but they did obtain an "attendance bonus", payable from 1 June, of 1s. per shift for adults and 6d. for others provided they worked a full working week. The proviso irritated the miners and was abandoned in September, the attendance bonus becoming a flat rate wage increase. Despite their wage advances since the outbreak of war the miners were being paid less than many factory workers and considering the dangers and discomfort in the pits they had reason to be much dissatisfied. The Miners' Federation therefore applied in May 1942 for a national wage increase of 4s. per shift for adults and 2s. for others and a national minimum wage of £4. 5s. per week for all adult mineworkers. This led to the Greene Board of Inquiry to investigate miners' wages. It recommended (18 June 1942) an addition of 2s. 6d. per shift for all adults and for all underground workers between 18 and 21, graduated additions for all underground workers under 18 and surface workers under 21 and a national minimum wage of 83s. per week for adult underground workers and 78s. for adult surface workers.

This award, estimated by the Board to cost £23 million, was unconditional, i.e. the increases were not dependent upon output or attendance. The immediate and substantial advance in wages, together with the establishment for the first time of a national minimum weekly wage for adult mineworkers, greatly reduced discontent in the coalfields.[108]

The Greene Board's third report (March 1943) proposed that a new National Conciliation Board should be created to replace the Joint Standing Consultative Committee which had been set up in 1936. This report was accepted by both sides of the industry and the new board quickly came into existence. It consisted of two bodies – a Joint National Negotiating Committee composed of equal numbers of Miners' Federation and Mining Association representatives and a National Reference Tribunal of three members unconnected with coal mining. If a national dispute could not be settled by the Negotiating Committee it would be referred to the National Tribunal which was empowered to make a final decision.[109]

The Tribunal had soon to make an important wages settlement. In December the miners put in a claim for a national minimum wage of £6 per week for underground workers and £5. 10s. for those on the surface. The Tribunal, with Lord Porter as Chairman, raised the national minimum weekly wage to £5 for adult underground workers and to £4. 10s. for surface workers, an increase of 17s. and 12s. respectively since the Greene award. These increases were necessary to placate the miners whose position in the table of industrial earnings had fallen from 23rd to 41st since June 1942 and to attract men to the pits where there was a serious shortage of labour.[110] The Porter award (22 January 1944) made no alteration to piecework rates and as a result reduced still further the pay differential between skilled and unskilled workers. Some pieceworkers gained nothing from the large increases in minimum wages and objected to such increases going to others.

A major revision of the industry's wage structure was now long overdue. A new national wages agreement to last for four years was accepted by the miners and colliery owners 20 April 1944. It was designed to increase productivity, reduce labour disputes and to reassure the miners by guaranteeing minimum wage rates for a long period. The ascertainments system was suspended for the duration of the agreement and the percentage additions were merged into the minimum basic wage. The flat rate additions to wages during the war had reduced the proportion of the pieceworker's earnings which depended upon his personal efforts and at the same time lessened the differential between his wages and those of daywage men. These factors contributed to a decline in productivity and to overcome this these additions (except the wartime cost-of-living bonus) were merged into the percentage increase in the tonnage or yardage piece rates in every district.* The former flat rate element in the pieceworker's wages could now be increased by additional effort on his part whereas the daywage worker was restricted to a fixed rate per shift. This, of course, would do something to restore the wage differential between the two kinds of workers.[111]

The abolition of the ascertainments system was welcomed by the miners who for years had believed they were unfairly treated by it; the minimum percentage addition was rarely improved upon and the profits from by-products, which were often high, did not come into the reckoning of district proceeds.[112]

During the war the miners' economic position was vastly improved. Their wage rates rose much faster than those in other industries and in no other industry was the guaranteed weekly minimum wage as high as the £5 Porter award. Between 1938 and 1944 the mineworkers rose from 81st to 14th in the list of earnings of about a hundred industries. Since the advance in their earnings exceeded the increase in the cost of living the miners obtained higher real wages. (Between 1939 and 1945 the cost of living index rose from 88 to 112·5 and the wage rates of mineworkers from 109 to 222).[113]

In this period the average earnings per shift of mineworkers in Cumberland were more than doubled:[114]

Year	*s.*	*d.*	*Year*	*s.*	*d.*
1939	10	11	1943	18	10¼
1940	12	3¾	1944	22	2
1941	14	4½	1945	24	1¼
1942	17	2½	1946	25	0¼

* For Cumberland the percentage addition which was merged into the consolidated wage was 30 and the percentage applicable to piece rates was 48. (R. P. Arnot, *The Miners in Crisis and War* (1961), 402.)

Between the wars there was little improvement in the lamentable accident record of the Cumberland coalfield. Many miners were killed by explosions of fire-damp, particularly in the Whitehaven pits, which resulted all too often from the carelessness of shot-firers:[115]

Date	*Pit*	*No. Killed*
26 April 1915	Brayton No. 4	7
6 December 1920	Wellington	3
11 July 1922	William	2
27 November 1922	St. Helens	6
5 September 1922	Haig	39
13 December 1927	Haig	4
12 February 1928	Haig	13
29 January 1931	Haig	27
9 December 1946	Harrington No. 10	15

The explosion at Brayton followed the firing of a shot in the Yard Band seam which ignited gas in the goaf. Eight men were severely burned, seven of whom died of their injuries in the next few weeks. The Inspector of Mines who investigated the accident reported that the explosive used, gelignite, was most unsuitable for the purpose, that the shot was fired next to a goaf the condition of which was unknown, and the attempt to bring down a rib of coal which was almost entirely undermined should not have been allowed. In his view, "had the two officials concerned, one a deputy and the other a shot-firer, possessed abilities equal to their responsibilities, the accident would not have happened."[116]

When three stone-drifters were killed at Wellington Pit by a gas explosion after blasting operations in 1920 the Inspector of Mines asserted that, "The judgement of both the shot-firer and the chargeman was at fault, first in drilling the hole at this point and then in firing it after it had been drilled. There can be no doubt that fire-damp was present if not at the face at any rate in the cavity above the shot, and that the shot, being badly placed and, therefore, overcharged through having so little work to do, ignited the gas and caused the explosion."[117] After a similar explosion at William Pit in July 1922 which killed two men it became evident that not enough had been done in testing for gas and the Divisional Inspector accordingly delivered a sharp warning to the Whitehaven Colliery Company, 24 August 1922,

> "Gentlemen – I am desired to communicate with you with reference to the explosion which took place last month in your William Pit and the explosion which occurred in your Wellington Pit in December 1920. In both cases it was the fortunate absence of coal dust which prevented a great disaster. Both cases also were due to incompetence or want of proper care on the part of men employed as shot-firers.
>
> I have therefore to point out to you that it would appear that sufficient care is not being taken in the choosing of men to carry out the responsible duties attached to the firing of shots. I have to suggest to you that the Managers of the respective collieries should call together all their foremen and shot-firers and instil into them the absolute necessity of carrying out to the full all the legal requirements, and especially those in regard to shot-firing."[118]

A fortnight after this warning thirty-nine men were killed at Haig Pit by an explosion of fire-damp ignited by the flame from a shot and propogated by coal dust. The shot-firer, who mishandled his task, was himself killed. It emerged in the subsequent inquiry that owing to the absence of a deputy on holiday the two remaining deputies in the district where the explosion took place were working twelve hour shifts instead of eight hours fixed as the normal maximum under the Coal Mines Regulation Act 1908 and the Coal Mines Act 1919. The explosion was made worse by the presence of coal dust which the management had mistakenly believed to be harmless because of its dampness.[119]

Haig Pit claimed four more victims 13 December 1927 in an undersea working $2\frac{1}{2}$ miles from the shaft. Fortunately the explosion occurred about 10 p.m. when the afternoon shift had ended and the night shift men were beginning to enter the pit and had not proceeded very far. Had the accident taken place an hour earlier or later at least 200 men would have been involved in the explosion and the fire which resulted from it. It proved impossible to put out the fire and it was decided to seal off two districts of the pit in the hope that by cutting off the air the fire would die out. For two months the pit was idle. In February 1928 preparations were made to attempt to recover the sealed area and to resume work at the pit. After, an inspection of the mine doors were made in some of the stoppings and air tests made after proceeding through the doorway of the main stopping. But 12 February thirteen members of a party of twenty-four engaged in exploration work were killed in a series of explosions. The dead included the works manager, an under-manager, an overman, six deputies and two Inspectors of Mines. Again the danger area of the pit had to be walled off.[120]

Three years later twenty-seven out of 169 men working underground in the same pit died in an explosion of fire-damp propogated by coal dust. According to the official report fire-damp was ignited either by a spark within the exploder or across the conductors of the firing cable or by a spark resulting from contact between an ejected Cardox cartridge and a steel tub.[121] The Cardox carbon dioxide cartridge, which obtained its blasting effect from the rapid gasification of liquid carbon dioxide, was approved in 1930 and brought into use at the Whitehaven pits in the same year.[122] After this explosion the Cardox cartridge was modified so that the discharge of the gas tended to pull the cartridge into the hole instead of firing it out.[123]

At William Pit, one of the most dangerous in the country, twelve men were killed in 1941. Since 1928 the Main Band seam had been sealed off and working concentrated in the Bannock band. From time to time fires occurred in the sealed off area and water was applied to the fire via holes bored through the sealing wall. An explosion took place 3 June 1941, not of fire-damp but of inflammable gas produced by the application of water to a considerable mass of glowing coal in the Main Band seam. After the explosion the Bannock Band seam was abandoned and sealed off as well as the Main Band seam and work was started in the Six Quarters seam.[124]

An explosion of fire-damp, 9 December 1946, killed fifteen men at Harrington No. 10 pit. These men had been sent to clear an accumulation of gas in No. 2 district of the Main Band about $3\frac{1}{2}$ miles from the shaft and presumably the accident occurred while they were thus engaged. But how the gas was ignited the official inquiry was unable to establish.[125]

Falls of roof, runaway tubs and haulage accidents were also responsible for many deaths and serious injuries. Cumberland had the highest accident rate in Great Britain in both 1922 and 1923; if the general accident rate for the country was 100 the percentage for Cumberland in 1922 was 117 and 110 in 1923.[126] It is a tragedy that many miners were killed or maimed unnecessarily; had the safety regulations been observed at all times and had the men and officials taken care necessary for survival in a highly dangerous industry it is beyond question that the list of casualities could have been greatly reduced.

COSTS AND PROFITS

From 1914 until the end of government control in 1921 production cost rose sharply in every coalfield. In the Cumberland coalfield, a high cost area, production costs per ton of coal disposable commercially were:-[127]

Year	*Wages Cost*		*Total Cost*		*Wages Cost as % of Total Cost*
	s.	*d.*	*s.*	*d.*	
1917[a]	15	2¼	20	4	74·69
1918	18	9	24	6¾	76·33
1919	27	3	34	2¼	79·70
1920	33	7¾	43	6½	77·27
1921[b]	24	4¾	35	4¼	69·00

[a] July-December [b] Excluding months of April-June

Between 1917 and 1920 both the wages cost and total cost per ton were more than doubled, with wages absorbing a greater proportion of total costs. Large wage advances awarded during the war and the immediate post-war years increased costs directly and, by reducing the incentive to maximum effort on the part of the miners, contributed to the same end indirectly. The backward state of the mines, particularly the neglect of road maintenance above and below ground, the lack of plant and a high rate of absenteeism reduced output and raised costs. There was an enormous increase in the prices of such materials as timber, explosives, wire ropes, oats and hay. In the words of Sir R. A. S. Redmayne, "It would be within the mark to say that for materials alone the cost of production per ton of coal raised had trebled by the end of 1919 as compared with the period immediately preceding the outbreak of war."[128] Since colliery owners were guaranteed profits during the period of control it may be that they were then less concerned to curb rising costs than they otherwise might have been.

During the years 1917-21 the financial record of the Cumberland coalfield was one of annual losses which reached a peak of 12s. 6¼d. in 1920:-[129]

Year	*Proceeds Per Ton*		*Losses Per Ton*	
	s.	*d.*	*s.*	*d.*
1917[a]	17	3½	3	0½
1918	20	10	3	8¾
1919	25	8¼	8	6
1920	31	0¼	12	6¼
1921[b]	27	7¾	7	8½

[a] July-December

[b] Excluding months of April-June. These figures do not include a subvention of about 3s. 1d. per ton July-Sept.

Fortunately for the colliery owners the agreement of 17 July 1917 (subsequently confirmed by Act of Parliament which became law 6 February 1918) guaranteed a standard profit for each undertaking during the period of control, the colliery owner being allowed to choose the profit earned in any one of the three years preceding the war as his standard profit.[130] The survival of some Cumberland pits depended upon the government subsidy and when this was withdrawn continuing heavy losses compelled their owners to abandon production, e.g. the Risehow Colliery, Coking & By-Product Co. at Alice Pit in 1921 and at Bertha Pit in 1923. Considering the annual losses it is surprising that so many pits in Cumberland continued working as long as they did, but the owners, buoyed up by Micawberish hopes of better times to come, were always most reluctant to surrender their collieries.*

After 1921 production costs in Cumberland per ton of coal disposable commercially fell notably; by 1925 they were less than half of those in 1920:-[131]

Year	*Wages Cost*		*Total Cost*		*Wages Cost as % of Total Cost*
	s.	*d.*	*s.*	*d.*	
1922	15	11½	22	10¼	69·82
1923	15	2½	22	1	68·86
1924	15	11	22	7¼	70·41
1925	15	2	21	5¼	70·74
1926[a]	15	2	21	1	71·93

a August 1925-April 1926 inclusive

Economies in production costs were achieved largely at the expense of the miners whose earnings were substantially reduced. The closure of some of the least efficient pits also helped to lower costs in the coalfield as a whole. Further, when the era of guaranteed profits came to an end colliery owners had to make an extra effort to hold down costs and increase production. Nevertheless, despite the significant reduction in costs the Cumberland coalfield suffered financial losses year after year:-[132]

Year	*Proceeds Per Ton*		*Losses Per Ton*		*Profit Per Ton*	
	s.	*d.*	*s.*	*d.*	*s.*	*d.*
1922	20	1¼	2	9		
1923	20	5¾	1	7¼		
1924	21	1½	1	5¾		
1925	18	2	0	11¼		
1926[a]	16	9¾			1	3½

[a] August 1925-April 1926 inclusive

	Subventions: Cumberland	*Average All Districts*
1925	2s. 6¾d.	1s. 1½d.
1926	5s. 6¾d.	2s. 8¾d.

* In the Cumberland coalfield losses amounted to £219,766 in 1922 and to £138,822 in 1923. (*Report of the Royal Commission on the Coal Industry (1925)*, Cmd 2600 (1926), I 108.)

But for these subventions (which in Cumberland were twice the national average) the financial position would have been much worse.

The dismal record of annual losses continued and it was not until 1938 that the Cumberland coalfield showed a profit, the first since pre-war days. (The profit of 1s. 3½d. per ton in 1926 was not a true working profit.) High production costs, always above the average for all districts, made it almost impossible for colliery owners in Cumberland to mine coal at a profit. In 1938, when a profit was made, production costs per ton in Cumberland were the highest of all the mining districts in the country.[133]

Year	*Cumberland*				*Average all Districts*			
	Costs per Ton d.c.		*Balance + Profit − Loss*		*Costs per Ton d.c.*		*Balance + Profit − Loss*	
	s.	*d.*	*s.*	*d.*	*s.*	*d.*	*s.*	*d.*
1927	16	10¾	−1	2¾	15	7	−0	5¾
1928	17	4	−2	3¼	14	2¼	−0	11
1929	16	10	−0	11½	13	6½	+0	4½
1930	17	3½	−1	7½	13	8¾	+0	4¼
1931	17	10½	−2	1¼	13	8¾	+0	3½
1932	16	1	−1	3	13	8	+0	2
1933	15	10	−1	4½	13	3¾	+0	2¾
1934	16	0½	−1	10½	12	11½	+0	5
1935	15	9½	−1	5¼	12	11¾	+0	6¼
1936	15	11	−0	0¾	13	8	+0	11½
1937	18	2¾	−0	2¼	14	7¾	+1	2¾
1938	20	2¾	+0	6	16	0¼	+1	4

To reduce labour charges, by far the biggest single item in production costs, a large scale programme of mechanization was necessary. In Cumberland this was difficult to carry out. Frequent faults and variations in the thickness of seams, together with a lack of capital, delayed the adoption of mechanized mining methods. Some progress was made but it was not enough.[134]

Year	*% Coal Mechanically Cut*	*% Coal Mechanically Conveyed*	*Output Per Manshift Overall Cwts.*
1925	9	—	15·41
1930	15	3	16·38
1936	65	18	18·01
1938	72	34	17·23

Other factors helped to inflate losses. Strikes throughout the coalfield in 1920, 1921, 1926 and 1931 impaired profits and at Whitehaven Colliery the position was made worse by a local strike in 1923 which lasted for seventeen weeks. Serious explosions caused considerable losses in output and, in addition, entailed much expenditure to repair the damage. This was particularly true of Whitehaven Colliery where major explosions took place in 1920, 1922, 1927, 1928, 1931 and 1941 and in the pits concerned production was stopped for several months. To meet marketing requirements Cumberland coal had to bear high cleansing and preparation costs. After the elimination of waste and allowing 5 per cent for mine consumption it was said (in 1931) that no more than about 80 per cent of the tonnage raised and weighed at the pithead was commerically disposable. Excessive absenteeism and the low output per manshift were also blamed for the coalfield's poor financial performance.[135]

In the Irish market Cumberland coal had to meet increasing competition from Scottish and Lancashire coal. Between 1932 and 1935 an import duty on coal imposed by the Irish Free State government further reduced the export of Cumberland coal. In the county itself growing quantities of coal were being imported – in the Carlisle area from Northumberland and in south Cumberland from Yorkshire.

A measure of the problems facing colliery owners in Cumberland is the reduction in the number of mines at work from thirty-five in 1920 to twenty-five in 1938. Some of these pits were old pits whose closure, compelled by the exhaustion of their coal reserves, was merely a matter of time, e.g. Allhallows in 1928 and Aspatria No. 4 in 1933. Others were kept going for years by their owners despite annual lossese e.g. Buckhill Pit which the Allerdale Coal Co. had to close in 1932 because it could no longer afford such a drain upon its resources. Some mining companies had insufficient capital to finance the working and development of their collieries. Both the Risehow Colliery Co. and the Oughterside Coal Co. (1928) went into liquidation, the former in 1928 and the latter in 1930. It was a lack of capital which caused the Whitehaven pits to cease production in 1935. Collieries belonging to the United Steel Cos. Ltd. continued to survive because they were backed by the enormous resources of a firm capable of absorbing financial losses better than the other colliery owners in the country.[136]

In bad times colliery operators sought a reduction in the tonnnage rents payable to the colliery owner. This was done by the Whitehaven Colliery Co. in 1924, the Moresby Coal Co. in 1925 and the St Helens Co. in 1928. Lord Leconfield during the period 1926-32 accepted reduced rents in respect of Buckhill Colliery, a lower rent from a struggling colliery being preferable to no rent from a closed colliery.[137] Records of the Cumberland Coal Owners' Association show the financial difficulties of the collieries controlled by members of the Association:

Year	*Gross Output Tons*	*Trading Loss Amount*	*(coal commercially disposable) Per Ton*	
		£	*s.*	*d.*
1927	2,716,555	134,803	1	$2\frac{3}{4}$
1928	2,391,048	217,074	2	$3\frac{1}{4}$
1929	2,172,895	83,851		$11\frac{3}{4}$
1930	2,224,496	140,906	1	$7\frac{3}{4}$

Losses in Cumberland for the quarter ended 30 September 1930 amounted to 1s. 7·83d. per ton raised and weighed at the pit top. This was 1s. more than in Scotland, 1s. 5·4d. more than in Yorkshire, 1s. 6·83d. more than in Durham and 1s. 9·36d. more than in Northumberland.[138]

Colliery owners in Cumberland were handicapped by having to pay royalty rents which were on average the highest in Great Britain. In 1920 the following tonnage rents were payable from different collieries:[139]

Seam	*Whitehaven*	*Moresby*	*Clifton*	*St. Helens*
Main	1s. 4d.	10½d.	9 d.	10½d.
Bannock or Ten Quarters	1s. 2d.	9¼d.	7½d.	9½d.
Other	1s. 0d.	8 d.	7 d.	8 d.

The owners of St Helens Colliery had also to pay a wayleave and shaftage rent of 2d. per ton on the undersea coal. Since the Lowthers had invested enormous amounts of capital in the development of Whitehaven Colliery before leasing it to others the tonnage rents there were always higher than in other royalties belonging to the family. The average cost of royalties per ton in the Cumberland coalfield amounted to 10·8d. in 1922 and to 11·27d. in 1923.[140]

At Whitehaven, in an attempt to help the owners of the colliery, tonnage rents were reduced after 1926:[141]

Seam	*1927*	*1928*	*1933*
Main	1s. 0d.	9d.	8d.
Bannock	10d.	7d.	6d.
Other	8d.	5d.	5d.

During the period 1924-1938 the average royalty per ton of coal in Great Britain varied from 5¾d. to 6¼d.[142]

The system of royalty payments was always a source of grievance. Colliery owners could be compelled to pay the royalty owner excessive rents and wayleave charges which meant increased costs and lower profits. When trade was bad the colliery owner could suffer financial losses, but the landowner continued to receive, without effort or risk on his part, his income from annual and tonnage rents. The miners strongly resented payments to landowners who, they felt, contributed nothing to the industry and enjoyed an unearned income from a gift of nature. To the miners royalty payments meant dearer coal, lower wages for themselves and, in a period of depression, unemployment.

The income from all the coal royalties belonging to Lord Lonsdale approached £50,000 in 1917 but declined in later years:[143]

Year	£	*Year*	£
1914	44,023	1919	34,021
1915	43,817	1920	33,747
1916	46,947	1921	28,574
1917	49,493	1922	34,221
1918	34,242		

His revenue from Whitehaven Colliery, the largest of his coal royalties, began to shrink fast after 1930; in 1936 it was less than half what it was in 1924:[144]

Year	£	s.	d.	Year	£	s.	d.
1924	22,192	7	9	1931	17,550	8	7
1925	23,831	2	0	1932	12,409	6	5
1926	24,847	7	4	1933	17,484	15	5
1927	22,764	11	2	1934	10,508	16	0
1928	24,310	14	8	1935	12,188	12	8
1929	24,384	5	8	1936	10,456	10	9
1930	23,719	8	4				

When output fell there was a parallel decline in tonnage rents and if there was at the same time a reduction in the amount of royalty per ton of output the income of the royalty owner was bound to suffer. Whilst the mineral owner, unlike the miner, was not wholly dependent for his livelihood upon the mining industry, his standard of living varied with the fortunes of the industry. Compared with the Lowthers the income of the Curwens from coal royalties was modest. Between 1920 and 1929 their royalties at Harrington, Flimby and St. Helens yielded a total of £35,045, an average of £3,504 per annum during these ten years.[145] In the twenty years 1914-1933 Lord Leconfield obtained a total of £210,259 in colliery rents, an average of £10,512 per annum. Of this total sum three-quarters came from one colliery, the Brayton Domain Colliery near Aspatria which was worked by Joseph Harris.[146]

In Cumberland the coal owners faced many inescapable problems – the restrictions imposed by geological difficulties, the declining output from ageing collieries and therefore, as more accessible coal was used up, the necessity to mine coal at higher costs and increasing competition for markets in an age of economic depression. Nevertheless it would appear that there was some lack of enterprise and initiative, an unwillingness to invest in new mines or in the modernization and reorganization of existing mines in an endeavour to maintain output. At Whitehaven annual output fell sharply after 1928:[147]

Year	Output Tons	Year	Output Tons
1927	881,239	1931	435,362
1928	744,986	1932	419,609
1929	612,693	1933	386,569
1930	603,978		

Such a decline in output suggests that criticisms by experienced mining engineers of the Whitehaven Colliery Co.'s sluggish handling of the colliery were valid. It may be noted that when work at William Pit was suspended in September 1928 the weekly output amounted to 3,550 tons. The pit resumed production in December 1929, but by February 1930 its output was not more than about 590 tons per week.[148] Under different owners both this pit and the colliery as a whole were later to yield much greater quantities of coal. Since there was a substantial decline in tonnage rents the fall in output was a matter of concern to the lessor who, in 1933, took the unusual step of compelling the company to surrender the colliery in the hope that under new management better results might be obtained.

National and local strikes in this period exemplify the bitter antagonism which then existed between both sides of the mining industry. Although it is impossible to quantify the effects of bad labour relations upon mining costs it is certain that the friction between employers and employees in the coalfields was prejudicial to profits. Neither side seemed able or willing to understand the viewpoint of the other; employers objected to wage increases and miners to wage cuts, even when circumstances made these necessary if not desirable. Negotiations when they took place were too often vitiated by an atmosphere of suspicion and hostility which made a peaceful settlement of disputes difficult. Both sides can be blamed for the national strikes which did tremendous harm to the industry and damaged both earnings and profits.

The art of man-management was notably lacking in British coalfields at this time. Too many colliery owners and managers still adopted towards their workmen a dictatorial attitude reminiscent of the nineteenth century. Colliery managers, highly qualified and professionally competent as many of them were, were not always skilled in human relations. The strike at Whitehaven in 1923, even if it could not have been averted, should not have dragged on as long as it did. Cavilling, the major cause of the strike, did not end with the ballot of Christmas 1924, but continued until 1937 when it was replaced by a rota system. Presumably the departure of the colliery manager in September 1924 had something to do with this. In 1931 it was the miners' leaders who were at fault when they brought about an unnecessary strike throughout the Cumberland coalfield, a strike which did not have the wholehearted support of the miners themselves.

During the Second World War the financial record of the Cumberland coalfield was one of annually increasing costs and losses. Between 1939 and 1945 total costs per ton rose from 20s. 10d. to 56s. 10d., an increase of 172·8 per cent, and working losses from 6·64d. to 10s. 10d. Production costs in the county were well above the national average and in 1945 were the highest in Great Britain.

Production Costs per Ton d.c.[149]

Year	*Cumberland*				*Great Britain*			
	Wages Cost		*Total Costs*		*Wages Cost*		*Total Costs*	
	s.	*d.*	*s.*	*d.*	*s.*	*d.*	*s.*	*d.*
1939	14	3	20	10	10	10	16	5
1940	17	0	24	9	12	9	18	11
1941	21	7	32	2	15	1	22	3
1942	26	3	37	5	18	1	25	6
1943	30	4	43	0	20	3	28	2
1944	35	8	50	8	23	8	33	3
1945	39	7	56	10	25	5	35	11

Wages costs, always the biggest item in mining costs, rose from 14s. 3d. to 39s. 7d. per ton, an increase of 177·7 per cent. Wages as a proportion of total production costs amounted to 68·24 per cent in 1939 and 69·64 per cent in 1945. Other factors contributed to rising costs, e.g. the higher prices of stores and timber and the growth of "Other Costs", i.e. salaries, management, insurance and office expenses.

As the output of coal declined year by year fixed charges had to be spread over a smaller output and the fall in output per manshift overall in Cumberland from 0·84 tons in 1939 to 0·67 tons in 1945 added to costs. Absenteeism in the coalfield increased from 10·9 per cent in 1943 to 14·6 per cent in 1945. Particularly serious was the growing absenteeism of faceworkers which in 1945 amounted to 18·1 per cent. The resulting decline in output was made worse by the fall in the output per manshift of the faceworkers, from 2·23 tons in 1939 to 2·09 tons in 1945.[150] The wages of oncost men (i.e. men above ground not directly engaged in cutting coal), like the fixed charges, had still to be paid on a reduced output. Annual wage increases together with annual decreases in productivity could only lead to continuous rises in production costs. In Cumberland, as in other coalfields, the substantial increases in wages during the war failed to stimulate improvements in output or productivity.

In the early years of the war annual losses in the Cumberland coalfield were as follows:-[151]

Year	*Proceeds per Ton*		*Loss per Ton*	
	s.	*d.*	*s.*	*d.*
1939	20	1·25	0	6·64
1940	23	7·36	0	10·34
1941	30	5·64	1	3·56

Throughout the war colliery owners in some coalfields found it impossible to meet the rising costs of production despite increases in the price of coal and their survival depended upon outside financial help. In 1940, under the War Emergency Assistance Scheme, the proceeds of a levy per ton of coal supply were used to assist those owners whose collieries had suffered from enemy action or from the loss of export markets. In 1941 a further levy was imposed to enable high cost producers to bear the cost of the guaranteed wage following the introduction of the Essential Work Order. An additional levy was introduced to subsidize "necessitous undertakings" whose output was essential to the war effort but which were in great financial difficulty.

Continuing increases in costs, particularly flat rate national wage increases, proved a burden to the least productive collieries. National price rises sufficient to compensate these undertakings would injure consumers and provide bonus profits for low cost producers. To deal with these problems the government instituted the Coal Charges Account in June 1942. It was decided to average "the rise in costs over the whole output of the industry, levying the funds to meet successive increases in costs by means of a national average charge per ton, and recouping the industry for the levy by national average price increases."[152] The payment of the guaranteed wage, the subsidy to necessitous undertakings and the cost of the Greene and later wage awards were met by the Account. Both the levy and the price of coal were increased, the price rise enabling colliery owners to meet the increased levy. From 1942 onwards any necessary increases in the price of coal were on a national basis and the proceeds were pooled, providing a fund from which subsidies in the shape of price allowances per ton of output could be paid to high cost producers to maintain their standard district credit balances. Certain districts covered their costs only by receiving substantial sums from the fund in excess of the levy they paid into it; in effect high cost coalfields were subsidized by low cost, profitable coalfields.

Cumberland was one of the seven districts which made no contributions to the Coal Charges Account since each year it recovered from the Account considerably more than it paid into it. (The other districts were: Durham, S. Wales and Monmouthshire, Lancashire and Cheshire, North Wales, Bristol and Somerset, and the Forest of Dean.) In fact Cumberland recoveries per ton between 1942 and 1945 were the highest of all mining districts in the country.[153]

Cumberland Transactions with the Coal Charges Account[154]

Year	*Proceeds Per Ton*		*Loss Per ton*		*Recoveries Per Ton*		*Adjusted Balance Per Ton +Profit −Loss*	
	s.	*d.*	*s.*	*d.*	*s.*	*d.*	*s.*	*d.*
1942	34	8	2	9	3	3	+0	6
1943	37	3	5	9	6	1	+0	4
1944	41	4	9	4	10	0	+0	8
1945	46	0	10	10	10	8	−0	2

THE COAL TRADE

Until 1930 over a quarter of Cumberland's coal output was exported, most of it going to Ireland. A feature of this trade, commented upon fifty years ago, was "the remarkable constancy of exports" which between 1922 and 1930 remained close to the pre-war level of 624,000 tons; in 1927 and 1928 coal exports exceeded those in 1913.

Cumberland Coal Exports[155]

	Thousand Tons	*As % of Cumberland Output*
1913	624	26·9
1922	571	26·0
1923	467	22·1
1924	567	25·9
1925	568	28·6
1926	301	30·8
1927	853	35·3
1928	766	35·8
1929	619	30·6
1930	606	30·0

A prolonged strike at Whitehaven Colliery in 1923 was largely responsible for the fall in output and exports in that year and the national strike for a bigger fall in 1926.

From 1930 Cumberland coal had to meet intense and growing competition in the Irish market from other British coalfields. This, together, with a prolonged strike in the Cumberland collieries in 1931 accounted for the sharp drop in the county's coal shipments in that year to 471,652 tons. In July 1932 the Irish Free State imposed a duty of five shillings per ton on imported British coal which reduced greatly the coal trade of this country. United Kingdom coal exports to the Irish Free State fell from 2,425,000 tons in 1931 to 1,040,000 tons in 1934.[156] Cumberland usually supplied about one-tenth of

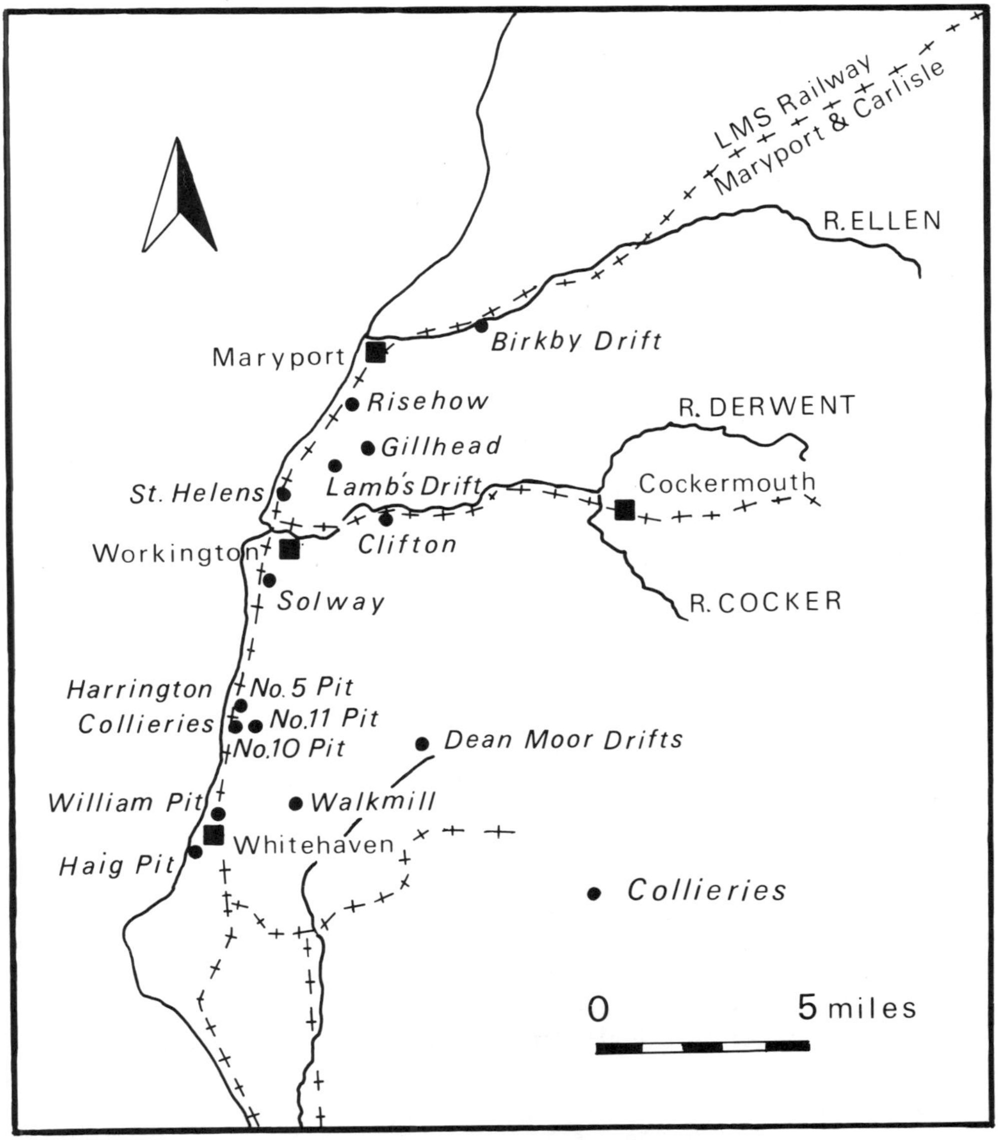

FIG. 6 – Coal Mines in Cumberland 1945.

these exports and this trade was seriously affected by the new tariff; in the next two years it was almost halved. Although the Irish import duty was removed in January 1935 and coal shipments to Ireland immediately increased total coal shipments from Cumberland in 1935 were lower than in the previous year. Worse was to come. The Whitehaven collieries closed in October 1935 and did not resume production until March 1937. As a result the coal trade from Whitehaven, which had been badly affected by the Irish tariff, ceased completely in 1936 and this is reflected in the following table:-

Cumberland Coal Shipments[157]

Year	*Total Shipments*	*Total Shipments to Ireland*	*Irish Free State only*
	Tons	*Tons*	*Tons*
1931	471,652	442,200	144,439
1932	494,090	447,779	142,170
1933	388,861	347,597	80,217
1934	405,731	316,200	83,739
1935	378,191	328,807	112,803
1936	181,489	161,701	25,745

With the re-opening of the Whitehaven pits the coal trade revived. From Whitehaven in the years ended 31 March 1938-40 146,601, 208,835 and 287,090 tons of coal were exported.[158]

Between the wars changes came in the relative positions of the various Cumberland ports. Until 1927 Maryport remained the leading port in Cumberland. For many years Maryport had been the chief port for the importation of iron ore but this came to an end in 1927 when iron manufacture ceased there and the Prince of Wales dock at Workington was opened. Similarly Maryport lost its position as the leading coal exporter when collieries in its neighbourhood closed. The trade of Maryport had been helped by its ability to accommodate vessels of up to 6,000 tons whereas Whitehaven could not handle vessels over 3,000 tons or Workington vessels over 2,000 tons. In 1927 most of the trade of the port was ruined by the opening of a new large dock at Workington. At Workington the Lonsdale Dock, opened in 1865, could not handle large vessels and ore for the iron and steel works had to be imported into Barrow and Maryport and then moved by railway at additional costs to Workington. To improve the accommodation at Workington and to reduce costs the United Steel Companies Ltd. therefore promoted the Workington Harbour and Dock Act in 1920 which led to the creation of the Prince of Wales Dock by extending and deepening the Lonsdale Dock. Work began in 1923 and was completed in 1927 at a cost of approximately £650,000. Whilst this work was in progress traffic was dealt with as far as practicable in the tidal harbour. The new dock could accommodate vessels up to 10,000 tons and largely because of this Workington took over from Maryport the position of leading port in Cumberland.[159] Maryport suffered not only from the cessation of iron ore imports but also from the diminution in its coal exports. A comparison of coal shipments from Workington and Maryport before and after 1927 illustrates the decline in the coal trade from the latter port:-[160]

Year	Workington Tons	Maryport Tons
1922	215,899	224,274
1923	160,871	318,489
1924	2,195	295,858
1925	2,026	263,582
1926	—	116,960
1927	2,702	270,657
1928	203,093	151,403
1929	143,480	140,822
1930	157,204	132,403

The small port of Harrington ceased to function in 1928 and its coal trade was taken over by Workington. Coal exports from Workington were affected to some extent by the fortunes of the local iron and steel industry and varied with the fluctuations in its demand for coal. Frequently only a small percentage of coal output in the Workington area was available for export.[161]

From 1930 until the outbreak of the second world war coal shipments from Maryport dwindled year by year, falling to 30,940 tons in 1938, the lowest figure for over a century.[162] The Senhouse Dock, which had been chiefly concerned with iron ore imports, remained idle for years and became in effect a large and free open air swimming pool for many of the town's inhabitants. In the years 1931-8 annual coal shipments from Workington ranged from 101,163 to 209,043 tons.[163] Silloth's share of the coal trade with Ireland was always comparatively small and in these years there was a marked reduction in the quantity of coal shipped from there:-[164]

Year	Tons	Year	Tons
1929	43,783	1933	6,696
1930	30,664	1934	11,759
1931	24,373	1935	12,205
1932	16,386	1936	12,741

The port of Maryport was saved from extinction by the outbreak of war in 1939 which led to a quick and extraordinary increase in coal exports. This followed not a sudden and dramatic outburst of mining activity in its hinterland but the arrival by rail of large supplies of North-East coal diverted to West Cumberland to avoid sinking by enemy vessels. Workington also shared in this windfall but Maryport benefited more from it simply because it was nearer to Newcastle. In the table below figures in brackets denote the amount of North-East coal included in total shipments from Maryport and Workington.

Coal Shipments[165]

Year	Maryport Tons		Workington Tons	
1939	34,602		156,303	
1940	83,395	(52,170)	241,380	(81,525)
1941	320,305	(294,834)	339,597	(249,677)
1942	467,405	(445,866)	318,041	(269,811)
1943	460,599	(450,745)	123,134	(94,717)
1944	518,600	(513,571)	330,275	(285,239)
1945	467,132	(464,841)	148,843	(131,959)
1946	476,880	(473,052)	22,935	

Maryport shipments 1944-6 are for year ending 30 September.

Notes and References

[1] *Report on Haig Pit explosion, 5 Sept 1922*, Cmd. 1796 (1923).
[2] *Report on William Pit explosion, 15 Aug. 1947*, Cmd. 7410 (1948).
[3] D/Lons/W, Colliery Output Books.
[4] D/Lons/W, Report on the Whitehaven Coalfield, 24 June 1933.
[5] D/Lons/W, Statement regarding the Priestman Whitehaven Collieries Limited.
[6] D/Lons/W, Report on the Priestman Whitehaven Collieries Limited, 16 Feb. 1935.
[7] *Ibid.*
[8] J. L. Carvel, *The Coltness Iron Company* (1948), 141-7.
[9] D/Lons/W, Coal Royalty Rent Account Books.
[10] *Ibid.*
[11] P. W. S. Andrews and S. A. Brunner, *Capital Development in Steel* (Oxford, 1951), 145.
[12] *Workington Star*, 20 May 1938.
[13] Information provided by the British Steel Corporation.
[14] D/Lons/W, Coal Royalty Rent Accounts.
[15] D/Lons/W, Output Books.
[16] *West Cumberland Times*, 6 Aug.-28 Sept. 1927.
[17] D/Lons/W, Output Books.
[18] D/Lec, Mineral Reports 1914-33.
[19] D/Lons/W, Letter books of T. S. Durham.
[20] D/Lons/W, Coal Royalty Rent Accounts, *Reports of Inspector of Mines*.
[21] D/Lons/W, Output Books.
[22] Andrews and Brunner, *op. cit.*, 244.
[23] *Report of Inspector of Mines*, 1948.
[24] D/Lec, Mineral Report 1915.
[25] D/Lec, Mineral Reports 1914-33.
[26] D/Lec, Mineral Reports 1914-24.
[27] D/Lec, Mineral Reports 1914-21.
[28] D/Lons/W, Letter Books of T. S. Durham. *Lists of Mines*, 1927-9.
[29] D/Lec, Mineral Reports 1926-7.
[30] *Lists of Mines*, 1926-9.
[31] *Report on Accident at Risehow Colliery, 26 Nov. 1957*, Cmnd. 396 (1958).
[32] Andrews and Brunner, *op. cit.*, 145.
[33] *Lists of Mines*, 1945-52. NCB, 1947-52.
[34] D/Lec, Mineral Reports 1914-33. *Lists of Mines* 1918-45. D/Lons/W, Letter Books of T. S. Durham.
[35] D/Lec, Mineral Reports 1913-33.
[36] D/Lec, Mineral Reports 1914-28.
[37] D/Lec, Mineral Reports 1924-30.
[38] *Lists of Mines*, 1930-7.
[39] *Reports of Inspectors of Mines*, 1914-18.
[40] Sir R. A. S. Redmayne, *The British Coal Mining Industry During the War* (Oxford, 1923), 94.
[41] *Reports of Inspectors of Mines*.
[42] *Ibid.*
[43] W. A. Lee, *Thirty Years in Coal* (1954), 14.
[44] *Reports of Inspectors of Mines*.
[45] G. H. J. Daysh, *West Cumberland: A Survey of Industrial Facilities* (Whitehaven, 1938), 157.
[46] *Reports of Chief Inspector of Mines*.
[47] Coal Mines Reorganisation Commission. *Colliery Amalgamations*, 30 quoted by Jewkes and Winterbottom, *op. cit.*, 95.
[48] Andrews and Brunner, *op. cit.*, 204.
[49] Jewkes and Winterbottom, *op. cit.*, 96.
[50] Records of the Cumberland Miners' Association.
[51] *Reports of Chief Inspector of Mines*.

[52] Andrews and Brunner, *op. cit.*, 203. Daysh, *op. cit.*, 134-5.
[53] Andrews and Brunner, *op. cit.*, 204.
[54] Daysh, *op. cit.*, 135.
[55] *Reports of Chief Inspector of Mines*, Ministry of Fuel and Power, *Statistical Digest from 1938*, Cmd. 6538 (1944); *Statistical Digest 1945*, Cmd. 6920 (1946); *Statistical Digest 1946 and 1947*, Cmd. 7548 (1948). *Colliery Year Book 1948.*
[56] *Annual Abstract of Statistics*, vol. 86, Table 138.
[57] Ministry of Fuel and Power, *Statistical Digest from 1938*, Cmd. 6538 (1944); *Statistical Digest 1946 and 1947*, Cmd. 7548 (1948). *Colliery Year Book 1948.*
[58] *Reports of Chief Inspector of Mines.* Ministry of Fuel and Power, *Statistical Digest 1944*, Cmd. 6639 (1945); *Statistical Digest 1945*, Cmd. 6920 (1946).
[59] D/Lec, Mineral Report 1919.
[60] *Reports of Chief Inspector of Mines.* Ministry of Fuel and Power, *Statistical Digests.*
[61] Ministry of Fuel and Power, *Supplement to Statistical Digest 1945* (1947).
[62] *Reports of Chief Inspector of Mines.*
[63] *Report of Chief Inspector of Mines*, 1938.
[64] Ministry of Fuel and Power, *Supplement to Statistical Digest 1945* (1947).
[65] Ministry of Fuel and Power, *Northumberland and Cumberland Coalfields: Regional Survey Report (Northern "A" Region) 1945.*
[66] *Reports of Chief Inspector of Mines.*
[67] *Reports of Inspectors of Mines.*
[68] *Colliery Year Book*, 1933, 1947. Political and Economic Planning, *Report on the British Coal Industry* (1936), 194.
[69] Jewkes and Winterbottom, *op. cit.*, 9.
[70] Ministry of Labour, *Reports of Investigations into Industrial Conditions in certain Depressed Areas*, Cmd. 4728 (1934), 11.
[71] Jewkes and Winterbottom, *op. cit.*, 30.
[72] *Census of Population 1951.*
[73] D/Lec, Mineral Reports 1932-3.
[74] Redmayne, *op. cit.*, 153-7, 181.
[75] Redmayne, *op. cit.*, 221.
[76] *Report of the Royal Commission on the Coal Industry 1919*, iii, 104.
[77] R. P. Arnot, *The Miners: Years of Struggle* (1953), 187.
[78] Redmayne, *op. cit.*, 218. G. D. H. Cole, *Labour in the Coal-Mining Industry 1914-1921* (Oxford, 1923) 165.
[79] F. A. Gibson, *Statistics of the Coal Mining Industry of the United Kingdom* (Cardiff, 1922), 146.
[80] Arnot, *op. cit.*, 245.
[81] *Workington Star*, 15 Oct.-5 Nov. 1920. Gibson, *op. cit.*, 146.
[82] *Workington Star*, 5 Nov. 1920.
[83] Lee, *op. cit.*, 14-15.
[84] *Workington Star*, 8 April 1921.
[85] Records of the Cumberland Miners' Association.
[86] *Whitehaven News*, 23 June 1921.
[87] *Report of the Royal Commission on the Coal Industry 1925*, Cmd. 2600 (1926), i, 131-4.
[88] *Reports of Chief Inspector of Mines.*
[89] *Report of the Royal Commission on the Coal Industry 1925*, Cmd. 2600 (1926), iii, 310.
[90] *Whitehaven News*, May-Nov. 1923. Cumberland Miners' Association, *Whitehaven Collieries Dispute: Minutes of Proceedings* (Carlisle, 1923), 371-8.
[91] *Report of Chief Inspector of Mines*, 1924, 110. *Report of the Royal Commission on the Coal Industry 1925*, Cmd. 2600 (1926) i, 133. Kirby, *op. cit.*, 68.
[92] *Report of Chief Inspector of Mines*, 1925, 8-11.
[93] Minute Books of the Cumberland Miners' Association, April-Dec. 1926.
[94] *West Cumberland Times*, July-Sept. 1927.
[95] *Records of the Cumberland Coal Owners' Association.*
[96] *Ibid.*

[97] *Whitehaven News*, 4-11 June 1931. Minute Books of the Cumberland Miners' Association, May-June 1931.
[98] *Whitehaven News*, June 1931. Minute Books of the Cumberland Miners' Association, June 1931.
[99] Minute Books of the Cumberland Miners' Association, July-Aug. 1931.
[100] *Report of Chief Inspector of Mines*, 1934.
[101] R. P. Arnot, *The Miners in Crisis and War* (1961), 164-74. Lee, *op. cit.*, 115.
[102] *Reports of Chief Inspector of Mines*, 1936-8.
[103] Records of the Cumberland Coal Owners' Association.
[104] *Reports of Chief Inspector of Mines.*
[105] *Ibid.*
[106] W. H. B. Court, *History of the Second World War: Coal* (1951), 139-40.
[107] Court, *op. cit.*, 232. Lee, *op. cit.*, 239-40.
[108] Court, *op. cit.*, 223-4. Lee, *op. cit.*, 136-8.
[109] Court, *op. cit.*, 234-5.
[110] Court, *op. cit.*, 254-6. Lee, *op. cit.*, 141-4.
[111] Court, *op. cit.*, 262-5.
[112] *Ibid.*
[113] *Ibid.*, 265-6, 327-8.
[114] Ministry of Fuel and Power, *Statistical Digests*, 1944-9.
[115] *Reports of Inspector of Mines.*
[116] *Report of Inspector of Mines* (Northern Division), 1915.
[117] *Ibid.*, 1920.
[118] *Report on explosion at Haig Pit, 5 Sept. 1922*, Cmd. 1796 (1923).
[119] *Ibid.*
[120] *Workington Star*, 16 Dec. 1927. *Whitehaven News*, 16 Feb.-8 March 1928.
[121] *Report on explosion at Haig Pit, 29 Jan. 1931*, Cmd. 3858 (1931).
[122] *Report of Chief Inspector of Mines*, 1930.
[123] *Report of the Royal Commission on Safety in Coal Mines, 1938*, Cmd. 5890 (1938), 324.
[124] *Report on explosion at William Pit, 3 June 1941*, Cmd. 6367 (1942).
[125] *Report on explosion at Harrington No. 10 Colliery, 9 Dec. 1946*, Cmd. 7222 (1947).
[126] *Report of the Chief Inspector of Mines*, 1923.
[127] *Reports of Chief Inspector of Mines.*
[128] Redmayne, *op. cit.*, 223.
[129] *Reports of Chief Inspector of Mines.*
[130] Redmayne, *op. cit.*, 94.
[131] *Reports of Chief Inspector of Mines.*
[132] *Ibid.*
[133] *Ibid.*, Records of Cumberland Coal Owners' Association. Mining Association of Great Britain, *Statistical Review of the Coal Industry 1933-4*. Miners' Federation of Great Britain, *Annual Volumes of Proceedings*, 1935-8.
[134] *Reports of Chief Inspector of Mines.*
[135] Reports of the Cumberland Coal Owners' Association.
[136] D/Lec, Mineral Reports.
[137] D/Lons/W, Coal Royalties Minutes 1924-8. D/Lec, Mineral Reports.
[138] Records of the Cumberland Coal Owners' Association.
[139] D/Lons/W, Analysis of leases of Lonsdale royalties.
[140] Mining Association of Great Britain, *Statistical Review of the Coal Industry 1925*.
[141] D/Lons/W, Lonsdale Coal Royalties, statement by T. S. Durham, 14 March 1930.
[142] Board of Trade, *Statistical Abstract for the United Kingdom*, 1940.
[143] D/Lons/W, Statement of Income derived from Coal Royalties.
[144] D/Lons/W, Coal Royalty Rent Accounts.
[145] D/Lons/W, Misc. Colliery Papers, "Statement showing Rents due to Mr. Curwen".
[146] D/Lec, Mineral Reports.
[147] D/Lons/W, Colliery Output Books.
[148] D/Lons/W, Lonsdale Coal Royalties Minutes, T. S. Durham to T. E. Forster, 25 Feb. 1930.
[149] Ministry of Fuel and Power, *Statistical Digest from 1938*, Cmd. 6538 (1944); *Statistical Digest 1946 and*

1947, Cmd. 7548 (1948). *Financial Position of the Coal Mining Industry: Coal Charges Account*, Cmd. 6617 (1945). *Colliery Year Book*, 1948.

150 Ministry of Fuel and Power, *Statistical Digest from 1938*, Cmd. 6538 (1944); *Supplement to Statistical Digest 1945*. *Colliery Year Book*, 1948.

151 Ministry of Fuel and Power, *Statistical Digest from 1938*, Cmd. 6538 (1944).

152 Court, *op. cit.*, 333-4.

153 Court, *op. cit.*, 334-6. *Financial Position of the Coal Mining Industry: Coal Charges Account*, Cmd. 6617 (1945). H. Wilson, *New Deal for Coal* (1945), 116-31.

154 *Coal Charges Account*. Court, *op. cit.*, 338-46.

155 Jewkes and Winterbottom, *op. cit.*, 98.

156 Political and Economic Planning, *Report on the British Coal Industry* (1936), 157.

157 Daysh, *op. cit.*, 157.

158 Whitehaven Harbour records.

159 Workington Harbour records.

160 Workington and Maryport Harbour records.

161 Daysh, *op. cit.*, 159.

162 Maryport Harbour records.

163 Workington Harbour records.

164 Daysh, *op. cit.*, 159.

165 Maryport and Workington Harbour records.

CHAPTER VII*

1947-1982/3

WEST CUMBERLAND

In 1947 the hopes and dreams of generations of miners became a reality when the coal mining industry was nationalized. At that time there were thirteen working pits in West Cumberland – Clifton, Siddick, William, Haig, Harrington No. 10, Harrington No. 11, Moresby, Solway, Risehow, Gillhead, Ellenbank, Dean Moor and Birkby.† Dean Moor was a licensed colliery not operated by the National Coal Board. Sadly, the reality in Cumberland proved harsh. Since nationalization the story of mining in this coalfield has been one of contraction. Of the twelve N.C.B. collieries which in 1947 produced 1,082,540 tons of saleable coal and employed 5,590 men only one still survives, Haig Pit which in 1979/80 had an output of 269,842 tons and employed 963 men.[1]

Because of the exhaustion of their coal reserves the smaller collieries had soon to abandon production. The Birkby & Allbright Colliery at Dearham Bridge closed in March 1950 and the Ellenbank Colliery, which only employed thirty persons and where working conditions were very bad, closed in 1952 after producing 10,208 tons of coal in the previous year. Dean Moor Colliery, an independent colliery which in 1950 employed fifty-six men, ended its long history in 1951 when the Moorside Drift No. 4 ceased production.[2]

Following a disastrous explosion in August 1947 William Pit, the oldest pit in the Whitehaven area, went out of production and coal getting was not resumed until 1948. Despite an annual production of 55-75,000 tons during the years 1950-53 the pit had to close 31 December 1954 because the workings had run into a solid rock fault which put an end to production. It had been unprofitable since 1947 and its 254 employees were transferred to neighbouring pits. Local miners were not seriously disturbed by this event since they believed, mistakenly as it turned out, that the remaining coal in William Pit would be worked from Harrington No. 10 Pit.[3] During the first decade of nationalization when both coal and colliers were urgently needed miners displaced by pit closures were quickly re-employed at other pits, but in the following decade such labour transfers became progressively more difficult.

Until 1956 the National Coal Board's policy was to obtain maximum output and even

* Many of the post-1947 statistics in this chapter and in the appendices are not available in published sources and I am greatly indebted to Mr L. R. Hoten, Area Statistician of the Western Area, for his considerable efforts in seeking out this information for me.

† There was a fourteenth pit, Micklam (or Harrington No. 5) Pit then owned by the United Steel Cos. Ltd. This pit, which was closed in March 1981, was primarily a fireclay pit and yielded very little coal.

unprofitable mines were kept going to ensure fuel supplies for British industries and to make expensive coal imports unnecessary. Coal prices rose rapidly, doubling within ten years. By 1957 there was no longer a seller's market for coal and an increased use of the alternative fuels gas, oil and electrictiy. The demand for coal was further decreased by notable improvements in fuel efficiency both in the home and in industry generally. It was now essential to hold down production costs in the mining industry and the N.C.B. embarked upon a programme of closing the older, uneconomic mines, reconstructing other pits, reducing the number of coal faces and the speedy adoption of mechanized mining. For a notoriously high cost district like the West Cumberland coalfield such a programme, if carried to its logical conclusion, would mean extinction. Soon the few remaining pits in the county became fewer as one after another their lack of profitability made their closure necessary.

In February 1959 the two collieries at Clifton and Gillhead were closed. Output at Clifton had fallen from 161,790 tons in 1950 to 73,556 tons in 1958 and at Gillhead from 55,239 tons to 22,466 tons in the same period. The decline in output, falling productivity and a high rate of absenteeism made their continued survival impossible. Of the 556 men employed at these pits (442 at Clifton and 114 at Gillhead) 420 became redundant.[4] Two years later the Walkmill or Moresby Pit had to close because of the exhaustion of its coal seams; production there had diminished annually from 81,297 tons in 1953 to 51,105 tons in 1960. Nearly two-thirds of the 170 men employed at the pit when it closed were transferred to the Haig, Lowca and Solway collieries and about forty retained for salvage and safety work.[5] At that time St. Helens Colliery, Siddick, which employed about 550 men, was threatened with closure because it was working at a loss. Fortunately for the men the colliery's life was saved by a bargain made with the National Coal Board. According to the Cumberland Development Council the colliery "has been reprieved until the end of 1964, but this has only been achieved by the National Executive of the National Union of Mineworkers consenting to the re-opening of opencast mining in the County, and is subject to the proviso that there is no serious deterioration in the normal operational results. The Council view the future of the coal mining industry with anxiety having regard to the declared policy of the Chairman of the National Coal Board to run mines on purely economic lines and to close those which operate at a loss."[6]

In June 1963 Harrington No. 11 Pit ceased production after exhausting its reserves of coal and running at a loss for several years. In the 1950's it had yielded 30-35,000 tons of coal per annum but in the early 1960's only 20-25,000 tons per annum. Work for the seventy men displaced was made available in other parts of the coalfield.[7] Two more collieries closed in 1966 because they had run out of coal – Risehow in April and St. Helens in July. Except for the year 1957, when Risehow Pit produced 96,866 tons of coal, each pit had an annual output exceeding 100,000 tons from 1950 to March 1966. The 450 employees at Risehow Colliery were tranferred to other pits. The closure of St. Helens Colliery was particularly unfortunate in that the miners there had succeeded in raising the overall output per manshift to 31·5 cwts. in 1964/5 and to 34·3 cwts. in 1965/6, the best performance in the coalfield. Unfortunately a geological fault suddenly put an end to production on the face that was being worked and since the coal reserves of this mine were too far away to be worked economically there was no option but to close the colliery. Its 510 workmen were re-employed in the Harrington No. 10, Solway and

Haig collieries, the last three active pits in the coalfield. For some of these miners this was their third move since Vesting Day from a closed pit to a working pit.[8]

Harrington No. 10 Pit was the next colliery to fall under the axe. In the last six years of its existence this pit had an annual output of over 100,000 tons but its productivity was far below what was acceptable to the National Coal Board. Overall output per manshift fell from 21·2 cwts. in 1963/4 to 13·14 cwts. in 1966/7, rising slightly in 1967/8 to 14·3 cwts. In August 1967 the Coal Board announced its decision to close the pit in December. Maurice Rowe, leader of the Cumberland miners, aided by local M.P.s, strove desperately with both the N.C.B. and the government to keep the pit going. Following intervention by the Prime Minister the demise of the pit, along with that of fifteen others, was postponed – the government agreeing to bear the costs incurred by the National Coal Board in deferring the closure of these collieries. Because of the unemployment situation in the county Harrington No. 10 was granted a further reprieve by the Minister of Power but in July 1968 production finally ceased.[9] Local miners had hoped and expected that the Harrington No. 10 Pit would be used to enable coal to be extracted from the reserves of the old William Pit. In 1966 a connection was made betweem the workings in the Six Quarters seam at the Harrington Pit and the main roads in the William Pit area which improved ventilation in the former pit by making use of the William Pit shaft as an additional downcast.[10] But with its dismal productivity record in its last years Harrington No. 10 had to close. There were now only two working pits in West Cumberland and therefore few vacancies for the 560 men employed at the time of its closure. Of these only sixteen were transferred to the Haig and Solway pits, sixty were employed at the Harrington washery, 120 were retained for temporary salvage work and 364 became redundant. Most of the displaced miners had no desire to transfer to other coalfields and there was little prospect of employment in other local industries.[11]

For many years after 1947 a tremendous effort was made by the N.C.B. to develop Solway Colliery and to win coal by using the most modern techniques. A programme of reorganization improved the layout underground, provided better haulage facilities and paved the way for mechanized mining. With the introduction of power loaders, underground locomotives and skip winding the extraction and movement of coal were accelerated. By the end of 1960 the Board had invested £1,569,718 in the reconstruction and development of the colliery.[12] It proved difficult and costly to exploit the undersea coal and work was hampered by an excess of gas and water. Frequent faults entailed long and expensive delays before the coal could be re-located and although Main Band coal was up to 12 feet thick some of the seams were too thin and dirty to make mining practicable. Only once, in 1967/8, did the annual output of Solway Colliery exceed 200,000 tons when 216,976 tons of saleable coal were raised. Severe faulting badly affected production in 1970 and from then until 1973 the longwall method of working was abandoned and coal getting was restricted to robbing the pillars left from previous workings. This was accompanied by a rapid decline in output and productivity. Output per manshift overall fell from 30·9 cwts. in 1968/9 to 15·7 cwts. in 1969/70 and to 15·1 cwts. in 1972/3. With such a record and with diminishing coal reserves the pit could not survive very long and 7 November 1969, with a loss then of £5 per ton, Solway Colliery was officially put in jeopardy. By the end of 1972 the pit was losing £10·61 per ton; by that time it had lost over £2½ million since November 1969. The pit closed in May 1973 and with it the associated preparation plant at Harrington which employed forty-five men. Of the 495 mineworkers at Solway Colliery more than 100 were offered jobs at the last remaining colliery in Cumberland, Haig Pit, Whitehaven and many of the rest were invited to transfer to other coalfields, particularly in the Midlands. About 160 of the Solway Pit miners were aged 55 or over and therefore eligible to benefit under the Redundant Mineworkers' Payments Scheme.[13]

In view of the experience of the past thirty years at Haig Pit the future of this colliery would appear uncertain. Output decreased from 417,371 tons in 1947 to 212,495 tons in 1977/8, the lowest figure since nationalization. In the next few years production increased, reaching 372,848 tonnes in 1980/1 and 471,170 tonnes in the following year. Concurrently there was a large reduction in the labour force, the number of men on the colliery books falling annually from 1,609 in 1966/7 to 793 in 1982/3. Coal production fluctuated widely during these years, reaching its highest level in 1967/8 when 474,507 tons of saleable coal were raised. Like all other collieries in the country Haig Pit suffered from the effects of the miners' strikes in 1972 and 1974. Productivity at Haig Pit since 1947 has been among the lowest of the Cumberland pits, ranging from an overall output per manshift of 16·6 cwts. in 1952 to 27·1 cwts in 1975/6, the colliery's best performance in that year comparing very badly with the Western Area's average figure of 44·6 cwts. Since then productivity at the colliery has improved, reaching 1·88 tonnes in 1980/1 and 2·41 tonnes in 1981/2.[14]

Until 1980 coal was worked in the Upper Metal Band seam by advancing longwall faces. The pit is fully mechanized with power loading machines and powered roof supports and in recent years more than 90 per cent of the output has been power loaded, the remainder being development coal won by heading machines. Working conditions have been considerably improved and the pit has become much safer since the introduction of methane drainage to remove the large quantities of gas emitted near the coal face. Although some of the men travel more than 4½ miles to their working places the journey has been eased by the provision of riding sets powered by electric rope haulage.[15]

Between March 1970 and March 1978 the National Coal Board invested £3,862,000 in Haig Pit. The reconstruction and improvement of surface facilities, begun in 1975 and completed in the following year, accounted for much of this expenditure. A new coal preparation plant was built to replace the old Ladysmith washery which was some distance from the colliery and a conveyor was provided to carry the coal from the new plant to Whitehaven harbour. In addition there were new stocking facilities and a new distribution depot to serve the local coal merchants. These changes have made it possible to reduce the surface area of the colliery and to reclaim and tidy up the coastal strip.[16]

Until the late 1970's Haig Pit produced coal suitable for the steel industry, about three-quarters of its output going to the British Steel Corporation's works at Workington and the remainder to the local domestic market. Because of the closure of so many pits in West Cumberland it was impossible to supply the needs of the Workington steelworks from local sources. Despite the decline in the iron and steel industry in recent years which reduced the demand for coal Haig Pit's output in 1977 is said to have been insufficient to feed the coke ovens at Workington and it was necessary to import some 6,000 tons of coal per week from Yorkshire and South Wales.[17] A change came in 1978 with the closure of the Bannock Band seam at Haig Pit for the coal produced at Haig Pit was no longer suitable for the Workington furnaces and coking coal had to be imported from the United States and from other N.C.B. mines.[18] Other markets had to be found for Haig coal and one was found at the new 2,000 mw Fiddlers Ferry power station near Warrington.

In the last thirty years the need to close unprofitable mines has converted the coal mining industry of West Cumberland, for many years of major importance in that area, to one of comparatively minor significance. As the number of working pits decreased coal production diminished; after 1958 the annual output never reached a million tons and after 1969 it was less than 500,000 tons. In 1977/8 the 212,495 tons raised at Haig Pit amounted to less than one-fifth of Cumberland's output in 1947. To some extent the loss of coal has been offset by the recent development of opencast mining in the county which now yields more coal than Haig Pit.

Cumberland Coal Production (Saleable Coal)[19]

Year	*Output (Deep-mined)* *Tons*	*Year*	*Output Tonnes*
1947	1,082,540	1977/8	212,495
1958	1,074,199	1978/9	264,838
1962	904,355	1979/80	269,842
1968/9	645,712	1980/1	372,848
1972/3	333,462	1981/2	471,170
1976/7	293,104	1982/3	305,880

The coalfield's highest output in post-war years was 1,221,205 tons in 1955 from nine collieries, but as pits closed in rapid succession production slumped. Other factors helped to reduce output. During the war miners had six days holiday per annum; from March 1975 the total of holidays and rest days came to twenty-seven days per man, excluding seven statutory holidays.[20] Each extension of the miners' holidays during these years meant a loss of coal production. A high rate of absenteeism, which reached a peak of 18·6 per cent in 1965/6, had a similar effect. Until very recently productivity in West Cumberland mines, whether measured in terms of annual output per man or output per manshift, lagged badly behind national standards.

Productivity[21]

	Output per man-year		*Output per Manshift Overall*	
	Cumberland	*Gt. Britain*	*Cumberland*	*Gt. Britain*
Year	*Tons*	*Tons*	*Cwts.*	*Cwts.*
1947	193·6	262·0	15·7	21·5
1953	204·3	294·9	17·4	24·6
1960	200·4	305·3	20·6	28·0
1967/8	275·9	413·8	25·3	39·0
1972/3	178·7	472	16·3	45·9
1976/7	253·1	440	24·5	43·6
1977/8	193·1	441 Tonnes	18·4	2·19 Tonnes
1978/9	256·8	448 ,,	1·19 Tonnes	2·24 ,,
1979/80	279·9	467 ,,	1·29 ,,	2·31 ,,
1980/1	397·4	476 ,,	1·88 ,,	2·32 ,,
1981/2	534·2	494 ,,	2·41 ,,	2·40 ,,
1982/3	388·0	504 ,,	1·78 ,,	2·44 ,,

Between 1947 and 1977 output per manshift doubled in British mines generally. In the first decade of nationalization output per manshift overall rose very slowly, from 21·5 cwts. in 1947 to 24·9 in 1957. After 1957 the urgent need to contain costs, particularly labour costs, intensified the movement towards mechanized mining, especially the use of power loaders, which in turn led to a rapid growth in productivity. In 1947 only 2·4 per cent of British coal was power-loaded; by 1957 the percentage had risen to 23 and to 93 in 1972/3. In West Cumberland the productivity record between 1947 and 1979/80 was vastly different, ranging from a minimum of 14·8 cwts. per manshift overall in 1948 to a maximum of 27·1 cwts. in 1975/6. The latter figure refers to the performance of one pit, Haig Pit, the only working pit in the coalfield; in the same year the national average output per manshift was 44·8 cwts. In Cumberland the National Coal Board acquired a coalfield that was subject to geological faults, where little investment had taken place between the wars and where some of the pits were old, badly planned and badly equipped. The fact that despite the growth of mechanized mining output per manshift until 1979/80 never exceeded 27·1 cwts. suggests than in Cumberland there were major difficulties not present to the same extent in many other coalfields. At Haig Pit, however, the most recent figures show a great improvement, the colliery achieving productivity records in 1980/1 and 1981/2, the output per manshift of 2·41 tonnes in the latter year equalling the national average for the first time since the coal mines were nationalized.[22]

Despite the annual reduction in manpower at Haig Pit, which followed the Board's early retirement and voluntary redundancy schemes, output has increased since 1978 and this has brought about a distinct improvement in productivity figures. These results depended to a large extent upon the full co-operation of the miners and of the union. In addition, the incentive bonus scheme has undoubtedly done much to increase output and reduce absenteeism. Other factors contributed to this end. A decision was taken in 1977 to concentrate workings in the more productive Upper Metal Band seam and to adopt the retreat mining system. The first retreat face began production in November 1980 and resulted in a doubling of face productivity. Skip winding was introduced in 1982 at a cost of £500,000. Six tonne capacity skips, wound from a depth of 1,200 feet, effectively increased mineral winding capacity by fifty per cent – to 300 tonnes per hour. The skips load and discharge automatically onto a surface conveyor system, eliminating the need for rolling stock either on the pit bank or at the pit bottom. Another new development in the same year was the introduction underground of a fleet of ten Free Steered Vehicles to replace rope haulage systems. These are rubber-tyred vehicles for the transportation of materials to working areas underground; they are not engaged in the movement of coal. This cost £1 million, but this was less than a new rope-hauled track system would have cost and operating costs will be much less. Haig Pit was the first deep mine in the country to use this method of moving materials by FSVs.*

At Haig Pit, where all the coal is obtained from undersea workings, the chief obstacle to production is the presence of severe, geological faults. Mining operations can be brought to a standstill until ways have been devised to overcome a fault and in any pit of this kind it is always difficult and expensive to construct level, underground roads. A local newspaper illustrates the handicap, "The ever-present problem of geological

* Information provided by Mr R. Crossley of the N.C.B. Western Area.

faulting has once again cropped up at Whitehaven's Haig Pit. Two new coal faces, opened a few months ago, have both developed problems and output has dropped by half, to only 2,500 tons a week . . . . A new district is being equipped at the moment, but this is unlikely to be ready for another twelve weeks."[23]

The distance the men have to travel to reach the face is a major hindrance. On average travelling time to the face and back is two hours and this reduces greatly the amount of time available for production. Given these conditions it seems unlikely that profitable working at Haig Pit is possible.

MINING METHODS

In 1947 the National Coal Board took over a mining industry which suffered from an excessive number of pits, many of them too small and impossible to work profitably. Technologically the industry was in a backward state – in too many pits mechanization had made insufficient progress, an extensive and expensive reorganization of underground roads was necessary to accelerate the movement of coal and both lighting and ventilation needed to be improved. In the West Cumberland coalfield to raise productivity and to lower costs would be more difficult than in most British coalfields and mining methods would have to be considerably improved.

Changes soon came. By the end of 1947 locomotives were in use underground at the Solway and Risehow collieries to move both men and coal.[24] To make locomotive haulage possible it had been essential to provide level roads and at Risehow further work was done in 1950 to construct level locomotive roadways to the thicker seams. In 1949 skip winding replaced the tub and cage system at Solway Pit. The coal was wound loose in large capacity skips which were filled from a huge storage bunker at the pit bottom and discharged on the surface. This was a swift operation which economized in labour. One man controlled the unloading of coal into the bunker and the charging of the skip and on the surface another man controlled the unloading of the skip on to a conveyor which carried the coal to the preparation plant.[25]

In the Cumberland mines where gas was abundant, especially in the Whitehaven district, ventilation had always been a major problem. Haig Pit had a history of disastrous gas explosions and to make it safer its entire ventilation system was completely reorganized in 1952. This provided the pit with two main intake and four main return airways stretching for over 2½ miles seawards and resulted in a much higher standard of ventilation.[26] Equally important was a methane drainage system begun in 1950 that was reminiscent of eighteenth century methods. A pipe line, fourteen inches in diameter, was installed in a return airway and the upcast shaft to pipe the methane from boreholes near the face to the surface where it was used to raise steam in the boilers.* In 1971 over 212 million cubic feet of the gas were utilized in the boiler plant. This gas drainage system was very successful; without it coal production would have been restricted owing to the high emission of gas near the face and the risk of another mining disaster would have been so much greater.[27] A similar, but much smaller, methane drainage system was brought into use at Harrington No. 10 Pit.[28]

* In 1953 fire-damp from Haig Pit was pumped to the Whitehaven gasworks.

An additional safety measure was the issue of general purpose methanometers to selected overmen and deputies in the Cumberland pits as a supplement to their flame safety lamps. The methanometer gave a direct reading of the fire-damp percentage and could detect the presence of the gas when the quantity was too small or the concentration too low for it to be detected by a flame lamp. Since it was provided with a probe it was possible to examine places normally out of reach.[29]

The improvement in the standard of ventilation at Haig Pit led to an important change in policy there in 1961 – the introduction of electricity into a working face instead of compressed air. This was done despite the misgiving of local miners who had always feared that the use of machinery powered by electricity could lead to sparks capable of precipitating a calamitous explosion in the gas-laden atmosphere of the pit.[30]

Another innovation at Haig Pit was the introduction of retreat mining i.e. driving headings into the coal towards the boundary of the area to be worked and then removing the coal from the boundary backwards towards the shaft on a longwall face. Retreat mining operations began on a new face in the Main Band seam about four miles out to sea in August 1971 and this method was later adopted in the Metal Band seam. Because of difficult roof control problems this system of coal extraction had to be abandoned in 1973.[31] Retreat mining, however, was resumed in the Upper Metal Band seam in 1980.

Part of the National Coal Board's programme of modernizing and reconstructing existing collieries was devoted to the improvement of underground layouts and haulage facilities. At Haig Pit in 1959 a cable belt conveyor, 2,480 yards long, was installed and work was begun on a feeder conveyor system designed to bring all the output of the mine to this conveyor.[32] The feeder conveyor belt system, which is about five miles in length, was completed in 1965. At Harrington No. 10 and St. Helens collieries conveyors were installed in 1964 to replace the endless rope haulage systems.[33]

In the past far too many miners had been killed by falls of roof and the N.C.B. quickly took steps to lessen this danger. In the early years of nationalization timber props were progressively replaced by rigid steel props and bars and these in turn by hydraulic yielding supports. With the growth of power loading there were rapid advances in the face and these meant that the advancing of roof supports by hand was much too slow. To remedy this self-advancing, powered roof supports were brought into use. These consisted of a cluster of hydraulic props (a 'chock') mounted on a steel platform with a canopy of steel cantilever beams. As the face moved forward the platform was propelled by the ram attached to it, the legs (i.e. the hydraulic props) being lowered and driven forward, either singly or in batches at the turn of a lever, and then raised vertically in their new support position. Haig Pit, where powered supports were introduced in 1957/8, was one of the collieries which obtained these supports early for trial purposes.[34] This new method enabled supports to be advanced immediately behind the cutting machine, giving prompt support to the newly exposed roof. There was a saving in time and physical effort and the men working at the face were much safer since they were in effect protected by a powerful, steel canopy which covered virtually the whole of the roof. It was now possible to abandon 'strip packing' (i.e. building by hand a pack or wall of stone to support the roof in the waste area) and to adopt total 'caving' i.e. allowing the waste to collapse as soon as the coal was extracted.[35]

Although power loaders had been introduced into British mines during the Second World War only 2·4 per cent of the total output was power-loaded in 1947. The usual method of extracting coal was to undercut the longwall face by machine and to bring down the coal with explosives, the coal then being hand-loaded into tubs or on to conveyors. More and better power loaders were needed to increase productivity and to reduce labour costs. After 1956 the National Coal Board embarked upon a vigorous programme of mechanized mining, particularly the widespread use of power loaders. As a result the percentage of output power-loaded was raised from 23·0 in 1957 to 85·7 in 1966/7 and to 93·0 in 1972/3. Of the power loaders adopted the most widely used was the Anderton shearer, a rotary drum armed with teeth. As it rotated at high speed it traversed the face shearing down the coal and directed it, by means of a plough attachment, on to an armoured conveyor. Power loaders of this type were introduced at Haig Pit about 1955. The armoured flexible conveyor was an important innovation. It was not only strong enough to bear the weight of the falling coal but also it could (because it was hinged) be "snaked" forward when the face advanced instead of having to be dimantled and reassembled further inbye. Coal could now be cut and loaded simultaneously and after travelling along the armoured conveyor be transferred to belt conveyors for the journey out of the colliery.[36]

Mechanized mining led to an enormous saving in manpower. Many fewer men were needed to cut, load and move coal and much less was demanded physically of the underground worker. This was accompanied by a striking improvement in the national overall output per manshift – from 21·5 cwts. in 1947 to 36·6 cwts. in 1966/7 and to 45·8 cwts. in 1972/3. In Cumberland the percentage of output power-loaded increased although for many years it was below the average for the industry as a whole; in the 1960's a major problem was that of finding machines which could cut and load coal in seams under three feet thick. In the following decade, however, over ninety per cent of Cumberland coal was power-loaded, a percentage closely approaching the national figure. But there was no large and immediate increase in productivity – overall output per manshift remained for many years among the lowest in Great Britain and it was not until 1981/2 that productivity at Haig Pit, the county's last deep mine, equalled that of the industry generally.[37]

THE MINERS

During the first twelve years of nationalization the number of employees in N.C.B. mines in West Cumberland changed very little, the maximum employed in any one year being 5,935 in 1953. After 1959, as more and more pits were abandoned, fewer men were required. Between January 1957 and January 1967 the size of the labour force fell from 5,641 to 3,058, a loss of 2,583 or 45·8 per cent; the national average loss of manpower for this decade was 41·6 per cent.[38] By 1977/8 there were only 1,100 miners at work in the coalfield, less than one-fifth of the total employed in 1947. Two years later this figure fell to 963; in 1982/3 it was no more than 793.[39]

Year	*No. of Employees*	*No. of Working Pits*
1947	5,590	12
1957	5,646	9
1967/8	2,906	3
1977/8	1,100	1
1978/9	1,031	1
1979/80	963	1

Women had been employed in the Cumberland coalfield since the eighteenth century, a practice which continued until recent times. In 1947 there were 130 women surface workers in the county, but in 1972 there were only two and both retired from Harrington No. 10 Colliery in June of that year. These two were employed as 'screen lasses', i.e. to pick out waste material from the coal carried through the washery, and they were the last women to work in collieries in the United Kingdom.[40]

The National Union of Mineworkers, which in 1945 replaced the Miners' Federation of Great Britain, has, with great skill, pertinacity and, on occasion, the conscious use of its vast economic power, been highly successful in improving the lot of its members. Since its foundation there have been few years in which this union has failed to squeeze some concession out of the National Coal Board. By an agreement made in 1947 (effective from 5 May) the normal working week for underground workers was to consist of five consecutive shifts of 7½ hours each plus one winding time and for surface workers 42½ hours, exclusive of meal times, worked in five consecutive shifts of 8½ hours. Those who worked five full shifts were to receive a bonus – daywagemen would receive an extra day's pay and pieceworkers sixteen per cent of their total earnings for the five shifts. Men who worked a Saturday shift would be paid at overtime rates for it, in addition to receiving the five day week bonus, if earned. It was hoped that the establishment of the 'five day week' would reduce the absenteeism of all mineworkers and increase the productivity of pieceworkers. This new system in effect gave the miners a large increase in wages and helped to place them among the most highly paid workers in the country.[41]

In the early years of nationalization the N.U.M. aimed at obtaining a seven hour day for underground workers and a forty hour week, including meal times, for surface workers. Its repeated applications for these terms were rejected and no progress was made until July 1960 when the National Reference Tribunal in its award reduced the hours of underground workers by one hour per week and those of surface workers from 42½ to 41½ hours per week exclusive of meal times. These arrangements came into effect 9 January 1961.[42] The union continued to press its claims on behalf of the surface workers and in 1964 accepted the Coal Board's alternative offer of seven rest days each year, the equivalent of 1¼ hours per week. Five years later the disagreement over surface hours led to an unofficial stoppage which lasted a fortnight and involved 140 collieries. A settlement was reached when the N.U.M., after an individual ballot of its members, accepted the Board's offer of forty hours exclusive of meal times, the reduction in hours taking effect in December 1969.[43] The surface workers obtained their original objective in November 1970 when their hours were reduced to forty per week inclusive of meal times totalling twenty minutes per shift, but by this time the N.U.M. was claiming a further reduction to 38¾ hours.[44]

In the first ten years of its existence the chief aim of the National Coal Board was to increase output and to attract men to the mining industry wage increases were necessary. As a result the national minimum weekly wage for adult mineworkers rose very quickly:-[45]

	Underground			*Surface*		
	£	*s.*	*d.*	£	*s.*	*d.*
Jan. 1947	5	0	0	4	10	0
Nov. 1947	5	15	0	5	0	0
Oct. 1950	6	0	0	5	5	0
Jan. 1951	6	7	0	5	10	0
Nov. 1951	7	0	6	6	1	6
Feb. 1953	7	6	6	6	7	6
Jan. 1954	7	15	0	6	15	0

Before the war miners' wages were based upon local agreements made in each district and wage rates for the same job varied widely from district to district. The progressive rises in the national weekly minimum wages helped to reduce the differences in district rates after 1947 and also compressed the range of rates for the different classes of workmen in a district. By the end of 1951 miners were the most highly paid of industrial workers, their wages since 1938 having risen faster and further than those in any other industry.[46]

A rational system of wages that would provide the same payments for the same work in all districts had still to be devised and in an attempt to do this the National Coal Board in 1955 created a national wage structure for the 400,000 daywagemen employed in the industry. This was a difficult task since it involved reducing 400 jobs to thirteen grades (five for underground workers, five for surface workers and three for craftsmen) and fitting within each grade jobs of approximately equal work value. National standard wage rates were fixed for each of the thirteen grades with differentials of 1s. per shift between all but one of the ten grades of mineworkers and 2s. 6d. between each of the three grades of craftsmen. This scheme was accepted by the N.U.M. on the understanding that no man's wage rate should be reduced; in fact about ninety-five per cent had their wage rates increased by at least 1s. per shift. The new system led to about seventy per cent of daywagemen being paid national standard rates, the rest retaining the higher rates they were earning before the agreement was made.[47]

National Standard Rates Per Shift

Men Underground			*Surface Workers*					*Craftsmen*				
Grade	*s.*	*d.*	*Grade*	*Men*		*Women*		*Grade*	*Underground*		*Surface*	
				s.	*d.*	*s.*	*d.*		*s.*	*d.*	*s.*	*d.*
I	31	9	IA	31	1	—		I+	35	5	32	1
II	30	9	I	27	5	22	6	I	32	11	29	2
III	29	9	II	26	5	21	6	II	30	5	27	1
IV	28	9	III	25	5	20	6					
V	27	9	IV	24	5	19	6					

The rapid and substantial increase in miners' earnings in post-war years is shown in the following table:

***Average Earnings per Manshift**[48]

	Cumberland		*Great Britain*	
Year	*s.*	*d.*	*s.*	*d.*
1947	28	8	28	10
1948	32	6	33	2
1949	33	5	34	4
1950	34	8	35	6
1951	38	2	38	10
1952	42	10	43	2
1953	45	8	45	5
1954	48	3	47	9
1955	50	10	50	11
1956	53	2	55	8
1957	56	3	60	0
1958	57	10	62	5
1959	59	0	63	11
1960	60	5	65	10
1961	64	5	69	7
1962	66	5	72	5

* Including the value of allowances in kind.

In 1948 average earnings per manshift in Cumberland were 32.6d., three times the 1938 average of 10s. 9d., and because their increases in earnings exceeded the rise in the cost of living the miners' real wages were substantially improved. In the first decade of nationalization earnings were doubled, but in the second decade, although wages continued to rise they did so at a much lower rate. After 1957 the fall in the demand for coal made it essential to check the rise in costs and prices and both the N.C.B. and the N.U.M. were aware that the large increases in wages of earlier years were no longer possible. Thus the average earnings per manshift (including the value of allowances) of all mineworkers rose from £3 in 1957 to £4.10s. in 1966/7, an increase of only fifty per cent.

Having established a national wages system for daywagemen the next step for the National Coal Board was to to do the same for pieceworkers. This called for the ending of the long tradition of piecework payments and the introduction of a daywage system for faceworkers. With the growth of mechanized mining payments based upon output became increasingly anachronistic as productivity came to depend more and more upon the use of machines than upon physical effort. The multiplicity of piece rates, which not only varied from district to district but also from one working face to another, required constant revision and renegotiation in every colliery and caused frequent disputes and stoppages. By an agreement between the Board and the National Union of Mineworkers in June 1966 a uniform, national shift rate for power-loading teams was to be established by the end of 1971. Initially the standard shift rate in each Wages District would remain but these separate shift rates would disappear gradually until a powerloader man's wage would be the same wherever he worked. At the end of 1966/7 55 per cent of all mechanized faces were being worked under the new agreement; in the next two years this percentage rose to 80 and then to 91 and by June 1971 the national wage structure was complete.[49] Faceworkers in Cumberland started working under the terms of the National Power Loading Agreement in June 1966. In 1967 their standard rate of payment was £3. 15s. per shift or £22. 10s. per week.[50]

This change to a daywage system was a major advance. Wages could be more effectively controlled, changes in shift rates could be negotiated on a national level instead of the old, colliery level and the number of disputes over wages reduced. The faceworker had the security of regular earnings in place of the fluctuating income under the former piecework system and if he moved to another face or another district he suffered no loss of wages. It now became easier for the Board to reduce the disproportionate gap between the earnings of faceworkers and those of other mineworkers. Between 1967/8 and 1970/1 the wage increases awarded to daywagemen exceeded those of faceworkers. With the cessation of local wage bargaining it has become possible for the first time to identify the wages of all pieceworkers; formerly this was impossible because these wages varied with the terms of the men's contracts.

Pursuing its policy of narrowing the gap between the earnings of ordinary daywagemen and those of power-loader men the Board, in 1967, increased the wages of power-loading teams by 7d. to 3s. 8d. per shift. Daywagemen, irrespective of grade, received an additional 12s. 6d. per week which was larger than the average increase awarded to power-loader men. The same pattern was followed in succeeding years:-[51]

Wage Increases per Week

Year	*Daywagemen*	*Faceworkers*
1968	15s.	3s. 6d. – 15s.
1969	27s. 6d.	24s.*
1970	£2·80 – £3	£2·37½

*Faceworkers in the Nottinghamshire and Kent Wages Districts (already the most highly paid) received increases of 20s. and 9s. per week respectively.

Despite the unprecedented contraction of the mining industry, from 958 producing collieries in 1947 to 317 in March 1969 and from an average of 711,000 wage-earners to 336,300 in the same period, relations between the N.U.M. and the N.C.B. were extremely good. That such a transformation was achieved smoothly says much for both sides. The continuing improvements in wages and working conditions since the beginning of nationalization removed many possible sources of grievance and the machinery set up in 1947 to to settle disputes at pit, district and national levels prevented serious trouble. It was sound policy on the part of the National Coal Board to keep the miners informed of future developments, to explain why changes were necessary and to minimize the adverse effects of necessary changes, e.g. the provision of jobs and houses for miners in other areas when their pits were closed.

Peace in the coalfields was broken in 1969 by an unofficial strike in some districts which lasted a fortnight and which was in support of a demand by surface workers for a reduction in hours of work. In the following year the Annual Conference of the N.U.M. passed a resolution to claim minimum wages of £20 on the surface, £22 underground and £30 for men conditioned to the National Power-loading Agreement. This was accompanied by another resolution, passed by a small majority, calling for strike action if the claim was not met in full. Wage increases offered by the Coal Board to meet this claim were rejected by the National Executive Council of the N.U.M. which recommended strike action to a ballot of members. Although the result showed a majority in favour of strike action it was not the two-thirds majority of those voting required by the rules of the Union before an official strike could take place. Negotiations were resumed and the National Coal Board raised their offer of a wage increase to £3 per week for the lowest grade of surface and underground workers. The day after this offer was made an unofficial strike began in Yorkshire which spread to other collieries in Wales and Scotland and lasted almost four weeks. In a second ballot of the miners a majority accepted the £3 per week increase so that the unofficial action had accomplished nothing for the men. It had, however, cost the industry nearly £13·5 million.[52]

The miners in Cumberland did not join in the unofficial strikes of 1969 and 1970. Considering the number of pit closures since 1947 and the reduction of manpower in the mines, in a district where employment opportunities for adult males were always difficult to find, a more restive attitude on their part would have been understandable. Although their leaders, Tom Stephenson, secretary of the Cumberland branch of the N.U.M. and his successor Maurice Rowe, strove vigorously, if unsuccessfully, to keep the pits open they were able to secure the peaceful acceptance by their members of the many changes which had taken place in the coalfield.

By 1971 it was becoming clear that a major conflict in the industry was approaching. Miners in high-wage areas resented the introduction of the daywage system which eliminated the wage differentials which had formerly existed between the different coalfields. The minimum weekly wage of £30 for faceworkers, £19 for other underground workers and £18 for surface workers no longer seemed adequate. Because of their modest wage increases in the previous ten years the mineworkers' earnings were exceeded by those of men in other industries and the miners were determined to regain their position as the highest paid industrial workers. At their annual conference in July 1971 the N.U.M. accepted a resolution demanding a minimum wage of £26 for surface workers,

£28 for underground workers and £35 for faceworkers. Significantly, at this conference union rules were amended so that the majority required in a strike ballot was reduced from two-thirds to fifty-five per cent of those voting. The Coal Board's offers of higher wages were rejected, a ban on overtime was imposed from 1 November and a ballot of the miners in December showed a majority of 58·8 per cent in favour of a national strike. Repeatedly the Board tried to have the dispute referred to the industry's National Reference Tribunal for settlement but the N.U.M. would not agree and appeared adamant in its refusal to modify its demands. A complete national strike began 9 January 1972 which lasted until 27 February. In Cumberland the strike proceeded peacefully. Miners picketed the opencast disposal centre at Maryport in an attempt to prevent opencast coal being moved there from opencast sites at Dearham and Clifton. In addition members of the N.U.M. took no part in the continuance of safety measures in Cumberland's two pits. At the Solway and Haig pits this important work was done by N.C.B. officials and staff aided by overmen, deputies and shot-firers whose union did not participate in the strike. To keep the pits safe these men had to work long hours every day seven days per week.[53]

As coal stocks diminished it became necessary to conserve fuel supplies and a state of emergency was declared. It was vital to end the strike as quickly as possible and with this in mind the government set up a Court of Inquiry under Lord Wilberforce to investigate the dispute. The Court sat in public 15-16 February and its report, issued 18 February, made it clear that the miners' claim should be given exceptional treatment. The Wilberforce report took into account the nature of employment in the mines, changes in the cost of living and the fall in recent years of the miners' relative earnings position.

It recommended wage increases (back-dated to 1 November 1971) of £5, £6 and £4·50 per week for surface, underground and faceworkers respectively which would give them a new minimum weekly wage of £23, £25 and £34·50. It also recommended that adult rates should be paid at the age of 20 immediately, moving progressively to 19 years and then 18 years at yearly intervals and that the miners should have five additional days of annual holiday. These terms were accepted by the N.U.M. Executive Council and in a national ballot (23 February) 96½ per cent of the miners voted for acceptance. In Cumberland 1,246 mineworkerse voted for the award and only seventeen against. Work at the collieries was resumed 28 February.[54]

There was much public sympathy for the miners during this strike and a growing awareness of the merits of their case. At the same time it came as a shock to many people to realize how much power could be wielded by a large, determined and well-organized union and the pressure it could exert upon the national economy.

Just over a year later, from 1 April 1973, all mineworkers received a further wage increase of £2·29 per week and more was to follow.[55] In this year the cost of living rose very rapidly and the miners were determined not to suffer the consequences of inflation or a fall in their relative pay position. In this they were helped immeasurably by the outbreak of the Arab-Israeli war and the quadrupling of oil prices during the winter of 1973-4. The Conservative government which had tried to control inflation by adopting a statutory prices and incomes policy had now to face an enormous increase in the cost of oil imports with adverse effects upon the balance of payments and the value of sterling.

The miners in putting forward their claim for higher wages refused to accept the government's limitations on wage increases and from 1 November 1973 again imposed a ban upon overtime working. They knew that oil was no longer a cheap alternative to coal and their intransigence was strengthened by the knowledge that the coal supplies which they controlled were more essential than ever to the nation's industries. All the pay offers made by the National Coal Board were rejected and in February 1974 eighty-one per cent of the miners voted in favour of a national strike which began 11 February and lasted until 10 March. In Cumberland 775 mineworkers (eighty-eight per cent of Haig Pit's N.U.M. members) voted for the strike and 105 against.

Again the Whitehaven miners, in accordance with their leaders' instructions, kept strictly within the law in their support of the strike. Picketing at the Carlisle and Barrow power stations to discourage the entry of fuel supplies was peaceful and orderly. At Haig Pit the task of ensuring that underground workings remained safe and clear fell to the manager and his staff as it had in 1972 and it was because of their efforts that work could be resumed immediately when the strike ended.[56]

The government declared a state of emergency, placed restrictions on heating and lighting to conserve coal and electricity and put British industry on a three-day week. In a final bid to end the struggle with the miners the prime minister, Edward Heath, called a general election in the hope of obtaining a popular mandate for his policy. In the event the election (28 February) led to his defeat and the return of a Labour government headed by Harold Wilson.

During the strike the Pay Board was asked to produce a report on miners' pay relativities and its report[57] when presented to Parliament showed that between 1948 and 1968 the average weekly earnings of mineworkers exceeded those of manual workers in manufacturing industries but by 1973 the wide gap in earnings had been greatly reduced.* Following its publication the National Coal Board made large concessions and reached a wages agreement with the National Union of Mineworkers 6 March. This agreement, which gave the miners substantial increases, was quickly accepted by Michael Foot, the new Secretary of State for Employment. The basic wage was raised to £32, £36 and £45 for surface workers, underground workers and faceworkers respectively. The miners returned to work 11 March.[58]

Not only did this strike cost the Coal Board about £68 million it also did great damage to the national economy by crippling industrial production. The two national strikes differed significantly from miners' strikes in earlier years. Formerly it had been possible to starve pitmen into surrender but the age of the omnipotent employer had now come to an end and power seemed to have been transferred to the men. After the 1974 strike public sympathy for the miners seemed to wane. It was not merely irritation at the inconveniences and discomfort resulting from the strike, there was also an uneasy fear that a new kind of "overmighty subjects" had emerged capable of overthrowing an elected government. In addition every increase in miners' wages led inevitably to higher coal prices which were passed on by industrial consumers – particularly the electricity and steel industries, the Coal Board's two biggest customers – to their customers. For the householder coal and electricity mounted in price year by year.

* The Pay Board Report shows that in 1969 and 1970 the average earnings of these industrial workers exceeded those of the mineworkers.

The continuing inflation of recent years and the fact that every rise in the price of oil strengthened the bargaining position of the mineworkers made it almost impossible for the Coal Board to resist demands for further wage increases. Wages soared rapidly after 1971 as exemplified in the following of minimum weekly wage rates.

Year	*Faceworkers*	*Others Underground*	*Surface Workers*
	£	£	£
1971*	30·00	19·00	18·00
Nov. 1971	34·50	25·00	23·00
March 1973	36·79	27·29	25·29
March 1974	45·00	36·00	32·00
March 1975	61·00	47·00	41·00
March 1976	67·00	53·00	47·00
March 1977	70·35	55·65	49·50
March 1978	78·45	61·95	54·85
March 1979	84·95	68·45	61·35
March 1980	101·95	82·15	73·65
1 Jan. 1981	111·95	90·20	80·85
2 Nov. 1981	121·55	97·95	87·80
1 Nov. 1982	130·30	105·00	94·10

*Before the Wilberforce settlement.

Within seven years the minimum wages of faceworkers more than doubled and those of other mineworkers trebled. Between the wage settlements of March 1974 and March 1975 "threshold" rises of £4·40 per week were paid to all miners to compensate for higher living costs. Annual wage settlements in subsequent years were conditioned by the government's policy on incomes and inflation. Thus in March 1976 there was a general increase of £6 per week, which represented an average increase of about eleven per cent on mineworkers' wage rates, and in the following year the increase was five per cent of gross earnings up to a maximum of £4 per week. In March 1978 the increase was limited to ten per cent and twelve months later there was a flat rate increase of £6·50 for all mineworkers. In addition the rate paid for achieving standard performance under the incentive scheme was raised from £23·50 to £26·50 per week.[59]

Higher wages and increased mechanization did not halt the decline in output, the production of saleable mined coal falling from 139·78 million tons in 1969/70 to 127·02 million tons in 1972/3 and to 114·99 million tons in 1974/5. In Cumberland output fell from 477,162 tons in 1969/70 to 302,370 tons in 1974/5. An incentive scheme devised by the Coal Board in an attempt to keep up production was rejected by a national ballot of the miners in October 1974. But in the following March they accepted a production bonus scheme based upon total national output. This proved to be ineffective. Only one bonus was paid under this scheme. This was during the first quarter of 1975/6 and was based upon results in the last quarter of the previous financial year. No further production bonus payments were made during 1975/6 because the national output at no time reached the base level necessary; in fact the output of 112·44 million tons in 1975/6 was less than the output of the previous year. The National Coal Board realized that a successful incentive scheme related to national output was impossible and that what was needed was one close to the point of production. The individual miner probably believed he could not affect national results but he could influence production in his own pit.[60]

A former Director of the South Notts. Area of the N.C.B. has argued cogently that the decline in production and productivity can be attributed to the daywage system which contains no element of incentive. "So long as a man attends at the pit for a shift, he is paid a fixed amount of money, no matter how much or how little work he actually does. The rates paid are national ones and apply equally to all coalfields, regardless of production or productivity." In view of what happened later his proposals to deal with the situation are worth noting, "As a suggestion, why should not 75 per cent of a man's wages be guaranteed as a day-wage, with the balance being an incentive reward for diligent and industrious work. Let there be fair and adequate rewards for those who earn them. The incentive for the producer must obviously be higher than that for the 'back-up' man."[61]

As output continued to fall the N.C.B. tried again, in 1977, to produce an acceptable and realistic incentive scheme. This involved a modification of the daywage system by the introduction of a piece rate element for additional output and a return to local wage bargaining. Under this scheme there would be a locally assessed target output figure for each pit and faceworkers would be paid a productivity bonus on all output above 75 per cent of the target figure, the bonus amounting to £23·50 for reaching 100 per cent. Other skilled workers underground would receive 65 per cent of the bonus payable, others underground 50 per cent and surface workers 40 per cent. Since there would be no upper limit output in excess of the standard performance would lead to higher bonus payments. In this way part of the faceworker's earnings would depend directly upon output i.e. upon his personal efforts.[62] In a national ballot of the miners the scheme was rejected by 110,634 votes to 87,901; in Cumberland the miners at Haig Pit voted 362 for and 393 against the proposal.[63] In less than a month the Cumbrian miners reversed their position and voted in favour of negotiating their own incentive scheme at local level. Despite the result of the national ballot the executive of the N.U.M. decided (8 December) to allow locally negotiated productivity agreements and by the end of March agreements had been reached for the whole industry. The incentive scheme proved successful; in its first full year of operation output per manshift at the face rose by nearly eight per cent.[64]

Those who objected to these arrangements asserted that the pursuit of higher output and earnings would lead to the neglect of safety precautions in the pits and create envy and rivalry between miners in different collieries, or, in the words of Arthur Scargill, the militant leader of the Yorkshire miners, "put man against man, pit aganst pit and area against area."

With the adoption of the incentive scheme came large additions to the miners' pay packets. At Haig Pit the first production bonus in January 1978 gave faceworkers £27 extra, other underground workers £13·50 and surface workers £11. These payments were made after the deduction of tax; the payment to faceworkers before tax was £40.[65] Nationally the average weekly production bonus amounted to about £22·50 for face and development workers and about £14·50 for colliery workers as a whole.[66] In 1980/1 the incentive pay for achieving standard performance was increased by £3·50 to £30 per week. Since its inception the incentive scheme has contributed to the greater efficiency of the industry by helping to bring about higher productivity at the face. Higher productivity brought its reward. Average payments under the incentive scheme in 1980/

1 were about £37·50 for face and development workers and about £22·50 for colliery workers as a whole;[67] in the following year these payments were about £40 and £25 per week respectively.[68] The rise in the average weekly earnings (including allowances) of mineworkers from £127·02 in 1979/80 to £167·21 in 1981/2 reflects their gain from these productivity arrangements.[69]

In recent years wage negotiations have become more difficult because of the militant attitude adopted by some of the miners' leaders who have put forward unrealistic wage claims, adamantly opposed the closure of unprofitable collieries and who have been too ready to invoke the strike weapon to enforce their claims. The wage negotiations in the latter half of 1980 were marred by the strong opposition of extremist leaders like Arthur Scargill and Michael McGahey who urged the miners to reject the Board's proposals although a majority of the executive of the N.U.M. had recommended their acceptance. In the November ballot the mineworkers, influenced by the views of their moderate president, Joe Gormley, voted to accept the new pay deal. In this ballot the Cumberland miners voted 515 for and 236 against acceptance.[70] This agreement, effective from 1 January 1981, was only for a period of ten months. It was followed early in 1981 by a series of widespread, unofficial strikes because the National Coal Board, in a plan put to the miners, proposed to increase the efficiency of the industry by accelerating the closure of certain high-cost collieries. A national stoppage seemed likely. The government, which at first had adopted the attitude that this was a dispute for the industry itself to settle, soon realized that a confrontation with the miners might lead to a national strike like that which had helped to bring down the Heath government in 1974. The Prime Minister, Mrs. Thatcher, had therefore to capitulate to the miners and agree to provide more money for the coal industry and accordingly the plan to close pits was abandoned 18 February.[71]

In a national ballot held in January 1982 the miners agreed to accept a new wages settlement, effective from 2 November 1981, despite the efforts of Arthur Scargill and other militants to persuade them to reject it. In July the National Union of Mineworkers put forward a demand for a general wage increase of £27 per week. Two months later the Coal Board's offer to raise the basic rates of surface workers of £87·80 by £6·30 and those of underground workers receiving between £97·95 and £121·55 by £7·05-£8·75 was rejected by the union's leaders. Arthur Scargill, the new president of the union, insisted on linking the pay issue with pit closures, claiming that to accept these proposals would mean supporting the government and the National Coal Board, risking pit closures, the loss of jobs and lower living standards. This was followed by an overtime ban which cost the miners an average of £40 per week in lost earnings. Scargill embarked upon an impassioned campaign among the miners to win support for a national strike for higher wages and to prevent further pit closures. But in the ballot held at the end of October the miners voted against a strike by 125,233 votes to 81,592; in Cumberland the miners at Haig Pit voted 371 to 209 against a strike. This meant that the miners would accept the new pay scales and the overtime ban was immediately lifted.[72]

This was a resounding defeat for Scargill and his fellow militants. Many miners had shown by their votes their unwillingness to be swayed by inflammatory hyperbole and their resentment at the loss of earnings during the period of the controversial overtime ban. Others felt they could not afford to go on strike with a consequent loss of income, particularly those burdened by large mortgage payments. There were also those who believed that the latest pay claim was impracticable, that the call for a national strike was untimely because there were some fifty million tonnes of coal stockpiled at the pits and power stations and that a prolonged strike could lose markets for their output and lead to the closure of more pits.

Improvements in the position of the miners were not restricted to changes in wages and hours of work. From April 1947 the men no longer had to pay their share of the operating costs of pithead baths, the National Coal Board taking over the whole of this responsibility.[73] Miners in Cumberland and Lancashire, unlike those in other coalfields, were not entitled to concessionary coal, a grievance which led to a strike of the Lancashire men in 1949.[74] In the following year the Board agreed to supply six tons of coal to each miner-household in both counties at 21s. per ton below the current retail price. As coal prices rose later the price of concessionary coal was adjusted to ensure that the miners still had cheap coal. In 1977 Cumbrian mineworkers received six tons of coal per annum at a delivery charge of £2·20 and retired miners four tons per annum at the same price.[75]

In 1947 miners were allowed one week's holiday and six statutory holidays. A second week's holiday was introduced in 1953 and a third week in 1975. The working year was further reduced by the adoption of "rest days" in 1964 when they were given four colliery rest days, i.e. days on which the colliery would be idle and three individual rest days, i.e. days on which coal production continues but a miner can take the day off by giving a week's notice. By 1975 miners could claim a total of three week's annual holiday, seven statutory holidays, eight individual rest days and four colliery rest days. Payment for these holidays is at the standard grade rate.[76]

Little had been done to provide compensation for the dependants of miners killed at work until the introduction of a regular system of compensation for fatal accidents in February 1950 by which a miner's widow received £150 for herself and £50 for each eligible child. From January 1976 the widow was entitled to an immediate payment of £1,000, followed by six annual payments of £250 for herself and £50 for each child.[77]

As more and more collieries were closed the National Coal Board had to tackle the growing problem of shedding superfluous manpower and for elderly miners redundancy was a serious threat. These redundant colliers, unemployed through no fault of their own, were unlikely to find work elsewhere because of their age, inclination and lack of experience outside mining. To help them the Redundant Mineworkers (Payments Scheme) Order was made 24 June 1968 by which miners aged 55 or over who became redundant between 17 July 1967 and 28 March 1971 were to receive for a period of three years, or, if shorter, until the age of 65, a weekly benefit designed to make up their unemployment benefit to approximately 90 per cent of their previous net earnings. Mineworkers retired at the age of sixty or over would receive an immediate miner's pension; other redundant mineworkers covered by the Scheme would receive the pension at the end of the three-year period.[78]

Retired as well as redundant miners needed support and for years the N.U.M. worked diligently to remove some of the fears of old age. To lessen the financial hardship that started with retirement a mineworker's pension to supplement the exiguous state pension was obviously desirable. A voluntary pension scheme was started 1 January 1952 and most miners were quick to join it, but from 1 January 1960 membership became a condition of employment. At first the weekly pension was £1 which rose in subsequent years:[79]

March 1968	£1·50	Sept. 1975	£4·60
April 1973	£3·00	Sept. 1976	£5·23
Sept. 1974	£3·60		

The doubling of the pension in 1973 would normally have entailed an increase of twenty-five pence per week in members' contributions but the Board granted an allowance of this amount to the mineworkers so that no extra cost fell on them. From April 1975 the pension scheme was related to earnings with the result that miners who continued working and whose earnings increased annually could look forward to higher pensions. Miners paid 5 per cent of their gross weekly earnings into the pension fund and the Coal Board paid a similar amount plus a substantial deficiency contribution. The size of the pension depends upon length of service and amount of earnings, the maximum pension being half of the miner's earnings after forty-five years in employment. On retirement the miner also receives a tax-free lump sum equal to three years' pension. The contributions of the miners and the Coal Board were raised from 5 to 5¼% per cent in 1979 with a further increase in the Board's deficiency contribution.[80]

Continuing its constant search for additional benefits the N.U.M. began to press for early retirement for its members and in 1976 was claiming retirement at sixty for the miners, with a further reduction in stages to fifty-five by June 1980. In a national ballot in December 78 per cent of the mineworkers voted in favour of using industrial action if necessary in support of this claim; in Cumberland 665 mineworkers (84 per cent) voted for and 128 against industrial action.[81] Negotiations between the Coal Board and the N.U.M. led to an agreement upon a voluntary scheme of early retirement, operative from 1 August 1977, by which miners with twenty years' service underground could opt for retirement at the age of sixty-two. They would receive a tax-free lump sum of £500 and, up to the age of sixty-five, weekly taxable payments making up State benefits for unemployment or sickness to slightly less than two-thirds of their average weekly gross earnings.[82] In another national ballot 55 per cent of the N.U.M. voted in favour of the scheme; at Haig Pit 478 miners voted in favour of the scheme (63 per cent) and 281 against it.[83] The age for early retirement was reduced to sixty-one in 1978 and to sixty in 1979. In the latter year weekly benefits were improved by raising the upper limit of earnings (the basis on which benefits were calculated) from £80 to £110. From April 1978 the scheme was extended to include men with between five and twenty years' service underground who had worked in the industry for twenty-five to thirty-five years.[84]

In 1947 the nation was shocked into a knowledge of the blood cost of coal when 104 out of 118 men working underground at William Pit, Whitehaven, were killed by an explosion of fire-damp. This disaster, next to the worst in the history of the Cumberland coalfield, was caused by the firing of a shot which ignited an accumulation of gas in the No. 2 dip longwall face, the flame of the explosion traversing several working faces, nearly 2,000 yards of roadway and nearly half a mile of the main intake haulage road. In view of the accident record of the Whitehaven pits safety standards in this area needed to be well above the average and the inspector who conducted the enquiry into this disaster felt obliged in his report[85] to make a number of specific recommendations:

1. The appointment of a full-time specialist mining engineer with an intimate knowledge of mine explosions and fire hazards for the collieries in the Whitehaven district.
2. A high standard of ventilation.

3. Adequate dust prevention and suppression measures.
4. The elimination as far as practicable of all possible sources of ignition.
5. Continuous and close supervision by responsible officials of all operations on all shifts.
6. The maintenance of a high standard of discipline. Although this was the last mining disaster* in the county fatal accidents and serious injuries continued to occur.

Falls of roof and gas explosions accounted for most of the victims, but, as had happened too frequently in the past, neglect of regulations and carelessness played a large part. At Risehow Colliery two men were killed in 1957 by an explosion of fire-damp issuing from a feeder in the Lower Three Quarters Main Drift where, because of inadequate ventilation, the percentage of gas had reached a dangerous level. The faulty condition of the shot-firing cables caused a spark which ignited the fire-damp. In his report[86] on this accident the inspector said, "It is with regret that I point out that there would have been no loss of life had there been compliance with two simple precautions attached to the Approval in regard to the use of the exploder, namely, 'The shot-firer's station shall be at least 150 yards from the shot-holes' and 'Before firing the shots the shot-officer shall see that all other persons are in proper places of safety not less than 200 yards from the shot-holes.' "

In 1957 five men were killed and twenty-three seriously injured in the Cumberland pits and in the following year seven were killed and seven seriously injured. The increase in fatalities was disturbing and should not have happened according to the investigating inspector who "continues to find that most of these accidents are avoidable."[87] In later years there was a welcome reduction in the number of fatal accidents; in the ten years 1959-68 fifteen men were killed and in the seven years 1972-8 only one man died in the pits. Undoubtedly the main cause of this improvement was the closure of collieries and as the years went by fewer men were at risk. Other factors helped. Improvements in ventilation and shot-firing methods lessened the danger of gas explosions, stronger roof supports meant reduced casualties from falls of roof and mechanized mining proved much safer than manhandling coal.

COSTS AND LOSSES

The following table shows the enormous increase since 1947 in the cost of mining coal both in Cumberland and in the industry generally.

	Cumberland		*Great Britain*	
Year	*Wages Cost Per Ton* s. d.	*Total Cost Per Ton* s. d.	*Wages Cost Per Ton* s. d.	*Total Cost Per Ton* s. d.
1947	34 3·6	56 3·4	26 11	41 3
1957	65 1·2	112 10·3	48 3	81 6
1966/7	73 8	163 3	48 2	98 5
1976/7	£11·05	£37·93	£8·84	£18·87
1977/8	£16·20	£54·84	£9·76	£21·75
1978/9	£15·63	£47·27	£11·18	£24·10
1979/80	£16·81	£48·03	£12·89	£28·57
1980/1	£14·08	£43·03	£14·99	£33·56
1981/2	£12·35	£39·62	£15·59	£36·60
1982/3	£17·14	£54·95	£16·18	£39·23

* The term "disaster" is restricted to mining accidents causing the death of ten or more persons.

In the decade 1947-57, when increasing output mattered more than costs, total costs doubled because of higher wages and the continuing operation of uneconomic pits. Costs in Cumberland rose by about 44 per cent in the next decade and between 1966/7 and 1976/7 they were more than quadrupled. Costs continued to rise, from £37·93 per ton in 1976/7 to £54·84 in 1977/8, an increase of almost 45 per cent in one year, and by the latter year they were nearly twenty times what they had been in 1947. In the next four years production costs fell below the peak figure of £54·84, but rose to £54·95 per ton in 1982/3.[88]

Large cost increases were incurred in every coalfield during this period, but costs in Cumberland were always among the highest; in 1966/7 they were the highest in the country and in 1976/7 twice the national average costs. Geological difficulties, the expensive reconstruction of old collieries, the low output per manshift, further diminished by the reduction of working hours per shift from 7½ to 7¼ in 1968, and annual wage increases were responsible for this situation. These wage awards were national, without regard to the productivity or profitability of individual districts and bore particularly heavily upon areas like Cumberland. Nor was the establishment of a daywage system conducive to high productivity so long as it contained no element of incentive.

Wages Cost as % of Total Cost

Year	*Cumberland* %	*Great Britain* %
1947	60·95	65·25
1957	57·68	59·21
1966/7	45·13	48·98
1976/7	29·12	46·84
1977/8	29·54	44·87
1978/9	33·06	46·39
1979/80	35·00	45·11
1980/1	32·72	44·00
1981/2	31·18	42·60
1982/3	31·19	41·30

In a labour intensive industry such as coal mining wage costs form a large proportion of total costs and in Cumberland in 1947 this proportion amounted to nearly 61 per cent. By 1957 total costs had doubled but the wages factor had fallen to just under 58 per cent. Despite the concurrent increase in miners' earnings the wages content of total costs fell sharply during the next twenty years, to 45·13 per cent in 1966/7 and to 29·12 per cent in 1976/7. This can be attributed to the closure of unprofitable pits and to the growth of mechanization, particularly the adoption of power loaders and powered supports.

Coal prices had, of course, to move upwards with production costs. From 46s. 9·3d. in 1947 the price of Cumberland coal more than doubled to 101s. 7·4d. per ton in 1957 and then rose to 123s. 2d. in 1966/7. Ten years later the price was £22·87 per ton and in 1977/8 £40. 10s, more than seventeen times the price in 1947.[89]

Despite these large price increases the Cumberland coalfield was run at a loss and since 1947 had only once shown an annual profit, in 1981/2. Losses have been especially high in recent years when only the Haig Pit was in production:

Year		Loss Amount £	Loss Per Ton s. d.	Loss Per Employee £
1947		514,679	9 6·1	92·07
1957		590,704	11 2·9	104·62
1966/7		1,325,000	40 1	415·23
1974/5		2,220,940	£7·35	1,761·25
1975/6		3,420,052	£9·80	2,819·49
1976/7		4,414,273	£15·06	3,811·98
1977/8		3,132,535	£14·74	2,847·75
1978/9		4,401,648	£16·62	4,269·29
1979/80		3,619,758	£13·42	3,758·84
1980/1		1,118,422	£3·00	1,192·35
1981/2	*	1,908,000	£4·05	2,163·27
1982/3		2,373,000	£7·75	2,992·43

* Profits are underlined

Between 1947 and 1959 annual losses ranged from £383,430 to £902,198 and then fell sharply in the next four years; in the fifteen months ending March 1964 losses amounted to only £56,000, or 11d. per ton. This improvement did not last long. In 1965/6 losses exceeded £1 million for the first time and from then onwards continued to increase; in the later 1970's Haig Pit lost between £2 million and £4 million per annum. In 1978/9 there was a deficit of £4,401,648 or £16·62 per ton. Three years later came a sudden reversal of fortune when Haig Colliery made a profit of £1,908,000, equal to £4·05 per ton. This improvement did not last for it was followed in 1982/3 by a loss of over £2 million.[90]

For many years the Cumberland coalfield has been maintained by profits made by other coalfields; it could not have survived by its own efforts. It is not, however, the only coalfield continuing to operate despite heavy, annual losses. This has given rise to much adverse criticism and to demands that unprofitable collieries or coalfields should be closed as soon as possible, demands that have become more urgent with the appearance in recent years of a large surplus of coal.

Clearly financial considerations have not been the sole criterion in determining the continuation of deep mining in Cumberland. No private colliery owner or colliery company would have continued to operate a loss-making enterprise for thirty-five years. But the National Coal Board recognized from the start that they "cannot be guided by business considerations alone" and that it would not always be in the national interest to close down unprofitable pits. In particular, "social costs" had to be taken into account, especially the unemployment that would follow such closures. Thus the Board, although obliged by the Nationalization Act to cover their costs, taking good years with bad, were prepared to allow the continuance of unprofitable pits, areas or coalfields and recoup themselves out of the profits made by others.[91] The Cumberland coalfield, like others in Great Britain, has for years been subsidized by the tax-payer and the Board's efforts to provide employment for miners have been costly.

The financial record of the National Coal Board in Cumberland is better than the statistics above suggest since they do not take into account the profits, said to be substantial, from opencast workings in the county. Unfortunately, it has been impossible to obtain any figures of the production costs and profits of opencast mining in Cumberland and hence a complete statement of the coalfield's financial performance cannot be made.

OPENCAST MINING

Opencast mining in this country began in 1942 and in 1952 responsibility for these operations was transferred to the National Coal Board. This is a method of winning coal at comparatively shallow depths and rarely goes beyond a depth of 300 feet. Since the Coal Mines General Regulations (1952) prohibit the deep mining of coal at depths of less than 150 feet such coal can only be removed by opencast workings.

Private firms holding contracts from the Board do the work. First the top soil and subsoil are removed and then the overburden is stripped by huge excavators to expose the coal seam. The coal is extracted by mechanical shovels or similar machines and the overburden is backfilled as the work proceeds. Much of the heavy expenditure inseparable from deep mining is avoided in opencast mines, e.g. the costs of shafts, roof supports, underground roads, electrification, ventilation, haulage and winding. At opencast sites coal production begins quickly compared with the long delay between sinking the shaft of a deep mine and the beginning of coal getting. With lower production costs than deep mines opencast workings have earned substantial profits and done much to offset the losses incurred by collieries using traditional methods. In addition, opencast mining can work seams which are too thin for deep mines which cannot work a seam of less than eighteen inches and the average quality of opencast coal is higher than that from deep mines.

This new development was not welcomed by miners who feared they might lose their jobs. Others objected to the loss of agricultural land for years at a time, to the noise of opencast operations and to landscapes being spoiled by the appearance of mountains of soil. By 1957 the "coal at any price" policy had to be abandoned and strenuous efforts made to lower production costs. It was therefore desirable to increase the output of opencast coal since it was the cheapest to produce. Opencast production, i.e. coaling, started later in Cumberland than in other coalfields, at Stubsgill, near Distington, and Broughton Moor in April 1958. Opencast output in Cumberland amounted to 91,176 tons in 1958 and 102,808 tons in the following year. Coaling ceased at Broughton Moor in June 1959 and at Stubsgill in March 1960. As a result only 19,584 tons of opencast coal were raised in 1960.

The fall in the demand for coal after 1957, caused largely by the advent of cheap fuel oil and economies in fuel consumption, threatened to create unemployment among miners and one of the many steps taken by the Board to meet this situation was to reduce the scale of opencast mining despite its high profitability. Opencast working in Cumberland was not resumed until 1964. This followed an agreement in 1962 between the National Union of Mineworkers and the National Coal Board, the former consenting to the resumption of opencast mining in the county and the latter, in return, agreeing to extend the life of Siddick Pit, then facing closure, until the end of 1964.[92] Presumably the Board believed opencast profits would exceed Siddick Pit losses and in Cumberland such profits were highly desirable to offset the continuing losses of deep mines.

Prospecting by the Opencast Executive went on over a wide area of West Cumberland in a search for new sites. Coaling began at the Blooming Heather site, near Dearham, in August 1964 and continued until June 1975. This site, 745 acres in extent, produced a total of 1,306,250 tonnes of coal. Between May 1966 and May 1967 81,984 tonnes of opencast coal were obtained at Wythemoor House, near Branthwaite. Land restoration is an essential part of opencast working and when coaling ceased at the various sites the land was restored to agriculture. Continuous opencast production calls for forward planning based upon the results of drilling operations to locate new sites which would be ready to yield coal when existing sites were worked out.

Coaling at Mabel Plantation, near Workington, began in April 1970 and continued until April 1979. This site, 240 acres in extent, produced 1,728,257 tonnes of coal. The National Coal Board's 1974 "Plan for Coal" aimed at increasing opencast output from ten to fifteen million tons per year; for Cumberland this would mean raising annual production from about 350,000 tons to about one million tons. To meet this need new sites had to be opened out. Production began in 1974 at Low Close, Little Broughton on a 255 acre site containing about a million tons of power station and industrial coal. This site was extended by the acquisition of adjoining land providing additional reserves of about 800,000 tons. A 417 acre site at Outgang near Branthwaite, estimated to contain about 520,000 tons of coal, began to produce coal in April 1977. Whilst this work was going on much valuable coal was won by removing the pillars of coal left as roof supports in old, shallow mines.

As opencast mining developed opposition in the county to its further development began to grow. An application by the Opencast Executive in November 1976 to obtain coal from a 541 acre site at Oughterside containing two million tons of coal led to a public inquiry in the following March. Many objectors opposed the scheme, basing their arguments on agricultural, amenity and environmental considerations. Their views were upheld by the government inspector who led the inquiry and he recommended against an authorization to work the site. His recommendation was rejected by the Secretary of State for Energy, Mr. Anthony Benn, who asserted that the site ought to be worked in the national interest and that the value of the coal and the employment opportunities provided outweighed any comparatively minor disadvantages that might arise. The site was authorized in March 1978 and coaling began in February 1980.[93]

An application in June 1978 to win coal by opencast operations at the 399 acre Moresby and Keekle site was followed by another public inquiry in March-April 1979.[94] Authorization to work this site, which is expected to provide 1·1 million tonnes of coal within seven years, was given in February 1980 and coaling began towards the end of the year.[95] Other sites were opened out. At Pica, near Distington, production started in 1981 on a 193 acre site containing 625,000 tonnes of coal. Early in the same year coaling began on a small site at Woodside, Broughton Moor containing 30,000 tonnes of coal and was completed in twelve months.

As new sites came into production the output of opencast coal rose quickly, from 108,725 tons in 1969 to 221,185 tons in 1970 and to 383,904 tons in 1971. From 1974 onwards the annual output of opencast coal exceeded that of Haig Pit and in 1982 was nearly 900,000 tons, the highest yet achieved.

Output of Cumberland Opencast Coal

Year	*Tons*	*Year*	*Tons*
1964	51,827	1974	361,276
1965	138,395	1975	388,526
1966	145,386	1976	376,991
1967	194,976	1977	374,206
1968	63,407	1978	406,796
1969	108,725	1979	304,060
1970	221,185	1980	549,551
1971	383,904	1981	887,989
1972	378,620	1982	891,882
1973	324,701		

Between 2 April 1958, when opencast production began in Cumberland, and 29 March 1980 a total of 4,547,663 tonnes of opencast coal was raised in the county. All the opencast coal is sent to a large disposal point at Maryport capable of handling up to a million tonnes per annum. There it is graded, blended and despatched by rail. It is sold to power stations, to the steelworks at Workington and other local industries and, until recently, a substantial proportion was shipped from Whitehaven to Ireland. It is unfortunate that trading ceased at the port of Maryport in 1961; had the port remained open to traffic the export of opencast coal would have been much easier.

Although opposition to opencast mining appears to be growing in Cumberland it must be said that this new departure in mining has contributed much to the economic life of the area. At a time when more coal was needed the steadily increasing output of opencast coal was all the more valuable as the output from deep mines declined. Regrettably, no information about profits is available but opencast mining is known to be highly profitable and in a county where employment opportunities for adult males are few and becoming fewer, it provides jobs for hundreds of them. Fortunately there are substantial coal reserves in the county that can be worked by opencast for many years to come and this should help to ease the employment situation.

Some doubts have been expressed about the efficacy of the measures taken to restore the land after opencast sites have been worked out. Nevertheless, this work has reclaimed much derelict land and vastly improved it by the removal of industrial scars which have marred the landscape for generations.*

THE COAL TRADE

With the post-war contraction of the mining industry in Cumberland there was less coal available for export and the coal trade suffered accordingly. Shipments of North East coal from the Cumberland ports continued until 1953, but on a rapidly diminishing scale. This was particularly unfortunate for Maryport which had benefited greatly from such shipments during the war years. The importance of North-East coal for the trade of Maryport and the effects of its withdrawal can be seen in the following figures.

Maryport Coal Shipments (Tons)[96]
Year ending 30 September

1947	374,902	(366,264)
1948	157,705	(134,900)
1949	75,642	(46,003)
1950	84,963	(30,622)
1951	87,355	(19,681)
1952	67,700	(4,133)
1953	24,953	(3,138)
1954	34,406	—

Figures in brackets denote N.E. coal included in total shipments.

* Much of the information and all the statistics in this section were provided by the NCB Opencast Executive, Northern Region, Newcastle-upon-Tyne.

Pit closures in the vicinity of Maryport also contributed to the decline of its trade. Birkby & Allbright Colliery closed in 1950, Ellenbank Colliery in 1952 and Gillhead Colliery in 1959. By 1958 total coal shipments from Maryport amounted to only 18,970 tons and in the following year fell to 8,409 tons. The beginning of opencast production in Cumberland in April 1958 generated a glimmer of hope for the port. In the year ending September 1959 1,758 tons of opencast coal were shipped from Maryport but six months later opencast mining ceased and was not resumed until 1964. This was too late for Maryport for trading stopped in 1960 and in the next year the port was closed to traffic. The death of the port ended an important chapter in the history of the coal trade for Maryport had for many years been the leading port in Cumberland for the export of coal. Now it can only be used by small fishing vessels.[97]

As happened at Maryport, the coal trade of Workington was badly hit when the export of North East coal came to an end and by the closure of neighbouring collieries. In post-war years annual shipments of deep-mined coal from Workington never reached 100,000 tons. After the war shipments of North East coal dwindled rapidly and ceased in 1952 when only 7,763 tons were exported. All the coal raised in the Workington area, however, was not available to the coal trade, much of it being retained for use in the local ironworks. As six nearby collieries closed between 1959 and 1973 there was even less coal for sale to the ships. As a result coal shipments from Workington fell from 36,000 tons in 1965 to between 1,000 and 3,000 tons between 1966 and 1973. From the closure of Solway Colliery in 1973 until 1979 the Workington coal trade was at a standstill; there was no coal locally either for shipping or for the ironworks.[98]

Until 1959 annual coal shipments from Whitehaven exceeded 200,000 tons and then fell year by year until the late 1960's when they amounted to less than 100,000 tons. There was a slight recovery in the coal trade between 1970 and 1973 but in 1974 only 70,516 of coal were shipped. Several factors accounted for the variations in the coal trade. The closure of William Pit in 1954 and of Walkmill Pit in 1961 lessened the quantity of coal available for export, as did the two national strikes in 1972 and 1974. For many years much of Cumberland's coal output was consumed by the Workington Iron & Steel Company and as collieries near Workington ceased production more coking coal had to be bought from Haig Pit. After 1973 this was the only deep mine in the county and most of its output went to the Workington steelworks. It is not surprising, therefore, that in 1977 Whitehaven coal shipments amounted to no more than 65,852 tons and some of this was opencast coal sent to Whitehaven from the opencast disposal point at Maryport.[99]

In 1978, however, the situation altered. Coal working ceased in the Bannock Band seam of Haig Pit and was concentrated in the Upper Metal Band seam where the coal was not suitable for the steel industry. As a result coking coal for the Workington steelworks had to be imported from other areas and Haig Pit had to ship more of its coal.[100] In the next three years there was an increase in Whitehaven coal shipments which reached 182,080 tonnes in 1980, the highest figure since 1960. But in 1981 this figure was nearly halved, only 95,804 tonnes being shipped.[101]

A change of policy by the National Coal Board led to the revival of the coal trade at Workington at the expense of Whitehaven. It was obviously more economical to move opencast coal from Maryport to Workington rather than to Whitehaven for shipment and at Workington vessels of up to 10,000 tons could be accommodated. After a gap of six years coal shipments from Workington were resumed in 1980 when 8,554 tonnes of coal were exported; in the following year 402,638 tonnes left the port for Ireland, Finland, Denmark, Germany and Spain. But the system of moving coal to the ships at Workington was much less efficient than it might have been. Coal from the Maryport disposal point was carried to the dock by lorries and loaded on to the ships by means of cranes and grabs originally designed to handle cargoes of pig iron. The construction of a new coal handling depot at the Prince of Wales Dock, Workington, begun in April 1982 and completed in August 1983 at a cost of £3·4 million, has brought about considerable improvement. Now the coal is transported by rail direct to the depot and then shifted by a system of conveyor belts from the trains to the ships. The new system, more efficient and economical than the old one, has helped to reduce traffic congestion on the busy Maryport-Workington road and the villagers of Flimby will no longer have to suffer from lorryborne coal dust. It is hoped that the adoption of these new methods will make it possible for Workington to export up to a million tonnes of coal each year. Opencast mining has been a godsend to the port of Workington and its future as *the* port for the export of coal seems assured. Coal shipments from Whitehaven ceased in January 1982 and all coal from Cumberland is now shipped from Workington.[102]

Thus after more than three centuries the port of Whitehaven has withdrawn from the coal trade, or perhaps more accurately, been removed from it. For the past twenty years coal exports from Whitehaven have been declining and coal has ceased to be the biggest item in the trade of the port. The fall in coal shipments has been offset by the import of phosphate for the chemical works of Marchon Products Ltd., situated to the south of the town. These imports began in 1954 and since 1963 the tonnage of imported phosphate has greatly exceeded that of coal exports; between 1979 and 1981 Whitehaven imported over 500,000 tonnes of phosphate rock a year.[103]

Notes and References

1 NCB, *Reports*.
2 Ministry of Fuel and Power, *Lists of Mines*, 1947-50. NCB, *Report* 1952, 21.
3 *Whitehaven News*, 23 Dec. 1954, 15 Sept. 1966.
4 *Cumberland Star*, 30 Jan., 13 Feb. 1959.
5 *Whitehaven News*, 30 Nov. 1961.
6 *Report of the Cumberland Development Council Ltd.* (Whitehaven, 1962).
7 *Whitehaven News*, 9 May, 20 June 1963.
8 *West Cumberland Times*, 16 July 1966.
9 NCB, *Report 1967/8*, 41. *Report of the Cumberland Development Council Ltd.* (Whitehaven, 1968).
10 *Report of Divisional Inspector of Mines*, 1966.
11 *West Cumberland Times and Star*, 27 July 1968.
12 NCB, *Reports 1947-60*.
13 Information provided by the NCB Western Area, Public Relations Department. *West Cumberland Times and Star*, 3 Feb. 1973.
14 NCB Western Area, Public Relations Department.
15 *Ibid.*
16 *Ibid.*
17 J. Y. Lancaster and D. R. Wattleworth, *The Iron and Steel Industry of West Cumberland* (Workington, 1977), 10.
18 *West Cumberland Times and Star*, 27 Sept. 1980.
19 NCB, *Reports 1947-1967*. Mr L. R. Hoten, Area Statistician, Western Area.

[20] NCB, *Report 1974/5*.
[21] NCB, *Reports 1947-1981/2*. NCB Western Area, Public Relations Department.
[22] *Ibid.*
[23] *West Cumberland Times and Star*, 6 May 1978.
[24] *Report of Inspector of Mines* (Northern Division), 1947.
[25] NCB, *Report 1950*, 18.
[26] *Report of Inspector of Mines* (Northern Division), 1952.
[27] *Reports of Inspector of Mines* (Northern Division), 1950-2, 1961, 1971. NCB, *Reports 1951-3*.
[28] *Report of Inspector of Mines* (Northumberland and Cumberland Division), 1957.
[29] *Reports of Inspector of Mines* (Northumberland and Cumberland Division), 1962-3.
[30] *Report of Inspector of Mines* (Northumberland and Cumberland Division), 1961.
[31] *Report of Inspector of Mines* (Northumberland Division), 1971. *West Cumberland Times and Star*, 26 Aug. 1971.
[32] *Report of Inspector of Mines* (Northumberland and Cumberland Division), 1959.
[33] *Report of Inspector of Mines* (North Western Division), 1964.
[34] NCB Western Area, Public Relations Department.
[35] Griffin, *op. cit.*, 115. Lord Robens, *Ten Year Stint* (1972), 96-7. I. Berkovitch, *Coal on the Switchback: The Coal Industry since Nationalization* (1977), 190.
[36] NCB, *Reports*. NCB Western Area, Public Relations Department. Robens, *op. cit.*, 96-7. G. L. Reid, K. Allen and D. J. Harris, *The Nationalized Fuel Industries* (1973), 33-4.
[37] NCB, *Reports*. *Report of Inspector of Mines* (Northumberland and Cumberland Division), 1963.
[38] E. M. Knight, *Men Leaving Mining; West Cumberland 1966-7. Report to the Ministry of Labour* (University of Newcastle-upon-Tyne, 1968).
[39] NCB, *Reports*. NCB Western Area, Public Relations Department.
[40] *West Cumberland Times and Star*, 26 June 1972.
[41] NCB, *Report 1948*, 33; *Report 1956*, 70.
[42] NCB, *Report 1958*, 36; *Report 1960*, 29-30.
[43] NCB, *Report 1969/70*, 35.
[44] NCB, *Report 1970/1*, 38.
[45] NCB, *Reports 1947-54*.
[46] NCB, *Report 1951*, 73-4.
[47] NCB, *Report 1955*, 38-9.
[48] NCB, *Reports 1947-62*.
[49] NCB, *Reports 1966/7-71/2*.
[50] N.U.M., *Annual Report 1967*.
[51] NCB, *Reports 1967/8-70/1*.
[52] NCB, *Report 1970/1*, 36-7.
[53] NCB, *Report 1971/2*, 32-3. *Whitehaven News*, 13 Jan.-3 Feb. 1972.
[54] NCB, *Report 1971/2*, 34-5. Records of the Cumberland Miners' Association.
[55] NCB, *Report 1972/3*, 6.
[56] *Whitehaven News*, 7 Feb.-14 Mar. 1974. Records of Cumberland Miners' Association.
[57] Pay Board, *Relative Pay of Mineworkers*, Cmnd. 5567 (1974).
[58] Records of the Cumberland Miners' Association.
[59] NCB, *Reports 1970/1-81/2*.
[60] NCB, *Reports 1969/70-75/6*. Records of Cumberland Miners' Association.
[61] N. R. Smith in the *Daily Telegraph*, 22 Dec. 1976.
[62] Records of Cumberland Miners Association. NCB, *Report 1977/8*.
[63] *Whitehaven News*, 3 Nov. 1977.
[64] NCB, *Report 1978/9*, 18.
[65] *West Cumberland Times and Star*, 14 Jan. 1978.
[66] NCB, *Report 1978/9*, 18.
[67] NCB, *Report 1980/1*, 22.
[68] NCB, *Report 1981/2*, 20.
[69] NCB, *Report 1981/2*, 32-3.
[70] Cumberland Miners' Association records. *Daily Telegraph*, 2 Dec. 1980.

[71] *Daily Telegraph*, Feb. 1981. NCB, *Report 1980/1*, 3.
[72] *Daily Telegraph*, 7 July, 24 Sept., 27 Oct., 1, 3 Nov. 1982.
[73] NCB, *Report 1947*, 121.
[74] NCB, *Report 1949*, 85.
[75] Information provided by Mr W. S. Proud, Area Agent of the Cumberland Miners' Association.
[76] *Ibid.*
[77] *Ibid.*
[78] NCB, *Report 1967/8*, 42.
[79] Information provided by Mr W. S. Proud.
[80] *Ibid.* NCB, *Report 1978/9*, 18.
[81] *Daily Telegraph*, 16 Dec. 1976.
[82] NCB, *Report 1976/7*, 14.
[83] *West Cumberland Times and Star*, 5 Feb. 1977.
[84] NCB, *Reports 1977/8-78/9*.
[85] *Report on explosion at William Pit, Whitehaven, 15 Aug. 1947*, Cmnd. 7410 (1948).
[86] *Report on explosion at Risehow Colliery, 26 Nov. 1957*, Cmd. 396 (1958).
[87] *Report of Inspector of Mines* (Northumberland and Cumberland Division), 1958.
[88] NCB, *Reports 1947-1979/80*. Mr L. Hoten, Area Statistician, Western Area and his successor, Mr G. W. Wood.
[89] *Ibid.*
[90] *Ibid.*
[91] NCB, *Report 1947*, 119.
[92] *Report of the Cumberland Development Council Ltd.* (Whitehaven, 1962), 1962.
[93] *West Cumberland Times and Star*, 29 Oct. 1977, 6 May 1978.
[94] *Ibid.* 14 April, 20 Oct. 1979.
[95] *Ibid.* 16 Aug., 22 Nov. 1980.
[96] Maryport Harbour records.
[97] *Ibid.*
[98] Workington Harbour records.
[99] Whitehaven Harbour records.
[100] *West Cumberland Times and Star*, 27 Sept. 1980.
[101] Whitehaven Harbour records.
[102] Workington Harbour records. *West Cumberland Times and Star*, 14 Nov. 1981, 15 Jan. 1983, 13 Aug. 1983.
[103] Whitehaven Harbour records.

CONCLUSION

For four centuries coal has been mined continuously in Cumberland. In the early days mining was on a small scale, undertaken by freeholders who worked outcrop coal or sank shallow pits. Large scale, efficient mining was beyond their means and impossible until the many small collieries in separate ownership were consolidated. This was first achieved in the Whitehaven area by the Lowthers who had the wealth and determination necessary to remove their competitors. Both Sir John and Sir James Lowther, aided by competent agents, not only exploited the coal deposits on their own estates but added to them by investing large sums in the acquisition of more coal bearing land. Their wealth enabled the Lowthers to build up a colliery empire which, under a succession of highly skilled colliery managers, was not only very profitable but also technically the most efficient in the country. Other Cumbrian landowners – the Christians, the Curwens and the Senhouses – followed the Lowther example in sinking more and more pits in their estates and applying the mining methods initiated at Whitehaven. But the landed gentry, like the small freeholders, had in turn to relinquish their grip on the coalfield in the late nineteenth and early twentieth centuries when industrialists took over most of the collieries. In 1947 the pits were nationalized and in less than forty years all, except one, were closed.

As the industry developed coal was won at greater depths and at coastal collieries further out to sea. There were improvements in drainage and ventilation but the main business of mining, the extraction of coal, changed very little until recent times. Since the war mechanization has revolutionized coal mining. Coal is removed from the face, not by haggers wielding picks and shovels, but by power loaders and underground haulage, once the work of trailers and pit ponies, has been taken over by conveyor belts. The physical effort demanded of underground workers has been further reduced by the adoption of self-advancing, powered roof supports. These innovations have brought an enormous increase in productivity. Improvements in lighting, dust suspension, shot-firing and ventilation have made mining much safer but it is still dirty, difficult and dangerous.

Changes in mining methods have been accompanied by a vast change in the status of the miners who now hold a position that would have been inconceivable fifty years ago. For centuries they were held in scant regard and to many they seemed to be barbarians and almost a race apart. When the demand for coal was high their wages compared favourably with those of other manual workers but in times of depression, unemployment and short-time working caused real suffering. Economically the miners were at the mercy of the employers; strikes, usually of brief duration, rarely succeeded and could be broken by evicting the pitmen from their homes. Today the miners are the the most highly-paid industrial workers and few would deny them this position. They are, therefore, much better fed, housed and clothed than in the 1930's, enjoy holidays with pay and have been freed from the former, oppressive fear of poverty in old age. On retirement the miner obtains a tax-free lump sum and a substantial pension in addition to his national insurance pension and, like all other citizens, has access to the services provided by the welfare state.

It cannot be said that the improvement in the position of mineworkers has proceeded from some form of enlightenment but rather that it followed the constant pressure exercised by the large and well-organized National Union of Mineworkers determined to obtain the best possible conditions for its members. In this aim the task of the union was made such easier by the large increases in oil prices since 1973. It is no longer possible for employers to exploit miners and it is difficult for the National Coal Board and the government to resist the demands of pitmen. For too long the miners were impotent in the face of hostile employers but there is now a feeling that they have too much power. In 1974 they toppled the Heath government and nearly brought the national economy to a standstill.

What matters most is how this power is used. Much, of course, depends upon the quality of the leadership, but, as the events of 1982-3 have shown, perhaps more important are the good sense and realism of the miners themselves. Aware of what they stand to lose the majority of them are reluctant to embark upon strikes – the old cry of "one out all out" has little force. The moderates do not support excessive wage claims or the absurd proposition that no pit should close until the last piece of coal has been removed. As the demand for coal declines and oil prices fall the bargaining position of the miners is weakened and they see that it would be foolish to price themselves out of jobs by striking in support of inflated wage demands or exhausted pits.

This attitude was clearly shown in March 1983 over the proposed closure of the Lewis Merthyr Colliery, near Pontypridd. This was an old colliery whose brittle roof and soft floor made mechanized mining difficult. Its productivity rate was much below the national average and heavy losses were incurred. Although the men at Lewis Merthyr were offered jobs at nearby collieries, travelling expenses and transfer allowances of up to £1,500, the South Wales miners went on strike in opposition to the closure and Arthur Scargill called an emergency meeting of the nationl executive of the N.U.M. in an attempt to obtain unanimous support for a national strike over threatened pit closures. Many miners, however, wanted a national ballot before a strike was called and in this way they were supported by the national executive whose members voted unanimously in favour of a national ballot to decide the issue. This was a rebuff for the militants and the national ballot of 8 March was a tremendous defeat, the miners voting 118,954 (61 per cent) to 76,540 (39 per cent) against strike action. In Cumberland the miners at Haig Pit voted 314 (58 per cent) to 226 (42 per cent) against a strike.*

Moderate miners resented an attempt to push them into a strike before a national vote had been taken. The majority opposed a national strike for much the same reasons as they had in October 1982. Highly paid miners had no desire to lose wages or possibly their jobs in the event of a prolonged strike. Nor was the time right – there were large stockpiles of coal, better weather was approaching and oil prices were falling. It was evident that excessive quantities of expensive coal had been produced, that the survival of uneconomic pits like Lewis Merthyr was largely responsible for the National Coal

* *Daily Telegraph*, 3-11 March 1983.

Board's huge deficit and that the government would have to find enormous sums to cover these losses, money that would be better invested in new pits. If production had to be restricted obviously the least efficient pits had to be closed. It is significant that even in the militant Yorkshire and Scottish coalfields less than fifty-five per cent of the miners (i.e. the minimum percentage needed to authorize a strike) voted in favour of a strike. Clearly the verbal froth of extremist union leaders did not appeal to the majority of mineworkers.

As coal mining developed in West Cumberland the coastal strip changed from an isolated, sparsely-populated agricultural area into a busy, industrial zone. Since the coal was mined primarily for export to Ireland shipbuilding grew up to provide the ships to carry the coal from the rapidly growing ports of Whitehaven, Workington and Maryport. Railways appeared to move coal from the pits to the ports. Local deposits of iron ore, coal and limestone made it possible for Workington to become an important centre for the manufacture of iron and steel. Unfortunately, drastic changes in the industrial situation have led to West Cumberland becoming once more a "depressed area". Shipbuilding disappeared some fifty years ago, iron ore mining ceased in 1980 and in the last three years thousands of employees have been paid off by the Workington Iron and Steel Company. The contraction of the coal industry has added to the high level of unemployment; once the coalfield employed ten thousand men, now it provides work for less than one thousand. Maryport has been worst hit by the industrial slump. Coal mining, like shipbuilding and iron manufacture, has come to an end, the harbour is silted up and the few, small factories on the industrial estate are having a struggle to survive.

Permanent production in an extractive industry is impossible. Pit closures necessarily followed the exhaustion of coal reserves and now Haig Pit is the only working pit in the county.* This has been distressing for the Cumbrian miners and also for their sons who have little prospect of finding employment in an occupation that is largely hereditary. The Cumberland coalfield has been dying for many years and but for nationalization mining would have ceased a long time ago. Under private ownership it could not have survived; it is an ailing industry kept alive by infusions of cash provided by the state.

If the survival of Haig Pit depends upon its ability to make profits then it must surely close for geological difficulties alone make profitability a remote possibility. Its closure would be most unfortunate. Hundreds of men would be added to an excessively large number of unemployed in an area where employment opportunities are becoming rarer day by day. It is true that Haig Pit has been losing money for years but to close it could be expensive – redundancy payments, earlier pensions for those who are eligible and unemployment benefit would have to be provided for the discharged workmen. In addition, industrial and domestic consumers would have to pay higher prices for coal imported from other coalfields if supplies of opencast coal were insufficient to meet local needs.

In contrast to deep mining in post-war years opencast mining in Cumberland has been a distinct success. There has been a steady increase in the annual output of opencast coal which in 1982 approached 900,000 tonnes. This compensated for the marked decline in

* Since completion of this text, the position has changed – production at the Haig ceased in August 1984 and the pit closed completely in March 1986.

the production of deep-mined coal and opencast profits could be offset against the National Coal Board's losses from the deep mines. The growth of opencast mining has brought much-needed employment to the county, the number of contractors' employees working on the opencast sites rising from 180 in 1977 to 480 in 1982. This new branch of the coal industry seems to have a bright future. Drilling operations have shown that there are large reserves of coal available capable of providing work for years for the men engaged in it.

Perhaps a final word may be said on behalf of the consumers who appear to have been neglected in most studies of the coal mining industry. Though they are said, collectively, to own the industry, they have little say in its running. Year after year these new colliery owners have to pay rising prices for their fuel and are obliged to pay the taxes necessary to subsidize the losses made by the industry. Industrialists complain of the burden imposed upon them by high energy costs, but they can at least try to pass on their cost increases to their customers. For the domestic consumer there appears to be no way of escape. Miners can strike successfully for higher wages but a consumers' strike for lower coal prices, if not impossible, would have little chance of success. For too many of the older and poorer members of society a reasonable standard of home heating in the winter months has become an unattainable luxury.

APPENDIX I

GLOSSARY

Adit. A road or tunnel driven into a hillside for the purpose of transport, drainage or ventilation.
After-damp. A suffocating gas left in a mine after an explosion of fire-damp.
Bank. The pit surface.
Banker. On reaching the surface a loaded corf was placed on a tram by the banker and then moved to the coal bank and emptied. The empty corf was then returned to the pitmouth.
Banksman. Surface overlooker. Responsible for keeping accounts of coal sent up by each hewer and for seeing that the baskets were full.
Bearmouth or day-hole. An opening in the side of a hill leading to the interior of a drift mine.
Bearmouth roader. Kept the road clear to facilitate movement in and out of the mine.
Blower. Escape of gas in coal mine.
Bord and Pillar. A system of working coal. Coal was excavated at intervals along the face – the excavated areas were the 'bords' – and pillars of coal were left untouched between the bords to support the roof.
Brattice. A vertical wooden partition which divided a pit shaft from top to bottom. Designed to assist ventilation, air passing down one side and up the other.
Cavilling. The allocation of working places periodically by the drawing of lots.
Chaldron. A measure of capacity. The Newcastle chaldron was equal to 53 cwts. and the imperial chaldron about 28 cwts.
Checkweighman. Employed by the miners themselves to check the weighing of coal sent up the pit.
Choke-damp (also known as *black damp, stythe*) Miner's term for carbonic acid gas. Could kill miners by suffocation.
Corf. Container for coal. The early corves were made from timber. Later, they were large, wicker baskets made from hazel rods.
Corver. Made and repaired corves.
Cribs. Wooden rings secured at intervals in a pit shaft from top to bottom. These cribs, joined together by close-joined lengths of timber, provided a complete shaft lining to prevent water flowing into the pit.
Drift mine. A mine driven from the surface. In a drift mine a shaft is not needed to reach the working places.
Fault. Break in continuity of strata caused by earth movement. As a result of this dislocation a coal seam may be depressed on one side of a fault (a downthrow) or raised up (an upthrow).
Filler. Filled the corves with coal cut down by the hewers.
Fire-damp. Miner's name for carburetted hydrogen. An inflammable gas, explosive when mixed in certain proportion with air.
Gatewatcher. Person employed to attend to the gateway of a bearmouth, regulating traffic in and out of the bearmouth.
Goaf or *Gob*. The waste area from which coal has been removed.

Guides or *conductors*. Two pairs of guide rails set vertically on opposite sides of the shaft to keep the cage steady. Slides fitted to the sides of the cage moved up and down the shaft in the channels between the pairs of guide rails.
Hagger. Hewer.
Hooker. Removed empty baskets from the winding rope and hooked on filled baskets.
Inbye. Towards the coal face. Conversely, *outbye*, movement from the face towards the shaft.
Level. A watercourse.
Odman. Employed to drive headings. This was very hard work since he had to hew against the cleat (i.e. the grain of the coal) and his coal output was therefore much less than that of the hewer working in the bords.
Rolley. A train or set of corves or tubs drawn by a horse to the pit bottom.
Slatepicker. Employed at the bank to pick out slate and refuse.
Sough. A drainage adit.
Stage whipper. In charge of pony drivers underground. Responsible for the movement of coal underground to the shaft and for keeping the roads clear.
Staple. An underground shaft connecting one seam with another.
Stopping. A wall built underground to seal off the air.
Sump. Pit or well for the reception of water at the foot of the pumps.
Thirl. To bore through.
Tracer. Worked with horses. A trace-horse draws in traces as distinct from a shaft horse. Where more than one horse was involved it was more practical that one should be placed in front of the other rather than side by side. Each horse was harnessed direct to the load with a supplementary harness between the first and second horse.
Trailer. Employed underground to haul loaded sledges or trams.
Trum. Four-wheeled carriage on which corves were placed.
Tub. A small waggon for transporting coal.
Tubbing. Shaft lining made of wood or iron.
Upcast. A shaft carrying the return air upwards and out of the pit.
Waste. Areas of a mine from which the coal has been excavated.

APPENDIX 2

Owners of the Lowther Whitehaven Estate

1630 Sir John Lowther of Lowther (1581-1637) bought the manor of St. Bees for £2,450. His son.
Sir John Lowther of Lowther (1605-1675), 1st bart granted the estate 5 Jan. 1639 to his brother.
Christopher Lowther (1611-1644) who was created a baronet in 1642. Succeeded by his son.
Sir John Lowther (1642-1706) 2nd bart. He disinherited his elder son Christopher (1661-1731) the estate passing to his younger son.
Sir James Lowther (1673-1755) 4th bart.* He died without issue and bequeathed the Whitehaven estate to.
Sir William Lowther of Holker (1727-1756). Succeeded by.
Sir James Lowther of Maulds Meaburn (1736-1802). Created Earl of Lonsdale 1784. Died without issue, the earldom becoming extinct. Succeeded by.
Sir William Lowther of Swillington. Created Earl of Lonsdale 1807. Died 1844. Succeeded by his son.
William (1776-1872) 2nd Earl. Succeeded by his nephew.
Henry (1818-1876) 3rd Earl. Succeeded by his son.
St. George Henry (1855-1882) 4th Earl. Succeeded by his brother.
Hugh Cecil (1857-1944) 5th Earl. Succeeded by his brother.
Lancelot Edward (1867-1953) 6th Earl. Succeeded by his grandson.
James Hugh William (b. 1922) 7th Earl.

* Christopher Lowther inherited the baronetcy in 1706 and died in 1731, the baronetcy passing to his younger brother James.

APPENDIX 3

Owners of the Egremont Cumberland Estate

1682-1748 Charles Seymour, Duke of Somerset (b. 1662), became owner of the estate on his marriage in 1682 to Elizabeth Percy (b. 1667, d. 1722), daughter of the 11th Earl of Northumberland. Succeeded by his son.

1748-1750 Algernon Seymour, Duke of Somerset (b. 1684). Created 1st Earl of Egremont in 1749. After 1750 the family estates were split, the Cumberland section passing to his nephew.

1750-1763 Sir Charles Wyndham (son of Catherine Seymour), 2nd Earl of Egremont (b. 1710).

1763-1837 George O'Brien Wyndham, 3rd Earl of Egremont (b. 1751). The estate passed by will on his death to his illegitimate sons; the earldom of Egremont, inherited by his brother's son George Francis Wyndham, became extinct in 1845. Succeeded by his second natural son.

1837-1860 General Sir Henry Wyndham. Held a life interest in the Cumberland estate under the will of his father. Succeeded by

1860-1869 George Wyndham (b. 1787), first natural son and adopted heir of the 3rd earl. Created 1st Baron Leconfield in 1859. Succeeded by his son.

1869-1901 Henry Wyndham, 2nd Baron Leconfield (b. 1830). Succeeded by his son.

1901-1952 Charles Henry Wyndham, 3rd Baron Leconfield (b. 1872). Succeeded by his brother.

1952-1963 Hugh Archibald Wyndham, 4th Baron Leconfield (b. 1877). Succeeded by his brother.

1963-1967 Edward Scawen Wyndham, 5th Baron Leconfield (b. 1883). Succeeded by his son.

1967-1972 John Edward Reginald Wyndham, 6th Baron Leconfield (b. 1920). Created 1st Baron Egremont 1963. Succeeded by his son.

1972 John Max Henry Scawen Wyndham, 7th Baron Leconfield. 2nd Baron Egremont. (b. 1948).

APPENDIX 4

The Curwens

- Sir Nicholas Curwen d. 1604
  - Sir Henry Curwen d. 1623
    - Sir Patricius Curwen d. 1664
    - Thomas Curwen d. 1672
    - Eldred Curwen d. 1673
      - Henry Curwen d. 1725
- Thomas Curwen of Sella Park 1590-1653
  - Darcy Curwen of Sella Park 1643-1722
    - Henry Curwen 1680-1727
      - Henry Curwen 1728-1778
        - Isabella Curwen 1765-1820 = John Christian (Curwen)* 1756-1828
          - Henry Curwen 1783-1860
            - Edward Stanley Curwen 1810-1875
              - Henry Fraser Curwen 1834-1900
                - Alan Delancy Curwen 1869-1930
                  - Isabel Mary Curwen 1897-1967 m. Frederick Selby Chance
    - Eldred Curwen 1692-1745
      - Jane Curwen m. John Christian (1719-1767)
        - John Christian (Curwen)* 1756-1828 = Isabella Curwen 1765-1820

* John Christian (1756-1828) married Isabella Curwen in 1782. He adopted the family name of Curwen in 1790. His first wife, Margaret Taubman, whom he married in 1775, died in 1778.

APPENDIX 5

Colliery Statistics

Lowther Collieries

a. Output (Year from Lady Day to Lady Day)

	Whitehaven Collieries										Lattera Colliery	
Year	*Howgill*		*Greenbank*		*Yard Band*		*Knockmurton*		*Total*			
	Tons	*Lo.*	*Tons*	*Lo.*	*Tons*	*Lo.*	*Tons*	*Lo.*	*Tons*	*Lo.*	*Tons*	*Lo.*
1695	15,196	4	2,321	0					17,517	4	1,387	4
1696	13,705	0	1,782	4					15.487	4	1,199	0
1697	18,763	0	1,864	0					20,627	0	1,312	4
1698	17,495	2	1,802	7					19,298	1	855	0
1699	16,021	4	1,002	0	1,211	3			18,234	7	2,130	0
1700	18,655	1			2,699	5	4,347	5	25,702	3	1,584	0
1701	15,580	4	2,993	5	13,406	4			31,980	5	2,430	0
1702	20,591	6	2,343	6	7,804	5			30,740	1	3,126	2
1703	32,253	6	2,227	4	1,855	3			36,336	5	4,533	0
1704	28,362	1	2,877	0					31,239	1	2,035	7
1705	27,540	6	3,055	0					30,595	6	2,830	4
1706	23,942	7	2,471	5					26,414	4	4,114	4
1707	33,713	6	791	0					34,504	6	4,599	4
1708	34,208	1							34,208	1	3,150	0

Year	Whitehaven Collieries: Howgill Tons	Lo.	Whingill Tons	Lo.	Total Tons	Lo.	Parton Collieries (Lattera, Gunerdine, Birketts, Quarterbanks) Tons	Lo.
1709	37,534	2			37,534	2		
1710	40,232	4			40,232	4		
1711	35,968	1			35,968	1		
1712	38,951	2			38,951	2		
1713	42,114	4	257	0	42,371	4		
1714	48,148	1	4,792	0	52,940	1	2,347	5
1715	42,680	4	9,869	5	52,550	1	6,697	2
1716	45,778	2	13,836	0	59,614	2	6,751	2
1717	54,739	7	10,973	0	65,712	7	7,017	4
1718	48,526	1	10,366	5	58,892	6	6,172	0
1719	50,222	7	12,452	5	62,675	4	3,914	3
1720	41,883	3	9,694	1	51,577	4	5,297	0
1721	41,883	5	9,193	1	51,076	6	5,256	0
1722	40,693	4	4,043	5	44,737	1	4,931	4
1723	48,109	3	3,174	5	51,284	0	3,948	0
1724	43,648	2	4,200	1	47,848	3	4,982	4
1725	56,381	5	8,377	4	64,759	1	6,652	1
1726	66,033	2	15,049	7	81,083	1	8,155	4
1727	50,848	6	17,849	1	68,697	7	6,857	2
1728	54,930	2	17,217	5	72,147	7	5,531	2
1729	55,215	2	15,484	2	70,699	4	5,596	4
1730	61,385	0	20,501	2	81,886	2	5,094	2
1731	69,776	0	17,674	3	87,450	3	5,103	6
1732	70,337	3	13,907	2	84,244	5	3,321	2
1733	71,298	0	14,014	6	85,312	6	4,447	4
1734	81,407	1	14,765	5	96,172	6	3,219	7
1735	86,035	6	7,312	7	93,348	5	2,693	4
1736	95,606	4	6,299	0	101,905	4	2,858	4
1737	84,030	2	6,588	6	90,619	0	2,459	4
1738	74,537	1	8,537	0	83,074	1		
1739	77,144	2	8,884	4	86,028	6		
1740	88,801	3	8,419	1	97,220	4		
1741	82,524	5	7,328	0	89,852	5		
1742	76,722	2	5,966	4	82,688	6		
1743	80,846	3	4,452	2	85,298	5		
1744	82,091	1	5,122	2	87,213	3		
1745	95,775	5	6,615	0	102,390	5		
1746	105,288	1	9,461	6	114,749	7		
1747	91,078	0	11,733	2	102,811	2		
1748	92,544	1	12,656	0	105,200	1		
1749	91,586	7	11,644	6	103,231	5		
1750	108,209	7	17,918	2	126,128	1		

b. Coal Shipments

	Whitehaven Collieries										Lattera Colliery	
Year	*Howgill*		*Greenbank*		*Yard Band*		*Knockmurton*		*Total*		*Tons*	*Lo.*
	Tons	*Lo.*	*Tons*	*Lo.*	*Tons*	*Lo.*	*Tons*	*Lo.*	*Tons*	*Lo.*		
1695	15,671	0	1,208	0					16,879	0	761	0
1696	16,271	4	554	0					16,825	4	769	4
1697	21,757	0	1,809	4					23,566	4	726	0
1698	18,205	3	1,638	4					19,843	7	511	0
1699	17,277	4	2,061	7	1,362	1			20,701	4	1,787	0
1700	17,460	4			2,619	7	5,208	4	25,288	7	1,077	4
1701	15,041	4	1,953	7	13,470	6			30,466	1	2,189	3
1702	21,013	0	1,106	3	5,484	7			27,604	2	2,579	3
1703	29,470	5	893	4	1,184	7			31,549	0	3,949	0
1704	24,171	6	1,125	1					25,296	7	1,808	3
1705	24,780	6	1,113	5					25,894	3	2,652	1
1706	20,415	4	1,140	0					21,555	4	3,372	3
1707	29,623	5	548	2					30,171	7	4,099	2
1708	29,210	0							29,210	0	3,256	2

Year	Whitehaven Collieries						Parton Collieries (Lattera, Gunerdine, Birketts, Quarterbanks)	
	Howgill		*Whingill*		*Total*			
	Tons	*Lo.*	*Tons*	*Lo.*	*Tons*	*Lo.*	*Tons*	*Lo.*
1709	32,813	3			32,813	3		
1710	35,129	4			35,129	4		
1711	30,244	0			30,244	0		
1712	36,155	4			36,155	4		
1713	41,456	4	129	2	41,585	6		
1714	40,165	0	3,550	2	43,715	2	1,083	4
1715	44,252	0	7,188	0	51,440	0	6,684	5
1716	41,689	4	11,348	3	53,037	7	6,201	0
1717	53,528	0	7,569	5	61,097	5	6,527	4
1718	49,012	7	7,070	3	56,083	2	4,944	2
1719	48,511	1	8,304	5	56,815	6	1,498	0
1720	40,154	1	6,405	4	46,559	5	3,798	5
1721	40,789	6	9,088	7	49,878	5	5,078	4
1722	43,518	1	3,639	5	47,157	6	5,878	2
1723	46,117	1	2,842	7	48,960	0	3,193	0
1724	47,259	4	4,155	6	51,415	2	5,447	4
1725	58,590	4	7,820	2	66,410	6	5,726	1
1726	63,788	0	11,459	6	75,247	6	5,328	3
1727	60,317	3	18,273	3	78,590	6	5,899	7
1728	51,220	7	12,410	0	63,630	7	6,514	3
1729	58,239	3	15,493	0	73,732	3	5,805	4
1730	62,968	4	18,259	6	81,228	2	5,375	5
1731	67,250	2	17,858	6	85,109	0	4,704	2
1732	69,698	5	11,611	4	81,310	1	3,235	0
1733	66,633	0	12,277	7	78,910	7	4,330	5
1734	84,143	3	10,591	0	94,734	3	3,187	6
1735	88,243	4	7,114	5	95,358	1	2,579	2
1736	81,893	0	5,053	2	86,946	2	2,712	7
1737	80,897	0	3,416	7	84,313	7	2,214	4
1738	78,255	0	4,071	7	82,326	7		
1739	90,775	0	4,631	1	95,406	1		
1740	92,032	0	4,511	3	96,543	3		
1741	76,590	0	2,764	5	79,354	5		
1742	76,899	0	1,841	5	78,740	5		
1743	94,906	0	1,452	5	96,358	5		
1744	87,233	0	1,690	4	88,923	4		
1745	91,250	0	2,699	1	93,949	1		
1746	95,240	0	5,074	7	100,314	7		
1747	97,970	0	5,828	1	103,798	1		
1748	91,416	0	5,498	5	96,914	5		
1749	107,943	0	4,840	2	112,783	2		
1750	106,812	0	10,806	3	117,618	3		

c. Local Sales and Fire Coal

Year	Whitehaven Collieries										Lattera Colliery	
	Howgill		*Greenbank*		*Yard Band*		*Knockmurton*		*Total*			
	Tons	*Lo.*	*Tons*	*Lo.*	*Tons*	*Lo.*	*Tons*	*Lo.*	*Tons*	*Lo.*	*Tons*	*Lo.*
1695	1,752	3	458	6					2,211	1	262	4
1696	2,527	1	536	2					3,063	3	236	2
1697	2,435	0	245	2					2,680	2	77	7
1698	2,791	3	422	1					3,213	4	359	2
1699	2,158	1	106	6	89	3			2,354	2	827	1
1700	1,406	5			118	5	139	7	1,665	1	88	2
1701	1,175	2	573	4	228	7			1,977	5	88	4
1702	796	5	1,397	3	2,421	6			4,615	6	93	4
1703	3,457	3	1,250	2	734	0			5,441	5	197	5
1704	4,456	7	1,387	6					5,844	5	109	0
1705	3,261	5	1,299	1					4,560	6	99	6
1706	3,326	2	958	2					4,284	4	629	2
1707	3,608	6	400	0					4,008	6	251	4
1708	4,828	0							4,828	0	142	1

Year	Whitehaven Collieries: Howgill Tons	Howgill Lo.	Whingill Tons	Whingill Lo.	Total Tons	Total Lo.	Parton Collieries (Lattera, Gunerdine, Birketts, Quarterbanks) Tons	Lo.
1709	4,717	4			4,717	4		
1710	4,233	5			4,233	5		
1711	3,661	3			3,661	3		
1712	3,306	2			3,306	2		
1713	3,096	2	47	6	3,144	0		
1714	3,074	4	392	2	3,466	6	371	4
1715	2,868	7	1,715	2	4,584	1	672	2
1716	3,705	1	2,290	4	5,995	5	530	3
1717	4,104	2	3,394	6	7,499	0	403	6
1718	3,343	5	3,218	0	6,561	5	958	4
1719	3,608	1	3,709	3	7,317	4	2,069	0
1720	3,317	3	3,403	7	6,721	2	578	4
1721	2,665	5	1,806	4	4,472	1	169	7
1722	2,802	6	499	2	3,302	0	313	5
1723	2,025	2	193	0	2,218	2	219	2
1724	2,561	0	284	3	2,845	3	275	6
1725	2,216	2	682	3	2,898	5	276	0
1726	2,186	3	1,147	1	3,333	4	335	0
1727	2,080	3	1,606	1	3,686	4	363	6
1728	1,997	1	1,263	6	3,260	7	272	3
1729	2,525	5	1,501	5	4,027	2	187	6
1730	2,610	6	2,377	3	4,988	1	174	1
1731	3,032	3	1,754	2	4,786	5	147	1
1732	2,611	2	1,380	1	3,991	3	128	5
1733	2,076	1	690	1	2,766	2	130	0
1734	3,297	7	1,698	1	4,966	0	98	1
1735	6,116	2	2,240	0	8,356	2	126	0
1736	8,147	6	2,395	2	10,543	0	185	0
1737	7,655	1	3,481	0	11,136	1	197	1
1738	5,974	5	5,260	7	11,235	4		
1739	5,097	6	5,300	2	10,398	0		
1740	6,693	0	4,766	7	11,459	7		
1741	7,357	1	5,231	3	12,588	4		
1742	6,974	6	4,172	0	11,146	6		
1743	7,350	1	3,768	4	11,118	5		
1744	9,253	3	4,227	0	13,480	3		
1745	8,386	1	4,916	2	13,302	3		
1746	8,203	3	4,750	4	12,953	7		
1747	8,357	3	6,271	5	14,629	0		
1748	8,576	7	7,597	1	16,174	0		
1749	8,936	5	7,520	6	16,457	3		
1750	9,617	3	8,277	0	17,894	3		

d. Costs and Profits

Colliery Profits 1666-1694

Year	£	*s.*	*d.*	
27 Jan. 1666-22 March 1667	428	11	8	
1667	261	5	0½	
1668	328	1	7¾	
1669	457	13	7¼	
1670	370	12	4	
1671	450	9	3	
1672	239	14	4¼	
1673	401	15	1	
1674	698	7	10¾	
2 April 1675-18 Aug. 1675	274	9	7½	
Year ending:				
16 Aug. 1676	433	16	2½	
8 Aug. 1677	350	3	6	
8 Aug. 1677-19 Feb. 1679	756	16	7	
19 Feb. 1679-18 Feb. 1680	352	10	7	
18 Feb. 1680-18 Aug. 1680	435	14	4½	
Year ending:				
17 Aug. 1681	892	12	1	
16 Aug. 1682	367	6	10	
12 Sept. 1683	343	16	0	
10 Sept. 1684	1,531	1	0	
Sept. 1685	1,457	17	6¾	
8 Sept. 1686	1,441	2	9¼	
7 Sept. 1687	1,268	5	4¾	
5 Sept. 1688	1,320	5	10	
5 Sept. 1689	295	18	9½	
4 Sept. 1690	170	17	8¾	Loss
2 Sept. 1691	624	3	7½	
31 Aug. 1692	820	18	0¼	
Half year ending 1 March 1693	282	14	1½	
1693	894	11	4¼	
1694	648	10	10½	

	Whitehaven Collieries									Lattera Colliery									Total Profits		
	Cost			Receipts			Balance			Cost			Receipts			Losses					
	£	s.	d.	£	s.	d.	£	s.	d.	£	s.	d.	£	s.	d.	£	s.	d.	£	s.	d.
1695	1,918	12	5	2,473	15	1	555	2	8	140	13	0	12	16	6	127	16	6	427	6	2
1696	1,912	14	11½	2,602	10	6	689	15	6½	136	3	6	7	8	5½	128	15	0½	561	0	6
1697	2,416	6	1	3,447	4	10	1,030	18	9	160	19	1	4	15	6	156	3	7	874	15	2
1698	2,249	13	1½	3,100	8	0	850	14	10½	175	1	6	12	10	8	162	10	10	688	4	0½
1699	2,640	3	2½	3,466	12	4½	826	9	2½	291	9	5	32	19	7	258	9	10	567	19	4½
1700	2,829	2	5	3,738	2	10	909	0	5	227	6	5½	3	0	5	224	6	0½	684	14	4½
1701	3,413	3	2	4,399	19	10	986	16	8	457	10	10	3	0	10	454	10	0	532	6	8
1702	2,957	18	1	4,147	19	1½	1,190	1	0½	519	6	8	6	5	0	513	1	8	676	19	4½
1703	3,276	0	11½	5,092	8	10	1,816	7	10½	525	0	11¼	33	0	0½	492	0	10¾	1,324	6	11¾
1704	3,095	8	3	4,218	9	5	1,123	1	2	355	1	3	9	1	8	345	19	7	777	1	7
1705	2,883	0	2	4,104	6	9½	1,221	6	7½	417	8	1	7	16	0½	409	12	0½	811	14	7
1706	2,606	0	0	3,852	11	8	1,246	11	8	454	2	6½	52	8	6½	401	14	0	844	17	8
1707	2,911	8	5½	4,998	19	8	2,087	11	2½	486	16	7	20	19	2	465	17	5	1,621	13	9½
1708	3,052	0	4	4,724	0	2	1,671	19	10½	377	10	4½	11	16	10½	365	13	6	1,306	6	4

Howgill Colliery

	Cost £	s.	d.	Receipts £	s.	d.	Balance £	s.	d.
1709	3,558	3	3¾	4,816	10	0	1,258	6	8¼
1710	3,847	12	6	4,894	14	1	1,047	1	7
1711	3,582	6	0¾	4,710	18	0	1,128	11	11¼
1712	3,799	18	5¼	5,422	19	0	1,623	0	6¾
1713	4,259	16	11¼	5,844	5	4	1,584	8	4¾
1714	4,477	12	1	6,610	19	0½	2,133	6	11½
1715	4,006	2	7¾	7,837	4	5	3,831	1	9¼
1716	4,751	16	3¾	8,657	11	10	3,905	15	6¼
1717	5,170	6	8	9,326	0	2	4,155	13	6
1718	4,383	5	5½	8,660	7	0½	4,277	1	7
1719	4,338	14	7½	8,589	3	11½	4,250	9	4
1720	3,845	7	4½	7,782	10	7½	3,937	3	3
1721	3,683	0	5	7,760	1	0	4,077	0	7
1722	4,115	14	3½	7,626	11	11	3,510	17	7½
1723	4,593	9	11	7,873	15	7	3,280	5	8
1724	4,601	5	7¼	8,754	3	3½	4,152	17	8¼
1725	5,655	8	9¼	9,716	1	1½	4,060	12	4¼
1726	7,071	14	1¾	12,702	7	6½	5,630	13	4¾
1727	6,883	6	10	12,518	15	5½	5,635	8	7½
1728	6,909	6	2¼	11,721	8	5¾	4,812	2	3½
1729	7,915	7	7⅛	12,705	17	3½	4,790	9	8⅜
1730	8,949	15	7⅜	12,792	19	6	3,843	3	10⅝
1731	9,887	1	9⅞	13,990	4	4	4,103	2	6⅛
1732	10,852	5	0¼	14,421	2	1	3,568	17	0¾
1733	10,192	4	11	12,912	1	7	2,719	16	9
1734	9,866	10	2⅝	15,233	18	0	5,367	7	9⅜
1735	9,753	19	1	16,362	19	4⅛	6,609	0	3⅛
1736	10,387	13	8½	16,332	0	2	5,944	6	5½
1737	10,204	4	10½	14,519	5	0	4,315	0	1½
1738	9,207	11	5¾	15,358	8	9	6,150	17	3¼
1739	10,923	2	4	16,716	9	7	5,793	7	3
1740	10,365	6	0¼	17,537	7	1	7,172	1	0¾
1741	10,127	0	10	14,173	0	8	4,045	19	10
1742	9,857	6	10½	14,341	1	8	4,483	14	9½
1743	10,088	2	4½	17,034	4	2	6,946	1	9½
1744	9,732	4	4½	17,145	9	7	7,413	5	2½
1745	10,509	10	4	16,832	5	5	6,322	15	1
1746	11,185	4	10½	18,549	0	10	7,363	15	11½
1747	10,759	18	5¾	19,266	16	8	8,506	18	2¼
1748	10,121	13	3	17,159	12	1	7,037	18	10
1749	10,523	0	0¾	19,401	17	6	8,878	17	5¼
1750	12,706	7	8¼	20,339	12	1	7,633	4	4¾

Whingill Colliery

	Cost			*Receipts*			*Loss*		
	£	*s.*	*d.*	£	*s.*	*d.*	£	*s.*	*d.*
1713	27	9	6½	3	3	8	24	5	10½
1714	381	13	6	31	8	9½	350	4	8½
1715	851	0	2	144	16	11½	706	3	2½
1716	992	14	8¼	229	1	0	763	13	8¼
1717	762	13	4	339	9	6	423	3	10
1718	786	11	8½	321	16	0	464	15	8½
1719	908	1	0	370	18	9	537	2	3
1720	716	10	7½	350	0	10	366	9	9½
1721	994	1	10	193	16	8	800	5	2
1722	579	16	11½	52	10	1	527	6	10½
1723	387	12	2½	20	5	1½	367	7	1
1724	604	17	1½	29	15	6	575	1	7½
1725	877	9	0½	73	18	5¾	803	10	6¾
1726	1,166	4	5	124	5	5¼	1,041	18	11¾
1727	1,581	5	3½	173	19	11¼	1,407	5	4¼
1728	1,541	7	1⅜	141	13	3¾	1,399	13	9⅝
1729	1,696	10	7⅞	175	3	9½	1,521	6	10⅜
1730	1,918	10	5¼	277	7	2½	1,641	3	2¾
1731	1,896	6	0¼	204	13	3	1,691	12	9¼
1732	1,548	12	1¾	161	0	3½	1,387	11	10¼
1733	1,646	2	8½	80	10	3½	1,565	12	5
1734	1,388	1	2¾	197	12	11½	1,190	8	3¼
1735	781	7	5⅜	261	6	8	520	0	9⅜
1736	645	14	8	279	8	11	366	5	9
1737	588	3	9½	406	2	4	182	1	5½
1738	668	17	11½	613	15	4½	55	2	7
1739	812	9	0	618	7	3	194	1	9
1740	806	11	10¾	556	2	9	250	9	1¾
1741	705	12	8½	610	6	6½	95	6	2
1742	623	17	10	486	14	8	137	3	2
1743	466	10	3¼	439	13	2	26	17	1¼
1744	538	7	6½	493	3	0	45	4	6½
1745	771	2	11½	573	11	3	197	11	8½
1746	1,036	9	4¼	554	4	2½	482	5	1¾
1747	1,306	16	5½	731	13	9½	575	2	8
1748	1,204	6	10½	886	6	7½	318	0	3
1749	1,243	13	4½	877	8	5	366	4	11½
1750	1,615	10	5½	965	13	0	649	17	5½

Howgill and Whingill Collieries

	Total Cost			*Cost per Ton*		*Total Profit*			*Profit per Ton*	
	£	*s.*	*d.*	*s.*	*d.*	£	*s.*	*d.*	*s.*	*d.*
1709	3,558	3	3¾	1	10·7	1,258	6	8¼		8·0
1710	3,847	12	6	1	10·9	1,047	1	7		6·2
1711	3,582	6	0¾	1	11·9	1,128	11	11¼		7·5
1712	3,799	18	5¼	1	11·4	1,623	0	6¾		10·0
1713	4,287	6	5¾	2	0·2	1,560	2	6¼		8·8
1714	4,859	5	7	1	10·0	1,783	2	3		8·0
1715	4,857	2	9¾	1	10·1	3,124	18	6¾	1	2·2
1716	5,744	11	0	1	11·1	3,142	1	10	1	0·6
1717	5,933	0	0	1	9·6	3,732	9	8	1	1·6
1718	5,169	17	2	1	9·0	3,812	5	10½	1	3·5
1719	5,246	15	7½	1	8·0	3,713	7	1	1	2·2
1720	4,561	18	0	1	9·2	3,570	13	5½	1	4·6
1721	4,677	2	3	1	9·9	3,276	15	5	1	3·3
1722	4,695	11	3	2	1·1	2,983	10	9	1	4·0
1723	4,981	2	1½	1	11·3	2,912	18	7	1	1·6
1724	5,206	2	8¾	2	2·1	3,577	16	0¾	1	5·9
1725	6,532	17	9¾	2	0·2	3,257	1	9½	1	0·0
1726	8,237	18	6¾	2	0·3	4,588	14	5	1	1·5
1727	8,464	12	1½	2	5·5	4,228	3	3¼	1	2·7
1728	8,450	13	3⅝	2	4·1	3,412	8	5⅞		11·3
1729	9,611	18	3	2	8·6	3,269	2	10		11·0
1730	10,868	6	0⅝	2	7·8	2,202	0	7⅞		6·4
1731	11,783	7	10⅛	2	8·3	2,411	9	8⅞		6·6
1732	12,400	17	2	2	11·3	2,181	5	2½		6·2
1733	11,838	7	7½	2	9·3	1,154	4	3		3·2
1734	11,254	11	5⅜	2	4·0	4,176	19	6⅛		10·4
1735	10,355	6	6⅜	2	3·0	6,088	19	5¾	1	3·6
1736	11,033	8	4½	2	1·9	5,578	0	8½	1	1·1
1737	10,792	8	8	2	4·5	4,132	18	8		10·9
1738	9,876	9	5¼	2	4·5	6,095	14	8¼	1	5·6
1739	11,735	11	4	2	8·7	5,599	5	6	1	3·6
1740	11,171	17	11	2	3·5	6,921	11	11	1	5·0
1741	10,832	13	6½	2	4·9	3,950	13	8		10·5
1742	10,481	4	8½	2	6·4	4,346	11	7½	1	0·6
1743	10,554	12	7¾	2	5·6	6,919	4	8¼	1	7·4
1744	10,270	11	11	2	4·2	7,368	0	8	1	8·2
1745	11,280	13	3½	2	2·4	6,125	3	4½	1	2·3
1746	12,221	14	2¾	2	1·5	6,881	10	9¾	1	2·3
1747	12,066	14	11¼	2	4·1	7,931	15	6¼	1	6·5
1748	11,326	0	1½	2	1·8	6,719	18	7	1	3·3
1749	11,766	13	5¼	2	3·3	8,512	12	5¾	1	7·7
1750	14,321	18	1¾	2	3·2	6,983	6	11¼	1	1·2

	Parton Collieries										Dearham Colliery			
	Cost			*Receipts*			*Balance*				*Profit/Loss*			
	£	*s.*	*d.*	£	*s.*	*d.*	£	*s.*	*d.*		£	*s.*	*d.*	
1714	320	19	7	123	16	7	197	3	0	Loss				
1715	899	9	8	1,080	3	6	180	13	10					
1716	757	14	9¼	969	2	8¼	211	7	11					
1717	807	8	2¼	908	19	10	101	11	7¾					
1718	768	8	2¾	966	11	6¾	198	3	4					
1719	468	17	4¾	395	4	9½	73	12	7¼	Loss				
1720	657	4	11⅛	558	7	11⅛	98	17	0	Loss				
1721	759	7	1¾	811	5	1	51	17	11¼					
1722	591	14	8½	874	8	0¼	282	13	3¾					
1723	338	1	9⅝	492	4	10⅜	154	3	0¾		130	5	4	
1724	489	11	6	895	11	10¾	406	0	4¾		67	8	5	
1725	1037	2	4⅞	1,141	0	11¾	103	18	6⅞		122	0	4½	
1726	749	6	6¼	839	11	11	90	5	4¾		87	18	0	
1727	883	11	2½	1,149	9	5¾	265	18	3¼		11	9	7¼	Loss
1728	938	1	4¾	1,283	15	11¼	345	14	6½		5	16	0	
1729	796	1	10¾	914	4	1¾	118	2	3		14	14	6¾	
1730	881	7	8¾	859	5	8	22	2	0¾	Loss	12	14	10½	
1731	757	0	6¾	755	11	0½	1	9	6¼	Loss	33	10	4¼	
1732	556	9	7½	545	15	7	10	14	0½	Loss	18	0	3¼	
1733	617	8	5¼	657	12	1	40	3	7¾		38	7	8¼	
1734	510	16	8½	552	16	7	41	19	10½		19	13	8	
1735	444	12	8½	402	5	7¼	42	7	1¼	Loss	49	8	0¼	
1736	493	8	5¼	469	7	6¾	24	0	10½	Loss	14	10	2¾	
1737	472	16	3½	369	7	2½	103	9	1	Loss	16	5	9¾	
1738											50	1	7	
1739											95	1	3¼	
1740											87	3	3½	
1741											85	18	0¼	
1742											40	11	9½	
1743											80	7	0	
1744											26	3	10¾	
1745											128	19	8¾	
1746											110	0	2¼	
1747											126	9	7¾	
1748											137	8	3	
1749											76	15	6¼	
1750											142	18	0½	
1751											19	14	10½	
1752											2	2	5	
1753											37	7	5¾	Loss
1754											31	0	2	Loss
1755											267	19	11¾	
1756											100	7	7¾	

* Dearham Colliery year is from Michaelmas to Michaelmas.

Scalegill Colliery Profits

	£	s.	d.
1737	259	10	1¼
1738	146	4	9½
1739	568	4	5½
1740	338	18	10
1741	563	19	0½

Sources: D/Lons/W Estate Account Books 1666-1699; Colliery Abstract Books 1695-1750; Dearham Colliery Accounts 1722-1756; Miscellaneous Colliery Papers, profit of Scalegill Colliery. N.B. The above colliery statistics of the Lowther collieries since 1695, although compiled from the same sources, differ very often from those given by J. V. Beckett in his *Coal & Tobacco: The Lowthers and the Economic Development of West Cumberland, 1660-1760* (Cambridge, 1981), 229-237.

Whitehaven Collieries 1757-1888

Year	*Output Tons*	*Shipments Tons*	*Country Sale Tons*	*Profits* £	s.	d.
1757		115,303				
1758		117,075				
1759	80,245	101,513	1,359			
1760	80,177	98,100	1,212			
1761		123,310				
1771				9,304	3	0
1775	115,337	129,846	1,987	9,218	0	11¼
1776	104,766	125,991	2,117	9,913	5	10
1777	97,074	121,707	1,677			
1778	88,887	120,141	1,555			
1779	94,153	108,654	1,406			
1780	100,991	114,012	1,472			
1781	89,700	116,022	1,327			
1782	83,950	112,224	1,285			
1783	98,060	127,182	1,291			
		Waggons				
1784		65,874				
1785		75,879				
1786		54,827				
1787		67,601				
1788		67,863				
1789		74,554				
1790		64,032				
1796		58,128				
1797		52,810				
1798		47,555				
1799		44,732				
1800		45,033	*Local Sales**			
1801		45,314				
1802		47,906	*Waggons*			
1803		76,864	13,511	24,882	4	4¾
1804		74,127	15,501	26,657	9	8
1805		69,585	17,698	22,663	13	5¼
1806		74,238	14,144	31,611	1	7
1807		75,457	9,288	26,889	18	10½
1808		100,883	9,709	48,495	16	2¼
1809		76,114				
1810		100,121				
1811		105,217				
1812		98,283		40,455	12	1
1813		89,040		45,979	2	7
1814		110,193		58,687	5	11¾
1815		107,100		54,704	16	8¾
1816		92,097		51,575	12	6
1817		93,926		45,909	7	4¾
1818		76,660		30,231	4	7¼

* Local Sales i.e. Sales to town and country

Year	*Output*	*Shipments Waggons*	*Profits* £	s.	d.
1819		87,174	35,523	13	3¾
1820		91,531			
1821		87,024			
1822		90,335	30,790	14	11
1823		93,169	28,484	11	7
1824		87,070			
1825		90,732			
1826		99,859			
1827		84,436			
1828		82,436			
1829		87,471			
1830		94,416			
1831		90,908			
1832		96,434	27,318	14	3½
1833		93,697	27,630	10	5½
1834		94,540	21,526	9	6
1835		99,149	25,779	6	5½
1836		102,546	29,069	14	9½
1837		111,427	30,895	10	4½
1838		102,833	22,927	0	3
1839		102,350	20,461	17	10
1840		104,179	22,221	14	6½
1841		99,032			
	Waggons	*Tons*			
1842	108,950	205,518	15,564	9	7½
1843	95,328	205,800	11,361	5	11½
1844	107,651	215,182	16,311	13	1½
1845	111,460	231,893	19,516	19	7½
1846	118,337	220,870	36,307	18	11
1847	115,561	217,576	33,002	6	7
1850		223,816			
	Tons				
1866	255,505				
1867	276,408				
1869			19,925	10	3
1870			15,873	15	8¼
1871			11,727	9	9½
1872			26,723	13	6½
1873	227,714		75,285	10	10
1874	262,641		48,991	9	6
1875	294,617		26,342	9	10
1876	281,968		11,054	17	8½
1877	281,695		6,168	10	1
1878	251,414		7,667	4	0½
1879	345,444		12,643	14	6½
1880	347,244		13,785	8	3

Year	*Output*	*Shipments*	*Waggons*	*Profits*		
				£	*s.*	*d.*
1881	363,449			23,101	18	5½
1882	334,842					
1883	336,020					
1884	349,063			3,477	9	10
1885	379,189			2,359	5	7
1886	417,039					
1887	392,874					
1888	398,951					

1759-60 Output and Country Sale and all figures 1775-83 refer to Howgill Colliery only.
Profits 1869-81 for year ending 30 June.
Output 1881-83 for year ending June.

Sources: D/Lons/W Collieries 11, Report on Whitehaven Collieries 24 Nov. 1762. D/Lons/W Colliery Abstract Books; John Peile, Correspondence and Colliery Reports 1811-1847; G. B. Forster, Correspondence and Reports on Lonsdale Collieries 1880-1900; R. W. Moore, MS. History of Whitehaven Collieries.

Lonsdale Income from Coal

Year	£	*s.*	*d.*
1771	9,806	8	8
1812	42,872	14	2
1813	48,725	19	3½
1814	60,409	0	7
1815	56,427	2	2¾
1816	52,995	5	4¼
1817	48,190	5	0¾
1818	31,694	17	7¾
1819	36,988	12	3¾
1820	36,098	12	1¼
1821	34,239	15	5½
1823	29,570	9	9¼
1824	36,491	11	8
1825	37,558	8	1¼
1826	39,347	18	4¾
1827	36,416	5	2
1828	29,751	4	9
1829	35,054	0	5¼
1830	35,664	0	3½
1831	25,913	14	7
1832	32,113	10	0½
1833	35,299	6	6
1834	24,005	19	8
1835	29,549	3	11½
1836	38,368	19	3
1837	48,978	15	5½
1838	33,159	10	7½
1839	27,964	15	7
1841	22,883	8	3½
1842	24,233	11	0½
1843	18,805	12	4½
1844	24,344	5	7
1845	32,535	15	10
1846	48,356	17	11
1847	44,279	5	11

Year	£	*s.*	*d.*
1869	31,659	4	10½
1870	29,379	9	4¼
1871	25,120	7	7
1872	39,634	8	7½
1873	90,684	7	3
1874	63,933	0	6
1875	37,219	11	11½
1876	21,738	18	8½
1877	17,899	0	1
1878	18,435	3	3½
1879	22,579	7	11½
1880	23,772	11	5
1881	36,492	1	9½
1896	29,282	3	0
1900	46,707		
1901	53,237		
1902	43,699		
1903	44,285		
1904	41,353		
1905	42,142		
1906	39,017		
1907	39,378		
1908	42,776		
1909	42,791		
1910	38,111		
1911	33,266		
1912	40,949		
1913	48,737		
1914	44,023		
1915	43,817		
1916	46,947		
1917	49,493		
1918	34,242		
1919	34,021		
1920	33,747		
1921	28,574		
1922	34,221		

Years 1869-1881 ending 30 June.

Sources: D/Lons/W Rental 1771; John Peile, Colliery Reports and Correspondence; Statements of income; Unlisted papers. R. W. Moore, MS History of Whitehaven Collieries.

Whitehaven Colliery 1889-1946

Year ending August	*Output Tons*	*As Percentage of Cumberland Output %*	*Rent due to Lord Lonsdale* £	s.	d.
1889	364,157	20·93	16,967	15	8
1890	326,824	18·77	17,945	8	5
1891	330,776	19·80	18,963	13	2
1892	273,543	19·19	15,818	2	4
1893	366,055	20·16	15,634	16	5
1894	436,913	21·22	22,146	0	11
1895	494,002	26·22	21,351	13	4
1896	495,657	25·65	20,362	8	9
1897	578,238	29·10	22,112	2	9
1898	584,396	28·34	22,603	1	3
1899	584,700	27·83	25,555	1	6
1900	548,586	27·12	30,403	6	2
			Year ending December		
1901	560,391	26·57	34,253		
1902	556,690	25·53	28,157		
1903	600,226	27·19	28,690		
1904	539,706	25·46	26,501		
1905	462,774	21·51	27,890		
1906	562,605	25·29	25,130		
1907	599,133	26·58	24,834		
1908	592,697	27·76	27,785		
1909	612,364	26·51	27,634		
1910	568,872	26·15	23,825		
1911	454,250	19·78	18,998		
1912	524,918	24·07	25,483		
Year ending December					
1913	517,827	22·31	26,051		
1914	511,618	21·67	25,019		
1915	512,641	22·79	24,471		
1916	560,374	24·58	23,210		
1917	514,466	23·80	25,488		
1918	462,392	24·28	19,268		
1919	494,143	26·82	20,723		
1920	497,839	26·94	20,725		
1921	425,479	31·86	18,899		
1922	553,918	25·28	19,449		
1923	362,780	17·16	–		
			Year ending 31 July		
1924	583,233	26·64	22,192	7	9
1925	704,228	35·46	23,831	2	0
1926	349,433	35·78	24,847	7	4
1927	881,239	36·47	22,764	11	2

Year ending August	*Output Tons*	*As Percentage of Cumberland Output %*	*Rent due to Lord Lonsdale* £	*s.*	*d.*
1928	744,986	34·83	24,310	14	8
1929	612,693	30·31	24,384	5	8
1930	603,978	29·93	23,719	8	4
1931	435,362	28·77	17,550	8	7
1932	419,609	27·73	12,409	6	5
1933	386,569	27·20	17,484	15	5
1934	499,965	31·96	10,508	16	0
1935	419,388	27·27	12,188	12	8
1936	–	–	10,456	10	9
1937	285,151	21·23	–		
			Year ending 31 December		
1938	598,819	38·18	10,757	19	6
1939	641,241	40·27	11,686	17	8
1940	607,829	39·86	10,523	14	0
1941	427,761	35·36	7,861	10	9
1942	296,143				
1943	283,031				
1944	302,530				
1945	301,301				
1946	342,966				

Output figures 1942-6 are for Haig Pit only.

Sources: D/Lons/W Coal Royalty Rent Account Books 1871-1943; Output Books 1913-41; Lonsdale Coal Royalties Minutes 1904-30.

Curwen Collieries

Workington Collieries 1771-1802

Year	*Output Tons*	*Shipments Tons*	*Country Sale Tons*
1771	1,701		
1772	10,742		
1773	12,949		
1774	23,600		
1775	25,303		
1776	25,724		
1777	31,784		
1778	36,417		
1779	43,540		
1780	53,543		
1781	56,110	33,765	
1782	61,062	32,493	
1783	58,195	37,598	
1784	58,102	40,269	
1785	47,123	43,167	
1786	45,685		
1787	52,890	63,618	4,753
1788	55,166	60,930	4,879
1789	50,070	57,909	4,504
1790	48,699	65,148	4,020
1791	55,354	72,912	4,645
1792	46,648	59,313	6,438
1793	53,203	69,012	6,624
1794	51,775	66,585	5,693
1795	54,555	69,033	5,561
1796	50,064	61,608	4,800
1797	52,664	63,885	5,216
1798	52,952	69,204	3,975
1799	49,193	70,731	4,338
1800	63,844	97,445	4,526
1801	57,996	92,469	4,258
1802	65,309	104,940	5,518

Sources: Whitehaven Library MS 48, Workington Colliery Book 1799-1853. D/Cu 6/11, 6/12, Workington Colliery Accounts 1787-1802; Workington Collieries Coal Shipped Books.

Workington Collieries 1803-1861

Year	*Output* *Tons*	*Shipments* *Waggons*	*Profits* £	*s.*	*d.*	
1803		39,867				
1804		39,301				
1805		37,178				
1806		39,022	13,580	7	10½	
1807		33,941	6,561	18	9	
1808		35,098	1,097	9	10½	
1809		25,196	306	16	1	Loss
1810		25,444	3,053	3	6½	Loss
1811		32,623	6,012	15	10	
1812		26,947	1,190	11	1	
1813		26,287	1,462	19	2	Loss
1814		23,499				
1815		26,429				
1816	40,707	31,618				
1817		28,310				
1818		32,035	393			
1819		34,103	3,525			
1820		29,743	2,154			
1821		32,578	3,242	18	7½	
1822		32,402	2,805	12	8	
1823		27,343				
1824		29,145				
1825		27,707				
1826		14,342				
Year ending 30 June:						
1827		22,097				
1828		21,121				
1829	39,670	24,095				
1830		24,752				
1831		22,279				
1832		20,243				
1833		16,353				
1834		15,086				
Year ending 31 Dec.:						
1835		21,086				
1836		32,089				
1837		15,760				
1838		2,460				
1839		7,177				
1840		2,681				
		Tons				
1842		15,941	482	12	9	
1843		18,424				
1844		14,728				

Year	*Output* Tons	*Shipments* Waggons	*Profits* £ s. d.
1845		*15,308*	
1846		*21,772*	
1847		*19,572*	
		Waggons	
1849		10,356	
1850		11,119	
1851		11,821	
1852		7,647	
1853		4,243	
1854		2,894	
1855		1,885	
1856		2,625	
1857		2,015	
1858		2,824	
1859		2,239	
1860		2,825	
1861		1,905	

Sources: D/Cu 3/43, 3/44 Letter books of William Swinburn 1805-24. D/Cu Workington Collieries Coal Shipped Books 1807-61. D/Lons/W Customs House returns quoted by John Peile to Lord Lonsdale, 7 June 1841 and in his annual colliery reports 1842-7.
VCH 367.

Harrington Colliery 1783-1862

Year	*Output Tons*	*Shipments Waggons*	*Country Sale Tons*	*Profits* £	s.	d.
1783		12,516½	4,294			
1784	40,432	13,423	4,468	2,644	19	11¾
1785	42,781	14,389	5,494	2,827	9	9
1786	35,164	12,866	3,954	2,904	12	3
1787	38,206	15,134	4,007	3,542	10	1½
1788	43,680	17,374	1,729	4,097	7	10¾
1789	49,516	20,293	1,687			
1790		20,245				
1791		25,643				
1792		24,222		5,210	19	2
1793		29,294		6,508	2	9¾
1794		27,096		5,633	6	1
1795		29,387		6,587	5	11½
1796		27,674		7,177	5	10½
1797		27,050				
1800		25,961				
1801		18,791				
1802		16,345				
1803		20,912				
1804	41,420	20,510				
1805		17,375				
1806		19,571		6,466	14	3
1807		18,323		6,078	18	1¾
1808		24,252		6,782	17	3½
1809		16,219		3,452	17	4½
1810		24,359		4,531	7	5
1811		21,481		5,517	17	0
1812				4,496	14	1
1813				4,719	0	11½
1814		15,391				
1822		10,221		1,549	8	2½
1823		9,704				
1824		7,085				
1825		9,216				
1826 to 30 June		4,824				
Year ending 30 June:						
1827		9,345				
1828		7,371				
1829		8,843				
1830		9,213				
1831		6,708				
1832		5,925				
1833		8,483				
1834		8,291				
30 June-31 Dec.		3,889				

Year	*Output* *Tons*	*Shipments* *Waggons*	*Country Sale* *Tons*	*Profits* £ s. d.
Year ending 31 Dec:				
1835		7,883		
1836		10,071		
1837		14,116		
1838		13,506		
1839		14,148		
1840		14,442		
1841		10,415		
1842		8,142		
1843		6,954		
1844		7,095		
1845		10,018		
1846		9,174		
1847		8,742		
1848		6,131		
1849		6,007		
1850		7,494		
1851		5,540		
1852		5,248		
1853		5,380		
1854		5,016		
1855		3,278		
1856		3,956		
1857		3,117		
1858		2,719		
1859		3,202		
1860		3,130		
1861		1,391		
1862		1,128		

Sources: D/Cu Harrington Colliery Account Books. D/Cu Harrington Colliery Coal Shipped Books 1788-1862. D/Cu Letter Books of William Swinburn. *VCH* 368.

Harrington Colliery – Output and Royalty Payments 1868-1907

Year	Harrington Coal Wrought Tons	Harrington Fireclay Tons	Harrington Royalty £ s. d.	Harrington Undersea Coal Wrought Tons	Harrington Undersea Royalty £ s. d.
1868	33,642	494	739 6 11		
1869	22,299	2,697	608 9 9		
1870	19,296	4,882	999 13 1		
1871			1,338 14 1		
1872			1,098 6 0		
1873	47,528	5,334	1,104 5 11		
1874	64,256	3,923	1,112 19 0		
1875	78,681	4,837	1,555 6 10½		
1876	75,299	3,447	1,748 4 3		
1877	66,106	3,313	1,584 3 9	1,985	122 15 0
1878	54,479	3,038	1,334 2 11	13,016	249 2 10
1879	48,415	3,512	1,197 11 9	20,228	435 9 0
1880	45,000	3,797	1,178 7 2	19,745	452 10 0
1881	48,408	3,427	1,396 19 3	19,485	446 11 0
1882	60,626	4,158	1,584 7 3	24,719	566 9 8
1883	65,334	3,527	1,621 3 7	41,659	954 13 8
1884	48,187	4,219	1,407 3 10	52,567	1,204 13 4
1885	46,308	1,586	1,413 15 7	13,554	312 0 8
1886	66,508	3,967	1,958 5 9	10,186	232 0 8
1887	48,958	3,062	1,697 5 0	34,644	793 18 8
1888	36,009	4,410	1,611 8 4	58,016	1,329 11 0
1889	51,583	2,695	1,399 14 0	38,333	878 9 2
1890	46,778	3,263	1,294 6 11	51,613	1,182 16 0
1891	47,106	3,947	1,273 8 0	58,431	1,338 19 10
1892	40,150	2,325	1,167 1 9	35,889	822 9 6
1893	48,143	1,972	1,120 16 8	59,941	1,373 13 0
1894	57,655	3,959	1,433 8 3	49,631	1,137 7 10
1895	48,477	2,787	1,210 6 4	54,255	1,243 6 2
1896	45,470	2,673	1,139 17 0	55,139	1,263 9 9

Total Royalty Payments

Year	£	s.	d.	Year	£	s.	d.
1901	2,028	16	7	1905	1,399	5	6
1903	1,101	1	9	1906	1,363	1	11
1904	1,145	10	8	1907	1,216	18	1

Source: D/Cu Harrington Colliery Royalty Accounts.

Broughton Colliery 1756-1822

Year ending midsummer:	*Output Tons*	*Shipments Waggons*	*Country Sale Tons*	*Lord Egremont's Rent* £	*s.*	*d.*
1756	7,175			200	0	0
1757	5,149	3,836	1,816	200	0	0
1758	9,399	8,101	2,437	200	0	0
1759	9,354	9,990	3,009	200	0	0
1760	8,981	6,748	3,514	200	0	0
1761	10,891			200	0	0
1762	19,756			203	4	7½
1763	37,269			383	7	9¼
1764	23,439	*Tons*		241	2	4¼
1765	30,214	30,011		310	16	3
1766	30,079	33,505		309	8	5¾
1767	32,562	33,228		334	19	4¼
1768	38,019	39,882		391	2	0¾
1769	45,665			469	15	2¼
1770	38,155			392	10	0¾
1771	39,074			301	19	1½
1772	37,462			385	7	5¾
1773	38,212½			301	9	5½
1774	36,361½					
1775	36,604½					
1776	38,200			1,588	16	10½
1777	40,060½					
1778	40,053½					
1779	39,822					
1780	36,161			379	18	5
1781	37,513	*Waggons*				
1782	22,057½	6,608	7,824	717	6	0
1783	22,202½	7,076	6,345	742	14	0
				448	18	9
				June-Dec.		
				Year ending Dec.		
1784	26,630½	7,626	11,052	970	7	2½
1785	35,458½	10,926	13,845	1,443	15	2
1786	32,738½	10,206	11,899	932	0	9
1787	26,457	8,218	9,843	942	14	9
1788	23,166½	8,959	5,458	945	12	8
1789	24,132	8,872	6,076	854	4	3½
1790	19,753	7,756	4,027	738	6	6¼
Half year ending Christmas:						
1790	10,801	3,971	2,755			

Year ending Christmas:	*Output Tons*	*Shipments Waggons*	*Country Sale Tons*	*Lord Egremont's Rent* £	s.	d.
1791	22,376	7,540	6,245	786	12	8
1792	26,486	8,901	8,658	946	5	2¾
1793	29,681	10,231	7,475	1,007	13	3½
1794	31,350½	10,781	6,982	1,098	0	0
1795	33,910½	12,568	9,373	1,295	8	3
1796		12,482	9,532	1,289	6	11½
1797		13,716	7,516	1,379	16	6
1798		12,579	6,753	1,263	13	10½
Year ending Christmas:	*Output Tons*	*Shipments Waggons*	*Country Sale Tons*	*Lord Egremont's Rent* £	s.	d.
1799		13,171	9,824	1,357	11	9¼
1800		13,140	9,884	1,355	8	6½
1801		14,254	12,341	1,490	11	7½
1802		10,086	10,481	1,265	8	2
1803	22,882½	11,003	5,412	1,600	19	2
1804	23,965½	10,902	6,693	1,641	14	2
1805		6,618	7,947	1,158	14	7
1806		7,121	4,713	1,086	15	3¾
1807		5,365	6,648	1,000	0	0
1809		9,917	8,706	1,754	10	11¼
			Waggons			
1812	18,498	8,874	3,187			
1813		5,932	2,935	1,178	0	7½
1814		6,131	3,725	1,404	5	11
1815		6,629	3,288			
1816		6,218	3,508	1,319	11	7¾
1817		5,174	1,935			
1818		5,897	1,716	1,043	2	2
1819		6,624	1,674	1,102	3	6½
1820		5,160	1,699			
1821		5,030	2,541	1,048	11	1¾
1822		4,497	1,919	862	2	3½

Sources: Forster View Book of John Barnes. D/Cu Broughton Colliery Abstract Books 1756-95; Coals Shipped and Sold the Country 1781-1822; Coals Shipped Broughton and Unerigg 1782-96. D/Lec/65/6, 65/7 Broughton Colliery Accounts. D/Sen Abstract of Broughton Colliery Quarterly Accounts 1757-60; Broughton Colliery Shipments 1763-8.

Leconfield Colliery Rents

Year	£	*s.*	*d.*	*Year*	£	*s.*	*d.*
1853	3,766	11	3¾	1914	14,977	12	11
1854	4,335	13	4	1915	16,808	8	8
1855	3,840	19	10¾	1916	21,346	2	3
1856	3,444	7	1	1917	22,295	0	0
1857	3,430	18	3¼	1918	18,702	14	5
				1919	13,842	8	10
1860	3,610	10	6	1920	13,724	10	10
1861	4,328	8	2½	1921	9,280	13	4
1862	4,346	12	5	1922	12,555	15	0
1863	4,623	4	3½	1923	12,469	5	1
1864	4,553	19	3	1924	10,643	8	2
1865	5,916	7	3½	1925	8,194	18	10
				1926	3,670	18	9
1869	5,724	7	0	1927	4,822	0	9
1870	4,800	3	2	1928	5,930	2	5
1871	5,739	14	5	1929	4,836	11	10
1872	7,409	3	9	1930	4,163	4	2
1873	9,428	10	7	1931	3,646	18	10
1874	6,940	17	7	1932	4,315	14	9
				1933	4,032	17	4
1876	5,925	3	5				
1877	4,785	9	9				
1878	5,024	10	11				
1879	3,552	15	6				
1885	6,531	1	3				
1886	5,375	8	7				
1887	5,928	10	3				
1888	5,818	13	8				
1889	5,262	5	8				
1890	6,067	1	9				
1891	6,818	8	0				
1892	6,243	19	10				
1893	5,631	14	8				
1894	4,968	10	3				

Source: D/Lec Mineral Reports.

Coal Shipments from Cumberland 1757-1847

Year	*Whitehaven*		*Workington*	*Harrington*	*Maryport*
	Tons	*Loads*	*Tons*	*Waggons*	*Tons*
1757	115,303	7			
1758	117,075	5			
1759	101,513	5			
1760	98,100	0			
1761	123,310	5			
1766			47,207		34,733
1769			32,468		
1770			41,664		
1771			45,167		
1772			38,541		
1773			41,226		
1774			46,426		
1775			58,848		
1776			60,380		
1777			64,320		
1778			75,193		
1779			66,405		*Waggons*
1780			65,403		17,255
1781	*Waggons*		64,663		15,548
1782	56,424		62,835		
1783	65,732		72,079		
1784	65,874		72,799		
1785	75,879		79,368		
1786	54,827		80,769		21,202
			Waggons		
1787	67,601		28,838		
1788	67,863		31,253	15,041	
1789	74,554		35,387	20,332	
1790	64,032		38,194	20,246	
1791			43,117	24,998	
1792			45,743	24,337	
1793			37,584	28,546	
1794			43,041	25,821	
1795			44,867	29,387	
1796	58,128		44,140	27,674	
1797	52,810		45,169	26,726	
1798	47,555		47,250		
1799	44,732		48,543		
1800	45,033		45,009		
1801	45,314		47,970		
1802	47,906		42,778		
1803	76,864		46,733		
1804	74,127		41,269		
1805	69,585		47,272		
1806	74,238		45,797		
1807	75,457		41,011		
1808	100,883		38,726		
1809	76,114		32,322		
1810	100,121		26,563		

Year	Whitehaven Waggons	Workington Waggons	Harrington Waggons	Maryport Waggons
1811	105,217	25,742		
1812	98,283	29,279		
1813	89,040	29,659		
1814	110,193	25,112		
1815	107,100	23,378		
1816	92,097	26,695	7,705	6,169
1817	93,926	27,692	10,406	5,133
1818	76,660			
1819	87,174			
1820	91,531			
1821	87,024			
1822	90,335			
1823	93,169			
1824	87,070			
1825	90,732			
1826	99,859			
1827	84,436			
1828	82,436			
1829	87,471			
1830	94,416			
1831	90,908	34,702	6,326	29,466
1832	96,434	37,254	8,440	30,969
1833	93,600	31,951	9,703	27,595
1834	94,279	33,402	8,995	28,855
1835	98,247	39,350	9,488	27,646
1836	101,019	46,944	12,213	22,799
1837	105,904	29,571	21,621	31,694
1838	94,846	25,560	21,633	35,789
1839	94,665	27,485	21,197	44,170
1840	97,684	23,148	19,885	49,359
	Tons	*Tons*	*Tons*	*Tons*
1842	205,518	71,731	34,012	140,473
1843	205,800	73,954	28,757	143,147
1844	215,182	74,209	26,399	168,709
1845	231,893	80,094	31,305	190,006
1846	220,870	70,430	28,383	213,152
1847	217,576	60,021	28,154	183,384

Workington shipments 1766-90, year ending 28 February; 1791-95 year ending April; 1796-1816 year ending 1 May.

Sources: D/Lons/W Report on Whitehaven Collieries, 24 Nov. 1762. D/Lons/W Colliery Abstract Books; Letters of John Bateman and John Peile; Peile reports on Whitehaven Collieries. D/Cu Tonnage Day Books, 1766-1800; Coal Shipments from Workington Books, 1801-1816. D/Cu 5/99 Udale's Account of Harrington Harbour 1788-97. E. Hughes, *North Country Life in the Eighteenth Century: Cumberland & Westmorland 1700-1830* (Oxford, 1965), 162, 366.

Coastwise Coal Shipments from the Ports of:

Year	Whitehaven Tons	Workington Tons	Maryport Tons	Carlisle Tons	Total Tons
1819	242,605				242,605
1820	242,086				242,086
1821	186,424				186,424
1822	255,224				255,224
1823	284,521				284,521
1824	284,122				284,122
1825	274,146				274,146
1826	281,770				281,770
1827	295,664				295,664
1828	266,956				266,956
1829	335,985				335,985
1833	346,341				346,341
1834	354,524				354,524
1835	354,098				354,098
1836	403,332				403,332
1837	409,493				409,493
1838	394,798				394,798
1839	439,188				439,188
1840	423,379				423,379
1841	451,370				451,370
1842	301,247		141,391		442,638
1843	300,948		124,700		425,648
1844	299,407		162,038		461,445
1845	348,307		185,549		531,856
1846	326,668		218,838		545,506
1847	308,846		195,159		504,005
1848	295,565		183,239		478,804
1849	330,495		197,061		527,556
1850	236,808	113,650	190,640		541,098
1851	226,266	95,703	201,676		523,645
1852	222,208	85,014	230,012		537,234
1853	223,143	126,289	269,011		618,443
1854	215,544	117,811	304,197		637,552
1855	206,383	111,216	283,603		601,202
1856	208,722	129,410	295,466		633,598
1857	193,603	133,872	337,338		664,813
1858	180,346	133,231	329,459		643,036
1859	189,305	126,436	382,093		697,834
1860	191,079	134,339	386,372		711,790
1861	183,366	128,959	380,304		692,629
1862	197,550	144,294	382,725		724,569
1863	152,945	140,411	413,863		707,219
1864	147,727	136,982	438,954		723,663
1865	146,589	145,666	459,725		751,980
1866	147,099	137,453	431,239		715,791
1867	169,954	136,607	476,162		783,723
1868	143,694	106,886	406,969		657,549
1869	148,478	99,484	422,622		670,584
1870	148,585	62,320	459,407		670,312

Year	*Whitehaven Tons*	*Workington Tons*	*Maryport Tons*	*Carlisle Tons*	*Total Tons*
1871	130,962	64,684	404,205		599,851
1872	121,823	53,787	154,815		330,425
1873	89,033	35,324	240,104		364,461
1874	100,245	20,774	236,205		357,224
1875	128,590	28,637	234,293		391,520
1876	130,290	47,038	276,874	45,791	499,993
1877	133,281	47,318	294,957	31,994	507,550
1878	129,540	55,334	282,607	28,518	495,999
1879	217,245	53,718	292,426	21,170	584,559
1880	198,791	52,980	242,176	20,088	514,035
1881	214,642	37,235	229,960	30,428	512,265
1882	186,784	44,130	202,427	23,590	456,931
1883	166,478	48,217	209,885	25,882	450,462
1884	165,949	54,885	186,328	25,056	432,218
1885	183,599	82,801	195,921	26,020	488,341
1886	202,565	106,074	193,848	22,739	525,226
1887	198,493	80,441	172,717	18,854	470,505
1888	196,169	58,046	179,691	19,611	453,517
1889	148,796	52,907	158,220	20,445	380,368
1890	159,943	51,012	143,438	19,530	373,923
1891	167,907	89,501	435,412	15,625	708,445
1892	145,326	73,216	143,501	16,157	378,200
1893	197,839	82,868	207,887	21,297	509,891
1894	275,194	115,631	245,564	37,720	674,109
1895	245,279	91,409	190,503	25,674	552,865
1896	269,499	88,113	179,757	22,954	560,323
1897	299,980	105,991	158,856	26,514	591,341
1898	295,653	129,584	194,723	25,641	645,601
1899	280,692	132,514	208,621	24,102	645,929
1900	233,314	106,291	208,720	9,189	557,514
1901	253,271	114,053	236,098	8,307	611,729
1902	256,992	135,771	231,326	17,163	641,252
1903	267,288	134,115	229,059	13,728	644,190
1904	201,218	150,425	277,148	16,149	644,940
1905	231,300	78,580	287,889	26,614	624,383
1906	251,436	55,944	315,335	34,801	657,516
1907	254,781	60,768	319,535	34,625	669,709
1908	266,787	68,374	316,593	23,148	674,902
1909	268,834	70,542	314,318	7,318	661,012
1910	212,749	115,735	299,231	8,507	636,222
1911	223,109	110,565	294,538	11,837	640,049
1912	237,559	79,059	288,858	10,605	616,081
1913	226,441	86,481	286,341	9,441	608,704

Sources: R. Hunt, *Mineral Statistics of the United Kingdom*, 1854-1881. Mines Department, *Mineral Statistics of the United Kingdom*, 1882-1896; *Mines and Quarries General Report and Statistics 1899-1913. Reports of Inspectors of Mines*, 1873-1914. *Report of the Royal Commission on Coal in the United Kingdom*, 1871, iii, Appendix 71-86.

Exports of Coal from the Ports of:

Year	Whitehaven Tons	Workington Tons	Maryport Tons	Carlisle Tons	Total Tons
1819	13,431				13,431
1820	11,502				11,502
1821	11,012				11,012
1822	12,860				12,860
1823	13,108				13,108
1824	13,108				13,108
1825	13,322				13,322
1826	14,165				14,165
1827	11,682				11,682
1828	10,457				10,457
1829	13,159				13,159
1830	16,040				16,040
1831	15,869				15,869
1832	17,693				17,693
1833	17,187				17,187
1834	18,121				18,121
1835	19,606				19,606
1836	20,172				20,172
1837	28,190				28,190
1838	28,021				28,021
1839	22,616				22,616
1840	24,318				24,318
1841	23,478				23,478
1842	19,106		3,903		23,009
1843	17,666		2,469		20,135
1844	12,734		1,658		14,392
1845	1,084		90		1,174
1846	1,428		322		1,750
1847	2,423		1,176		3,599
1848	2,914		1,118		4,032
1849	2,094		1,000		3,094
1850	2,033	250	1,243		3,526
1851	3,159	140	1,753		5,052
1852	1,627	500	2,492		4,619
1853	2,072	1,267	2,310		5,649
1854	2,747	1,664	3,184		7,595
1855	2,341	1,210	2,503		6,054
1856	3,943	1,514	4,075		9,532
1857	3,575	2,757	3,534		9,866
1858	2,036	1,325	2,244		5,605
1859	1,487	675	7,103		9,265
1860	3,441	1,964	6,446		11,851
1861	2,598	2,124	7,825		12,547
1862	965	8,379	8,800		18,144
1863	769	1,344	8,036		10,149
1864	1,103	1,211	4,718		7,032

Year	*Whitehaven Tons*	*Workington Tons*	*Maryport Tons*	*Carlisle Tons*	*Total Tons*
1865	1,454	840	6,976		9,270
1866	2,115	130	4,328		6,573
1867	1,506	–	4,085		5,591
1868	1,132	1,446	8,086		10,664
1869	818	890	7,256		8,964
1870	1,060	27	6,602	–	7,689
1871	550	–	3,189	–	3,739
1872	150	191	2,947	–	3,288
1873	250	–	240	–	490
1874	325	–	2,394	–	2,719
1875	80	180	2,142	–	2,402
1876	71	624	4,328	493	5,516
1877	–	53	5,235	840	6,128
1878	75	–	5,500	1,262	6,837
1879	310	–	3,190	–	3,500
1880	65	309	3,484	–	3,858
1881	–	–	1,046	–	1,046
1882	250	810	3,157	–	4,217
1883	150	321	2,897	–	3,368
1884	670	–	2,042	–	2,712
1885	–	53	748	–	801
1886	–	–	4,352	–	4,352
1887	–	–	1,293	–	1,293
1888	–	–	31	–	31
1889	348	114	607	–	1,069
1890	340	–	162	–	502
1891	371	–	3,343	–	3,714
1892	136	–	352	–	488
1893	–	–	3,376	–	3,376
1894	300	–	1,396	–	1,696
1895	55	817	15	–	887
1896	1,226	150	448	–	1,824
1897	462	150	445	–	1,057
1898	366	450	1,343	–	2,159
1899	–	–	538	–	538
1900	450	–	206	–	656
1901	130	74	69	–	273
1902	85	–	985	401	1,471
1903	–	146	1,147	90	1,383
1904	–	74	1,729	–	1,803
1905	–	90	124	–	214
1906	100	230	60	220	610
1907	–	–	–	–	–
1908	10	–	2,092	–	2,102
1909	–	–	6,910	–	6,910

Year	*Whitehaven Tons*	*Workington Tons*	*Maryport Tons*	*Carlisle Tons*	*Total Tons*
1910	–	–	1,842	–	1,842
1911	–	–	–	–	–
1912	–	–	–	–	–
1913	–	–	–	–	–

Sources: R. Hunt, Mineral Statistics of the United Kingdom, 1854-1881. Mines Department, *Mineral Statistics of the United Kingdom*, 1882-1896. *Reports of Inspectors of Mines*, 1873-1914. *Report of the Royal Commission on Coal in the United Kingdom*, 1871, iii, Appendix 71-86.

Movement of Cumberland Coal

Year	*Shipped Tons*	*% Shipped*	*Moved by Rail Tons*
1819	256,036		
1820	253,588		
1821	197,436		
1822	268,084		
1823	297,629		
1824	297,230		
1825	287,468		
1826	295,935		
1827	307,346		
1828	277,413		
1829	349,144		
1833	363,528		
1834	372,645		
1835	373,704		
1836	423,504		
1837	437,683		
1838	422,819		
1839	461,804		
1840	447,697		
1841	474,848		
1842	465,647		
1843	445,783		
1844	475,837		
1845	533,030		
1846	547,256		
1847	507,604		
1848	482,836		
1849	530,650		
1850	544,624		
1851	528,697		
1852	541,853		
1853	624,092		
1854	645,147	74·98	24,346
1855	607,256	75·01	11,346
1856	643,130	70·37	64,316
1857	674,679	71·62	58,745
1858	648,641	70·49	
1859	707,099	67·86	
1860	723,641	61·79	
1861	705,176	56·16	
1862	742,713	55·83	
1863	717,368	54·04	
1864	730,695	52·91	
1865	761,250	53·19	
1866	722,364	48·46	
1867	788,314	52·11	

Year	*Shipped Tons*	*% Shipped*	*Moved by Rail Tons*
1868	688,213	48·49	
1869	679,548	48·16	
1870	678,001	48·14	
1871	603,590	42·32	
1872	333,713	18·69	
1873	364,951	20·88	
1874	359,943	32·65	
1875	393,922	32·11	371,795
1876	505,509	37·73	459,729
1877	513,678	33·88	488,635
1878	502,836	36·22	497,921
1879	588,059	40·30	519,796
1880	517,893	30·81	516,530
1881	513,311	29·01	552,262
1882	461,148	26·39	554,904
1883	453,830	25·54	546,101
1884	434,930	25·47	497,640
1885	489,142	28·22	491,469
1886	529,578	29·60	505,522
1887	471,798	26·26	476,493
1888	453,548	25·97	486,650
1889	381,437	21·92	505,414
1890	374,425	21·51	512,709
1891	712,159	42·65	478,823
1892	378,688	26·57	422,392
1893	513,267	28·27	673,980
1894	675,805	32·82	614,832
1895	553,752	29·39	592,305
1896	562,147	29·10	603,077
1897	592,398	29·82	652,126
1898	647,760	31·41	681,402
1899	646,467	30·77	677,322
1900	558,170	27·60	665,955
1901	612,002	29·02	683,640
1902	642,723	29·48	694,989
1903	645,573	29·24	701,457
1904	646,743	30·51	663,589
1905	624,597	29·03	699,861
1906	658,126	29·58	779,746
1907	669,709	29·71	770,979
1908	677,004	31·71	728,146
1909	667,922	28·92	845,478
1910	638,064	29·33	804,902
1911	640,049	27·87	850,498
1912	616,081	28·87	802,630
1913	608,704	26·23	808,934
1914			765,200

Sources: R. Hunt, *Mineral Statistics of the United Kingdom*, 1854-1881. Mines Department, *Mineral Statistics of the United Kingdom*, 1882-1896. *Reports of Inspectors of Mines*, 1873-1914. *Report of the Royal Commission on Coal in the United Kingdom*, 1871, iii, Appendix 71-86.

Cumberland Coal – Output and Value 1854-1946

Year	*Output Tons*	*Approximate Price per Ton at Mines* s.	d.	*Value* £
1854	887,000			
1855	809,546			
1856	913,891			
1857	942,018			
1858	920,137			
1859	1,041,890			
1860	1,171,052			
1861	1,255,644			
1862	1,330,287			
1863	1,327,368			
1864	1,380,795			
1865	1,431,047			
1866	1,490,481			
1867	1,512,514			
1868	1,378,026			
1869	1,410,808			
1870	1,408,235			
1871	1,423,661			
1872	1,784,774			
1873	1,164,532			
1874	1,036,274			
1875	1,171,341			
1876	1,392,046			
1877	1,447,223			
1878	1,393,514			
1879	1,525,039			
1880	1,680,841			
1881	1,769,213			
1882	1,747,317	5	6	480,512
1883	1,776,249	5	6	488,468
1884	1,707,582	5	1·86	440,129
1885	1,732,746	4	9·29	413,620
1886	1,788,981	4	7·44	413,253
1887	1,796,594	4	9½	430,434
1888	1,745,954	4	10·8	427,758
1889	1,739,491	5	9·2	501,553
1890	1,740,413	7	6¾	658,094
1891	1,669,756	7	1½	594,851
1892	1,424,749	6	6¾	467,495
1893	1,815,552	6	2	559,795
1894	2,058,867	6	0	617,660
1895	1,883,592	5	6·41	521,206
1896	1,931,771	5	5	523,188
1897	1,986,536	5	7·04	554,905

Year	*Output Tons*	*Approximate Price per Ton at Mines* s.	d.	*Value* £
1898	2,061,878	5	9·65	598,374
1899	2,100,418	6	9·24	710,991
1900	2,022,327	9	2·65	932,377
1901	2,108,360	8	1·74	858,630
1902	2,179,703	7	2·57	786,237
1903	2,207,476	6	9·63	750,818
1904	2,119,382	6	7·26	699,926
1905	2,151,060	6	6·02	699,274
1906	2,224,423	6	6·59	728,406
1907	2,253,790	8	0·14	902,831
1908	2,134,336	7	9·24	829,831
1909	2,309,370	7	2·05	828,005
1910	2,174,755	7	3·39	791,883
1911	2,296,254	7	2·04	823,207
1912	2,133,563	8	11·90	959,256
1913	2,320,120	9	4·25	1,085,123
1914	2,360,510	8	6·02	1,003,455
1915	2,248,810	11	9·48	1,325,638
1916	2,279,295	14	8·58	1,676,965
1917	2,161,550	15	1·00	1,630,149
1918	1,904,324	17	6·85	1,673,067
1919	1,842,153	20	8·98	1,911,102
1920	1,847,699	24	1·92	2,232,056
1921	1,335,120	22	10	1,521,270
1922	2,190,499	16	3	
1923	2,114,037	18	3	
1924	2,188,795	18	8	
1925	1,985,442	16	11	
1926	976,354	17	1	
1927	2,416,049	15	8	
1928	2,138,823	14	4	
1929	2,021,405	14	10	
1930	2,017,477	14	9	
1931	1,512,878	14	8	
1932	1,512,681	14	0	
1933	1,420,697	13	7	
1934	1,564,204	13	6	
1935	1,537,432	14	4·43	993,878
1936	1,189,948	15	10·25	831,854
1937	1,343,093	18	0·57	1,118,985
1938	1,568,175	20	8·74	1,446,390
1939	1,592,300	20	1·25	1,423,989
1940	1,524,600	23	7·36	1,581,933
1941	1,209,600	30	5·64	1,579,116
1942	1,106,600	34	3·33	1,625,008
1943	1,074,000	36	9·90	1,704,857

Year	*Output Tons*	*Approximate Price per Ton at Mines* s. d.	*Value* £
1944	1,049,500	40 9·57	1,893,248
1945	999,900	45 4·75	
1945	999,900	45 4·75	
1946	1,016,300	46 5·5	

Sources: R. Hunt, *Mineral Statistics of the United Kingdom*, 1854-1881. Mines Department, *Mineral Statistics of the United Kingdom*, 1882-96. *Reports of Inspectors of Mines*, 1897-1946. Ministry of Fuel and Power, *Statistical Digest from 1938*, Cmd. 6538. (1944). Mining Association, *Statistical Review of the Coal Industry 1933-34*. M.F.G.B., *Annual Volumes of Proceedings* 1935-38. *Colliery Year Book 1948*.

Cumberland Coal – Employees & Productivity 1874-1946

Year	*No. of Persons Employed*			*Total*	*Output per Employee*	
	Under Ground	*Above Ground*	*Clerks etc.*	*Employed*	*Cumberland Tons*	*U.K. Tons*
1874	4,251	1,427		5,678	182	234·9
1875	4,763	1,745		6,508	179	248·7
1876	4,401	1,623		6,024	231	260·6
1877	4,381	1,440		5,821	245	271·4
1878	4,253	1,525		5,778	241	278·9
1879	4,281	1,528		5,809	262	280·4
1880	4,407	1,537		5,944	282	303·0
1881	4,633	1,542		6,175	286·	311·1
1882	4,556	1,591		6,147	284	310·5
1883	4,736	1,632		6,368	278	317·9
1884	4,850	1,630		6,480	263	308·9
1885	5,051	1,729		6,780	255	306·0
1886	5,039	1,676		6,715	266	302·9
1887	5,252	1,637		6,889	260	308·0
1888	4,983	1,484		6,467	269	309·4
1889	5,032	1,412		6,444	269	305·0
1890	5,644	1,489		7,133	243	287·1
1891	6,286	1,499		7,785	214	277·6
1892	6,118	1,506		7,624	186	265·9
1893	5,981	1,512		7,493	242	240·5
1894	6,221	2,061		8,282	248	266·9
1895	5,980	1,747		7,727	243	270·8
1896	5,929	1,832		7,761	248	282·0
1897	5,933	1,859		7,792	254	290·7
1898	5,875	1,951		7,826	263	285·8
1899	5,894	2,003		7,897	265	301·8
1900	6,469	2,077		8,546	236	288·6
1901	6,694	2,190		8,884	237	271·5
1902	6,818	2,230		9,048	240	275·3
1903	7,022	2,305		9,327	236	273·5
1904	6,803	2,164		8,967	236	274·2
1905	6,816	2,210		9,026	238	275·0
1906	6,920	2,197		9,117	243	284·5
1907	7,170	2,311		9,481	237	284·7
1908	7,206	2,377		9,583	222	264·7
1909	7,521	2,476		9,997	231	260·1
1910	7,957	2,586		10,543	206	251·9
1911	7,905	2,673		10,578	217	254·7
1912	8,125	2,617		10,742	198	239·0
1913	8,369	2,667		11,036	210	254·8
1914	8,801	2,621		11,422	206	234·3
1919	8,562	2,793		11,355	162·2	192·8
1920	8,963	2,759		11,722	157·6	183·8

Year	No. of Persons Employed Under Ground	Above Ground	Clerks etc.	Total Employed	Output per Employee Cumberland Tons	U.K. Tons
1921	8,920	2,641		11,561	115·4	144·2
1922	8,822	2,674		11,496	190·5	217·3
1923	9,104	2,701		11,805	179·0	229·3
1924	9,272	2,608	164	12,044	181·7	220·0
1925				11,036	179·9	220·5
1926					n.a.	n.a.
1927	8,974	2,494	154	11,622	207·8	245·3
1928	8,158	2,164	167	10,489	203·9	252·9
1929	7,670	2,000	153	9,823	205·7	269·5
1930	7,568	2,032	152	9,752	206·8	261·8
1931	6,626	1,874	133	8,633	175·2	252·8
1932	5,448	1,496	126	7,070	213·9	254·7
1933	5,001	1,338	120	6,459	219·9	262·4
1934	5,367	1,437	118	6,922	225·9	280·0
1935	4,993	1,309	104	6,406	239·9	288·8
1936	3,770	1,095	97	4,962	239·8	297·8
1937	4,580	1,304	106	5,990	224·2	303·6
1938	5,033	1,511	114	6,658	235·5	287·0
1939				6,716	237·0	301·9
1940				6,471	235·6	299·4
1941				5,584	216·6	295·8
1942				5,521	200·4	287·1
1943				5,510	194·9	274·8
1944				5,622	186·6	259·2
1945				5,730	174·5	246·4
1946				5,772	176·0	260·1

Cumberland figures 1938-46 refer to wage earners.

Sources: *Reports of Inspectors of Mines*, 1874-1946. Ministry of Fuel & Power, *Statistical Digest 1945*, Cmd. 6920; *Statistical Digest 1948 and 1949*. Board of Trade, *Statistical Abstract for the United Kingdom 1924-38*. Vol. 83, 310-311.

Cumberland Coal – Costs and Balances Per Ton of Coal Disposable Commercially

Year	*Proceeds*		*Costs*		*Balance*	
	s.	*d.*	*s.*	*d.*	*s.*	*d.*
1917[a]	17	3½	20	4	− 3	0½
1918	20	10	24	6¾	− 3	8¾
1919	25	8¼	34	2¼	− 8	6
1920	31	0¼	43	6½	−12	6¼
1921[b]	27	7¾	35	4¼	− 7	8½
1922	20	1¼	22	10¼	− 2	9
1923	20	5¾	22	1	− 1	7¼
1924	21	1½	22	7¼	− 1	5¾
1925	18	2	21	5¼	− 0	11¼
1926[c]	16	9¾	21	1	+ 1	3½
1927	15	8	16	10¾	− 1	2¾
1928	15	0¾	17	4	− 2	3¼
1929	15	10½	16	10	− 0	11½
1930	15	8	17	3½	− 1	7½
1931	15	9¼	17	10½	− 2	1¼
1932	14	10	16	1	− 1	3
1933	14	5½	15	10	− 1	4½
1934	14	2	16	0½	− 1	10½
1935	14	4¼	15	9½	− 1	5¼
1936	15	10¼	15	11	− 0	0¾
1937	18	0½	18	2¾	− 0	2¼
1938	20	8·74	20	2·68	+ 0	6·06
1939	20	1·25	20	7·89	− 0	6·64
1940	23	7·36	24	5·70	− 0	10·34
1941	30	5·64	31	9·20	− 1	3·56
1942	34	8	37	5	− 2	9
1943	37	3	43	0	− 5	9
1944	41	4	50	8	− 9	4
1945	46	0	56	10	−10	10

[a] July-December
[b] Excluding months of April-June
[c] August 1925-April 1926 inclusive

Subventions:
1925 2s. 4d. per ton
1926 5s. 6¾d. per ton

Sources: Reports of Inspectors of Mines. Records of Cumberland Coal Owners' Association. Mining Association, *Statistical Review of the Coal Industry 1933-34*. Miners' Federation of Great Britain, *Annual Volumes of Proceedings*, 1935-38. Ministry of Fuel and Power, *Statistical Digest from 1938*, Cmd. 6538 (1944). *Financial Position of the Coal Mining Industry: Coal Charges Account*, Cmd. 6617 (1945). W. H. B. Court, *History of the Second World War: Coal* (1951), 185, 338-346.

Average Cash Earnings per Manshift (All ages) – Cumberland and Great Britain

Year	*Cumberland* *s.*	*d.*	*Great Britain* *s.*	*d.*
1914 (June)	6	2¼	6	5¾
1918[a]	12	4½		
1919	14	4		
1920	17	1	16	10¼
1921[b]	14	5¾	15	11
1922	9	10½	9	11¾
1923	9	4½	10	1
1924	10	3½	10	7¾
1925	10	8	10	6
1926[c]	10	7½	10	5
1927	9	0½	10	0¾
1928	8	8¾	9	3½
1929	8	7¾	9	2¾
1930	8	9¼	9	3½
1931	8	8¾	9	2¼
1932	8	9¼	9	2
1933	8	8	9	1½
1934	8	8	9	1¾
1935	8	10¼	9	3¼
1936	9	7¼	10	0¼
1937	9	11¾	10	8
1938	10	9¼	11	2¾
1939	10	11	11	7
1940	12	3¾	13	0
1941	14	4½	14	11
1942	17	2½	17	5
1943	18	10¼	19	1
1944	22	2	21	8
1945	24	1¼	23	1
1946	25	0¼	23	11

[a] July-December 1918.
[b] Average of quarters Jan.-March and Oct.-Dec.
[c] Aug. 1925 – April 1926 inclusive.

Sources: Reports of Inspectors of Mines, 1914-1938. Ministry of Fuel and Power, *Statistical Digest from 1938*, Cmd. 6538 (1944); *Statistical Digest 1948 and 1949*.

Main Provisions of Cumberland District Wages Agreements 1922-1938

Minimum Percentage Addition to Basis Rates

Dec. 1921 - April 1924	30
May 1924 - 1925	44·4
1926 - March 1927	35·0
April 1927 - July 1931	30
Aug. 1931 - 1936	22·50
1937 - 16 May 1938	27
17 May 1938	30

Hours of Work

Underground per Day		*Surface per Week*	
1927-30 Nov. 1930	8	1927-1938	49
1 Dec. 1930-1938	7½		

Subsistence Wages per Shift

	Men Aged 21 and over		*Men aged 18 and under 21*		*Women*	
	s.	*d.*	*s.*	*d.*	*s.*	*d.*
1914	6	10	6	6	3	8
1922 Jan.-Feb.	6	11	6	8	3	8
March-May	6	10	6	6		
June	6	8½	6	4		
1923 4 June	6	7½	6	3	3	7
1924 18 June	7	5½	7	0½	4	0½
1925	7	5½				
1926 Dec.	6	7½	6	3	3	7
1927 1 April	6	10	6	3	3	8
1928	6	10	6	8	3	8
1929	6	10	6	6	3	8
1930	6	10	6	6	3	8
1931	6	10	6	6	3	8
1932	6	10	6	6	3	8
1933	6	10	6	6	3	8
1934 11 June	7	1	6	8·85	3	9·61
1935	7	1	6	8·85	3	9·61
1936	7	1	6	8·85	3	9·61
1937	7	3½	6	10·25	3	10·61

From 16 October 1937 these rates were merged into the basis rates.

Source: Reports of Chief Inspector of Mines, 1921-38.

Cumberland & Westmorland – Coal Cutting Machines

Year	*No. of Mines at which Machines in use*	*No. of Machines*	*Quantity of Coal cut*	*% of Output cut by Machines*	
				Cumberland & Westmorland	*Great Britain*
			Tons		
1902		2	16,245		
1903		2	6,487		
1904		1	1,585		
1905		2	7,695		
1906		6	16,592		
1907		4	5,200		
1908		1	2,242		
1909		1	220		
1910		–	–		
1911		3	606		
1913		11	29,963	1	8·5
1921	7	18	74,054	6	14·1
1922	6	24	166,802	7	15·0
1923	7	32	217,615	12	17·2
1924	10	35	155,166	7	19·0
1925	9	35	181,165	9	20·0
1926	9	34	87,599	9	22·0
1927	10	38	253,964	10	23·0
1928	11	41	298,865	14	26·0
1929	11	41	220,875	11	27·9
1930	13	52	309,857	15	31·0
1931	11	51	304,622	20	35·0
1932	9	44	433,681	29	38·5
1933	11	54	490,556	35	42·4
1934	10	63	594,827	38	47·0
1935	10	70	725,535	47	51·0
1936	9	62	767,743	65	55·0
1937	12	76	898,146	67	57·0
1938	12	80	1,128,765	72	59·5
1939	12	77	1,131,454	71	61·5
1940	12	76	1,258,931	83	63·7
1941	12	70	749,647	62	65·7
1942	13	80	861,344	78	66·4
1943	13	88	802,008	75	69·0
1944	12	82	764,624	73	71·6
1945	12	87	901,375	76	72·2
1946			895,606	73	73·9

Sources: *Reports of Inspectors of Mines*, 1902-1938. Ministry of Fuel & Power, *Statistical Digest from 1938*, Cmd. 6538 (1944); *Statistical Digest 1944*, Cmd. 6639 (1945); *Statistical Digest 1945*, Cmd. 6920 (1946); *Statistical Digest 1946 and 1947*, Cmd. 7548 (1948); *Statistical Digest 1948 and 1949*.

Cumberland & Westmorland – Conveyors

Year	*No. of Mines at which Conveyors in Use*	*No. of Conveyors*	*Tonnage of Coal Mechanically Conveyed*	*% of Output Mechanically Conveyed*
1929	3	4	8,163	
1930	5	8	67,987	3
1931	4	8	57,434	4
1932	5	16	113,596	7
1933	5	19	123,305	9
1934	5	20	237,487	15
1935	4	22	262,874	17
1936	3	19	215,182	18
1937	5	36	392,883	29
1938	5	34	527,128	34
1939	5	40	606,382	38
1940	5	38	624,082	41
1941	6	29	472,214	39
1942	6	35	483,096	44
1943	7	58	529,944	49
1944	7	71	544,316	52
1945	7	100	659,765	55
1946			751,951	62

Sources: Reports of Inspector of Mines, 1929-1938. Ministry of Fuel and Power, *Statistical Digest from 1938*, Cmd. 6538 (1944); *Statistical Digest 1944*, Cmd. 6639 (1945); *Statistical Digest 1945*, Cmd. 6920 (1946); *Statistical Digest 1946 and 1947*, Cmd. 7548 (1948).

Cumberland & Westmorland – Washing Plants

Year	*No. of Washing Plants*	*% of Total Output Washed* *Cumberland & Westmorland*	*Great Britain*
1927	13	38·8	20·5
1928	13	61·4	25·4
1929	13	58·5	27·6
1930	12	65·1	29·8
1931	11	62·7	30·4
1932	11	58·6	34·3
1933	11	67·0	37·4
1934	9	65·5	39·6
1935	9	65·4	41·4
1936	9	65·9	43·1
1937	9	66·1	43·8
1938	8	64·9	45·4
1939			47·4
1944	7	62·9	45·8
1945	8	71·6	47·4
1946	7	70·0	47·4

Sources: Reports of Inspectors of Mines, 1927-1938. Ministry of Fuel and Power, *Statistical Digest 1945*, Cmd. 6920 (1946); *Statistical Digest 1946 and 1947*, Cmd. 7548 (1948); *Statistical Digest 1948 and 1949*.

Cumberland & Westmorland – Electrical Equipment

Year	*No. of Mines at which equipment is installed*	*No. of Electric Motors in use*	*Aggregate H.P. of Electric Motors Below Ground*	*Above Ground*	*Total H.P. Above & Below Ground*
1913			2,079	2,134	4,213
1920			3,624	3,887	7,511
1921	19		4,098	4,195	8,293
1922	22		4,501	4,922	9,423
1923	21		5,466	5,751	11,217
1924	22		6,430	5,563	11,993
1925	22	363	6,817	5,408	12,225
1926	22	388	7,144	6,103	13,247
1927	21	412	7,745	6,229	13,974
1928	20	359	7,349	5,317	12,666
1929	21	397	7,982	6,254	14,236
1930	21	444	9,099	6,522	15,621
1931	21	445	9,018	6,840	15,858
1932	22	440	9,188	7,021	16,209
1933	18	441	8,956	6,721	15,677
1934	15	421	7,574	7,225	14,799
1935	16	440	8,263	7,481	15,744
1936	18	468	8,749	8,356	17,105
1937	18	513	9,372	8,784	18,156
1938	20	555	9,740	10,929	20,669
1944	14	690	12,039	13,067	25,106
1945	15	791	14,079	12,566	26,645

Source: Reports of Chief Inspector of Mines.

Coal Shipments from:

Year	*Whitehaven* *Tons*	*Workington* *Tons*		*Maryport* *Tons*	
1914				270,239	
1915				233,498	
1916				197,671	
1917				146,143	
1918				202,613	
1919				215,072	
1920				200,747	
1921				107,704	
1922		215,899		224,274	
1923		160,871		318,489	
1924		2,195		295,858	
1925	273,192	2,026		263,582	
1926	143,584	–		116,960	
1927	160,806	2,702		270,657	
1928	382,408	203,093		151,403	
1929	327,546	143,480		140,822	
1930	198,272	157,204		132,403	
1931	183,310	154,382		113,740	
1932	173,073	101,163		98,971	
1933	161,196	133,720		68,116	
1934	153,566	209,043		44,831	
1935	196,909	178,709		56,327	
1936	103,270	151,122		42,996	
1937	–	121,674		32,942	
1938	146,601	190,500		30,940	
1939	208,835	156,303		34,602	
1940	287,090	241,380	(81,525)	83,395	(52,170)
1941	235,651	339,597	(249,677)	320,505	(294,834)
1942	176,685	318,041	(269,811)	467,405	(445,866)
1943	205,940	123,134	(94,717)	460,599	(450,745)
1944	178,698	330,275	(285,239)	518,600	(513,571)
1945	196,472	148,843	(131,959)	467,132	(464,841)
1946	169,715	22,935		476,880	(473,052)

Figures in brackets denote the amount of North-East coal included in total shipments from Workington and Maryport.

Whitehaven shipments are for year ending 31 March. Maryport shipments 1944-46 are for year ending 30 September.

Source: Harbour Offices at Whitehaven, Workington and Maryport.

Whitehaven Colliery 1947-1982/3

	Haig Pit			*William Pit*	
Year	*Output*	*Manpower*	*Output per Manshift Overall Cwts.*	*Output*	*Manpower*
	Tons			*Tons*	
1947	417,371	1,300	19·9	n.a.	440
1948	341,838		17·4	n.a.	
1949	289,453	1,368	17·2	n.a.	411
1950	296,374	1,420	16·9	55,033	364
1951	333,793	1,576	17·1	64,395	381
1952	347,575	1,645	16·6	75,699	394
1953	379,356	1,576	19·3	68,882	312
1954	398,650	1,574	20·4	66,992	274
1955	438,132	1,775	19·8	60	24
1956	442,100	1,794	20·8		
1957	396,968	1,789	18·3	Pit closed 1965.	
1958	401,084	1,740	20·3		
1959	336,089	1,643	18·0		
1960	386,204	1,533	21·8		
1961	409,858	1,506	23·8		
Year ending March:					
1963	402,442	1,475	23·8		
1964	388,524	1,500	22·6		
1965	378,436	1,591	21·9		
1966	304,533	1,561	19·0		
1967	322,298	1,609	19·0		
1968	474,507	1,589	26·5		
1969	428,331	1,487	26·0		
1970	372,364	1,449	23·7		
1971	271,958	1,385	18·39		
1972	290,936	1,382	22·30		
1973	241,321	1,319	16·8		
1974	273,863	1,273	22·4		
1975	302,370	1,261	22·3		
1976	348,883	1,213	27·1		
1977	293,104	1,158	24·5		
1978	212,495	1,100	18·4		
	Tonnes		*Tonnes*		
1979	264,838	1,031	1·19		
1980	269,842	963	1·29		
1981	372,848	938	1·88		
1982	471,170	882	2·41		
1983	305,880	793	1·78		

Sources: Public Relations Department, National Coal Board, Western Area, Stoke-on-Trent. Manpower figures 1949-61 from Statistics Department, National Coal Board, Hobart House, London. Iliffe Books Ltd., *The Colliery Year Book and Coal Trades Directory 1947*.

Cumberland N.C.B. Coal – Output & Value (Saleable Coal)

Year	*Output Tons*	*Pithead proceeds* £	*Proceeds Per Ton* *s. d.*
1947	1,082,540	2,531,614	46 9·3
1948	1,002,438	2,647,117	52 9·8
1949	1,077,675	2,934,291	54 5·5
1950	1,130,655	3,094,796	54 8·9
1951	1,204,403	3,612,815	59 11·9
1952	1,205,182	4,146,425	68 9·7
1953	1,212,710	4,424,805	72 11·7
1954	1,211,689	4,623,647	76 3·8
1955	1,221,205	5,005,553	81 11·7
1956	1,144,555	5,382,292	94 0·6
1957	1,051,105	5,340,650	101 7·4
1958	1,074,199	5,536,926	103 1·1
1959	937,250	4,786,291	102 1·6
1960	952,611	4,974,364	104 5·2
1961	945,144	5,201,674	110 0·9
1962	904,355	5,073,780	112 3
Fifteen months to March:			
1964	1,184,000	6,678,000	112 10
Year ending March:			
1965	874,287	4,786,000	109 6
1966	923,752	4,884,000	105 9
1967	661,905	4,077,000	123 2
1968	801,892		
1969	645,712		
1970	477,162		
1971	377,487		
1972	381,536		
1973	333,462		
1974	282,119		
1975	302,370	4,729,600	£15·64
1976	348,883	6,559,000	£18·80
1977	293,104	6,702,000	£22·87
1978	212,495	8,520,500	£40·10
	Tonnes		
1979	264,838	8,117,900	£30·65
1980	269,842	9,339,971	£34·61
1981	372,848	14,924,118	£40·03
1982	471,170	20,412,000	£43·32
1983	305,880	14,172,000	£46·33

Figures of proceeds 1967/8-1973/4 are not available.

Sources: N.C.B. *Reports and Accounts*. Mr L. R. Hoten, Area Statistician, Western Area, and his successor Mr G. W. Wood.

Cumberland N.C.B. Mines – Employees

Year	*No. of Employees* *Below Ground*	*Above Ground*	*Total*
1947			5,590
1948	4,246	1,327	5,573
1949	4,266	1,333	5,599
1950	4,157	1,323	5,480
1951	4,155	1,390	5,545
1952	4,290	1,370	5,660
1953	4,470	1,465	5,935
1954	4,480	1,295	5,775
1955	4,300	1,238	5,538
1956	4,380	1,257	5,637
1957	4,394	1,252	5,646
1958	4,394	1,129	5,523
1959	4,358	1,087	5,445
1960	3,818	935	4,753
1961			4,042
1962			3,701
1963	2,780	899	3,679
Year ending March:			
1964	2,794	899	3,693
1965	2,878	951	3,829
1966	2,851	977	3,828
1967	2,401	790	3,191
1968	2,187	719	2,906
1969	1,749	573	2,322
1970	1,675	683	2,358
1971	1,472	491	1,963
1972	1,481	480	1,961
1973	1,407	459	1,866
1974	1,054	344	1,398
1975	972	289	1,261
1976	957	256	1,213
1977	921	237	1,158
1978	894	206	1,100
1979	835	196	1,031
1980	783	180	963
1981	758	180	938
1982	710	172	882
1983	645	148	793

Source: Mr L. R. Hoten, Area Statistician, Western Area and his successor, Mr G. W. Wood.

Productivity in N.C.B. Mines – Cumberland & Great Britain

Year	*Output per Manshift Overall*		*Output per Man-year*	
	Cumberland Cwts.	*Gt. Britain Cwts.*	*Cumberland Tons*	*Gt. Britain Tons*
1947	15·7	21·5	193	263
1948	14·8	22·3	179	273
1949	15·8	23·4	192	282
1950	16·4	24·2	206	293
1951	16·6	24·5	217	303
1952	16·3	24·2	212	299
1953	17·4	24·6	204	296
1954	17·5	24·9	209	303
1955	23·8	24·7	220	299
1956	24·2	24·8	203	298
1957	24·0	24·9	186	296
1958	17·1	25·6	194	288
1959	17·9	26·9	172	294
1960	20·6	28·0	200	305
1961	21·8	28·9	233	316
1962	22·1	31·2	244	342
1963	22·3	32·9		359
Year ending March:				
1964	22·8	33·4	252	362
1965	21·3	34·8	228	373
1966	23·5	36·1	241	381
1967	19·7	36·6	207	391
1968	25·3	39·0	275	414
1969	25·3	42·5	278	454
1970	21·2	43·4	202	457
1971	17·7	44·1	192	463
1972	20·6	41·9	194	387
1973	16·3	45·8	178	472
1974	21·1	42·3	201	384
1975	22·3	45·0	239	466
1976	27·1	44·8	287	455
1977	24·5	43·6	253	440
1978	18·4	2·19 *Tonnes*	193	441 *Tonnes*
1979	1·19 *Tonnes*	2·24 *Tonnes*	256 *Tonnes*	448 *Tonnes*
1980	1·29 *Tonnes*	2·31 *Tonnes*	279 *Tonnes*	467 *Tonnes*
1981	1·88 *Tonnes*	2·32 *Tonnes*	397 *Tonnes*	476 *Tonnes*
1982	2·41 *Tonnes*	2·40 *Tonnes*	534 *Tonnes*	494 *Tonnes*
1983	1·78 *Tonnes*	2·44 *Tonnes*	388 *Tonnes*	504 *Tonnes*

Sources: N.C.B. *Reports and Accounts*. Mr L. R. Hoten, Area Statistician, Western Area and his successor, Mr G. W. Wood.

Wages Cost in N.C.B. Mines – Cumberland & Great Britain Per Ton of Saleable Coal

	Cumberland			Great Britain		
Year	*Wages Cost*		*As % of*	*Wages Cost*		*As % of*
	s.	*d.*	*Total Cost*	*s.*	*d.*	*Total Cost*
1947	34	3·6	60·95	26	11	65·25
1948	40	8·5	61·35	29	9	65·26
1949	38	5·1	61·80	29	4	65·18
1950	38	10·8	61·17	29	5	64·77
1951	42	11·3	61·66	31	9	64·57
1952	49	7·4	59·22	35	9	62·99
1953	48	11·2	57·65	36	11	62·50
1954	51	3·7	59·23	38	4	61·93
1955	52	10·9	58·73	41	2	61·30
1956	58	10·6	58·44	44	9	60·21
1957	65	1·2	57·68	48	3	59·21
1958	60	1·6	54·24	48	9	58·09
1959	59	3·8	51·23	47	5	57·52
1960	53	7·3	49·21	47	0	55·68
1961	54	0·0	47·78	48	0	54·29
1962	55	1	47·66	45	8	52·29
1963/4	55	1	48·39	45	0	51·36
1964/5	61	5	46·70	45	10	51·69
1965/6	58	8	44·90	46	7	50·32
1966/7	73	8	45·13	48	2	48·98
1967/8				46	10	48·75
1968/9				45	0	46·58
1969/70				46	10	45·00
1970/1				56	7	48·79
1971/2				71	2	45·46
1972/3				75	5	49·21
1973/4				91	7	47·21
	£			£		
1974/5	8·79		38·24	5·86		48·91
1975/6	9·27		32·40	7·89		49·09
1976/7	11·05		29·12	8·84		46·84
1977/8	16·20		29·54	9·76		44·87
1978/9	15·63		33·06	11·18		46·39
1979/80	16·81		35·00	12·89		45·11
1980/1	14·08		32·72	14·99		44·66
1981/2	12·35		31·18	15·59		42·59
1982/3	17·14		31,19	16·18		41·24

Cumberland figures 1967/8-1973/4 are not available.

Sources: N.C.B. *Reports and Accounts*, 1947-1981/2. Mr L. R. Hoten, Area Statistician, Western Area and his successor, Mr G. W. Wood.

Cumberland N.C.B. Mines – Costs & Losses

Year	Wages Cost Amount £	Wages Cost Per Ton s. d.	Total Cost Amount £	Total Cost Per Ton s. d.	Loss Amount £	Loss Per Ton s. d.	Loss per Employee £
1947	1,856,740	34 3·6	3,046,293	56 3·4	514,679	9 6·1	92·07
1948	2,040,239	40 8·5	3,325,343	66 4·2	678,226	13 6·4	121·69
1949	2,070,689	38 5·1	3,349,545	62 2·0	415,254	7 8·5	74·16
1950	2,199,090	38 10·8	3,594,799	63 7·0	500,003	8 10·1	91·24
1951	2,586,240	42 11·3	4,193,754	69 7·7	580,939	9 7·8	104·76
1952	2,989,890	49 7·4	5,048,623	83 9·4	902,198	14 11·7	159·39
1953	2,966,998	48 11·2	5,146,408	84 10·5	721,603	11 10·8	121·58
1954	3,108,307	51 3·7	5,247,734	86 7·4	624,087	10 3·6	108·06
1955	3,230,465	52 10·9	5,499,654	90 0·8	494,101	8 1·1	89·22
1956	3,369,799	58 10·6	5,765,722	100 9·0	383,430	6 8·4	68·02
1957	3,421,572	65 1·2	5,931,354	112 10·3	590,704	11 2·9	104·62
1958	3,229,634	60 1·6	5,953,761	110 10·2	416,835	7 9·1	75·47
1959	2,779,513	59 3·8	5,425,037	115 9·2	638,746	13 7·6	117·30
1960	2,553,275	55 7·3	5,187,600	108 10·9	213,236	4 5·7	44·86
1961	2,551,847	54 0·0	5,339,970	113 0·0	138,296	2 11·1	34·21
1962	2,488,315	55 1	5,220,634	115 6	146,854	3 3	39·67
Fifteen months to March:							
1964	3,259,000	55 1	6,734,000	113 9	56,000	11	15·16
Year ending March:							
1965	2,683,000	61 5	5,744,000	131 5	958,000	21 11	250·19
1966	2,709,000	58 8	6,033,000	130 8	1,149,000	24 11	300·15
1967	2,438,000	73 8	5,402,000	163 3	1,325,000	40 1	415·23
		£		£		£	
1975	2,658,219	8·79	6,950,540	22·99	2,220,940	7·35	1,761·25
1976	3,233,232	9·27	9,979,052	28·60	3,420,052	9·80	2,819·49
1977	3,237,911	11·05	11,116,273	37·93	4,414,273	15·06	3,811·98
1978	3,442,941	16·20	11,653,035	54·84	3,132,535	14·74	2,847·75
1979	4,138,361	15·63	12,519,548	47·27	4,401,648	16·62	4,269·29
1980	4,534,871	16·81	12,959,729	48·03	3,619,758	13·42	3,758·84
1981	5,249,666	14·08	16,042,208	43·03	1,118,422	3·00	1,192·35
1982	5,821,000	12·35	18,667,000	29·62	*1,908·000	4·05	2,163·27
1983	5,242,000	17·14	16,809,000	54·95	2,373,000	7·75	2,992·43

Cumberland figures 1967/8-1973/4 are not available. *Profits are underlined.

Sources: N.C.B. *Reports and Accounts*, 1947-1966/7.
Mr. L. R. Hoten, Area Statistician, Western Area and his successor Mr. G. W. Wood.

Cumberland Coal – Average Earnings (All Workers)

Year	*Per Manshift*		*Value of Allowances in kind*		*Total*		*Per Week*		*Value of Allowances in kind*		*Total*	
	s.	*d.*	*s.*	*d.*	*s.*	*d.*	*s.*	*d.*	*s.*	*d.*	*s.*	*d.*
1947	28	8·3			28	8·3	135	5			135	5
1948	32	6·3			32	6·3	149	11			149	11
1949	33	5·3			33	5·3	166	1			166	1
1950	34	6·9		1·5	34	8·4	176	5		8	177	1
1951	37	10·9		3·4	38	2·3	199	2	1	6	200	8
1952	42	4·7		5·7	42	10·4	220	4	2	6	222	10
1953	44	10·5		9·6	45	8·1	225	8	4	0	229	8
1954	47	3·5	1	0·4	48	3·9	245	5	5	4	250	9
1955	49	9·1	1	1·7	50	10·8	259	2	5	11	265	1
1956	51	9·8	1	5·1	53	2·9	267	3	7	5	274	8
1957	54	8·0	1	7·8	56	3·8	275	10	8	4	284	2
1958	55	10·7	1	11·9	57	10·6	264	3	9	5	273	8
1959	56	10·7	2	2·0	59	0·7	264	7	10	1	274	8
1960	58	1·4	2	4·4	60	5·8	277	2	11	3	288	5
1961	61	10·2	2	6·9	64	5·1	298	0	12	5	310	5
1962	63	9	2	8	66	5						
Fifteen months to March:												
1964	66	10	2	9	69	7						
Year ending March:												
	£											
1975	9·817											
1976	12·565											
1977	13·529											
1978	14·661											
1979	18·645											
1980	21·705											
1981	26·452											
1982	29·620											
1983	30·510											

Cumberland figures 1964/5-1973/4 are not available.

Sources: N.C.B. *Reports and Accounts*. Mr L. R. Hoten, Area Statistician, Western Area and his successor, Mr G. W. Wood.

Absenteeism in N.C.B. Mines – Cumberland & Great Britain

Year	*Cumberland* %	*Great Britain* %
1947	11·7	12·4
1948		11·6
1949		12·3
1950	9·4	12·0
1951		12·2
1952		12·0
1953		12·4
1954		12·2
1955	10·7	12·5
1956	11·3	12·9
1957	12·4	13·8
1958	13·1	14·1
1959	14·3	14·7
1960	14·3	14·7
1961	14·9	15·4
1962	14·2	15·4
Year ending March:		
1963	13·8	15·4
1964	13·1	15·9
1965	15·2	16·0
1966	18·6	18·0
1967	17·4	17·3
1968	17·3	18·2
1969	17·0	18·1
1970	18·1	18·3
1971	18·1	19·2
1972	16·2	17·7
1973	15·7	17·6
1974	14·6	17·9
1975	14·1	16·0
1976	14·2	16·7
1977	17·0	17·3
1978	17·2	17·6
1979	16·3	17·1
1980	12·5	14·8
1981	11·0	12·4
1982	10·1	11·4
1983	9·9	10·4

Sources: N.C.B. Reports and Accounts. Mr L. Hoten, Area Statistician, Western Area and his successor, Mr G. W. Wood.

Cumberland Collieries – Fatal Accidents

Year	*No. killed*	*Year*	*No. killed*	*Year*	*No. killed*
1851	14	1896	9	1941	20
1852	11	1897	16	1942	7
1853	19	1898	10	1943	9
1854	22	1899	11	1944	8
1855*	9	1900	22	1945	n.a.
1856	7	1901	8	1946	18
1857	6	1902	12	1947	109
1858	13	1903	16	1948	4
1859	9	1904	16	1949	3
1860	15	1905	15	1950	–
1861	21	1906	13	1951	5
1862	16	1907	20	1952	4
1863	8	1908	19	1953	6
1864	9	1909	8	1954	1
1865	9	1910	150	1955	5
1866	10	1911	21	1956	2
1867	13	1912	11	1957	5
1868	16	1913	10	1958	7
1869	11	1914	14	1959	1
1870	12			1960	5
1871	10			1961	1
1872	14	1921	7	1962	2
1873	25	1922	56	1963	1
1874	7	1923	10	1964	1
1875	12	1924	17	1965	1
1876	11	1925	14	1966	1
1877	7	1926	5	1967	–
1878	7	1927	13	1968	2
1879	10	1928	21	1969	–
1880	4	1929	6	1970	1
1881	14	1930	5	1971	1
1882	17	1931	33	1972	–
1883	13	1932	13	1973	–
1884	7	1933	13	1974	–
1885	11	1934	6	1975	1
1886	8	1935	3	1976	–
1887	8	1936	2	1977	–
1888	39	1937	11	1978	–
1889	11	1938	10	1979	1
1890	10	1939	10	1980	–
1891	13	1940	10	1981	–
1892	6			1982	–
1893	11			1983	–
1894	15				
1895	9				

* 30 June-31 December only.
Figures for 1921 and 1925-34 are for Cumberland and Westmorland.

Sources: Annual *Reports of Divisional Inspector of Mines*, 1851-1973. Records of the Cumberland Miners' Association. Health and Safety Executive.

Cumberland Collieries – Female Employees

Year	*No.*	*Year*	*No.*
1873	306	1906	335
1874	314	1907	328
1875	343	1908	346
1876	323	1909	384
1877	306	1910	430
1878	336	1911	420
1879	336	1912	448
1880	349	1913	393
1881	343	1914	404
1882	379		
1883	373	1919	368
1884	418	1920	310
1885	412	1921	254
1886	395	1922	312
1887	392	1923	343
1888	303	1924	357
1889	261	1925	368
1890	233	1926	n.a.
1891	249	1927	388
1892	284	1928	260
1893	273	1929	250
1894	277	1930	256
1895	260	1931	249
1896	262	1932	142
1897	252	1933	147
1898	264	1934	172
1899	277	1935	89
1900	279	1936	87
1901	330	1937	157
1902	319	1938	188
1903	332		
1904	366	1947	130
1905	373		
		1972	2
		1973	–

Source: Annual Reports of Divisional Inspector of Mines.

Cumberland Opencast Coal 1958-1982

Year	*Output Tons*	*No. of Employees**
1958	91,176	
1959	102,808	
1960	19,584	
1961	Nil	
1962	Nil	
1963	Nil	
1964	51,827	
1965	138,395	
1966	145,386	
1967	194,976	
1968	63,407	
1969	108,725	
1970	221,185	
1971	383,904	
1972	378,620	
1973	324,701	
1974	361,276	
1975	388,526	
1976	376,991	91
1977	374,206	180
1978	406,796	240
1979	304,060	260
1980	549,551	420
1981	887,989	440
1982	891,882	480

* i.e. Contractors' employees

Source: N.C.B., Opencast Executive, Northern Region, Newcastle-upon-Tyne.

Coal Shipments 1947-1982

Year	Whitehaven Tons	Workington Tons		Maryport Tons	
1947	220,432	13,151	(4,406)	374,902	(366,264)
1948	279,042	43,334	(8,865)	157,705	(134,900)
1949	202,771	49,118	(380)	75,642	(46,003)
1950	217,312	67,601		84,963	(30,622)
1951	266,342			87,355	(19,681)
1952	280,048	42,962	(7,763)	67,700	(4,133)
1953	252,338	38,710		24,953	(3,138)
1954	258,010			34,406	
1955	226,339			51,186	
1956	284,094			40,618	
1957	263,113			25,695	
1958	254,385			18,970	
1959	223,862			8,409	
1960	194,947				
1961	173,663	1,687		Trading ceased	
1962	155,464				
1963	158,650	6,861			
1964	120,197	36,300			
1965	116,307	35,900			
1966	82,871	1,400			
1967	87,404	300			
1968	90,431	2,940			
1969	77,482	2,102			
1970	109,939	2,026			
1971	155,374	1,018			
	Tonnes	*Tonnes*			
1972	125,572	1,031			
1973	132,021	1,119			
1974	70,516	–			
1975	86,831	–			
1976	72,510	–			
1977	65,852	–			
1978	139,716	–			
1979	176,076	–			
1980	182,080	8,554			
1981	95,804	402,638			
1982	–	430,322			

Figures in brackets are of N.E. coal included in total shipments. Whitehaven shipments 1947-1955 year ending 31 March. Maryport shipments year ending 30 September. Workington shipments 1947-1953 year ending 30 June; 1961-1963 year ending September. Figures of total shipments from Workington 1951, 1954-1960 and 1962 are not available.

Sources: Whitehaven, Workington and Maryport harbour records. National Ports Council, *Digest of Port Statistics* 1966-1979. Northern Economic Planning Council, *Regional Ports Survey 1969*.

NOTES

Abbreviations Used in Notes

Manuscript Sources:

a. At the Carlisle Record Office:-

Bateman	Letters of John Bateman, 1802-11. (In Lonsdale MSS.)
D/Ben	Benson MSS.
D/Cu	Curwen MSS.
D/Lec	Leconfield MSS.
D/Lons/W	Lonsdale MSS. (Whitehaven section)
D/Sen	Senhouse MSS.
Lowther	Letters of Sir James Lowther to John Spedding.
Peile	Letters of John Peile to Lord Lonsdale.
Spedding	Letters of John Spedding to Sir James Lowther.

b. At the North of England Institute of Mining and Mechanical Engineers, Newcastle-upon-Tyne:-

Brown	Letter Books of William Brown.
Buddle	John Buddle MSS.
Forster	T. E. Forster Collection. (This collection includes the view books of John Barnes and Matthias Dunn.)
Watson	Watson Collection

Printed Sources:

CSPD	*Calendar of State Papers, Domestic.*
CSPI	*Calendar of State Papers, Ireland.*
CTB	*Calendar of Treasury Books.*
CTB & P	*Calendar of Treasury Books and Papers.*
PCR	*Privy Council Register.*
CW1, CW2	*Transactions of the Cumberland and Westmorland Antiquarian and Archaeological Society.* Old Series, Vols. 1-16, 1875-1900. New Series, Vol. 1, 1901 to date.
Hutchinson	W. Hutchinson, *History of the County of Cumberland* 2 vols. (Carlisle, 1794).
NCB	National Coal Board, *Annual Report and Accounts.*
VCH	Victoria County History, *Cumberland* Vol. ii (1905).

Unless stated otherwise the works cited below were published in London.

BIBLIOGRAPHY

Manuscript Sources

BIRMINGHAM
Reference Library:
Boulton and Watt MSS.
Assay Office:
Boulton MSS.

CARLISLE
Record Office, The Castle, Carlisle:
Benson MSS.
Curwen MSS.
Leconfield MSS.
Lonsdale MSS.
Senhouse MSS.
Lutwidge Letter Books.
Records of the Cumberland Coal Owners' Association.
Maryport Harbour Records.
Whitehaven Harbour Records.
Workington Harbour Records.

LONDON
Public Record Office:
Records formerly held in the Custom Houses at Whitehaven, Workington and Maryport:-
Whitehaven Letter Books, Customs, 1703-1860.
Letter Books, Miscellaneous, 1724-1858.
Workington Letter Books, 1855-1913.
Maryport Letter Book, Customs, 1856-60.
Letter Book, Miscellaneous, 1847-60.

NEWCASTLE-UPON-TYNE
North of England Institute of Mining and Mechanical Engineers, Neville Street:
Bell Collection.
Letter Books of William Brown, 2 vols., 1749-56 and 1765.
John Buddle Collection.
T. E. Forster Collection.
Watson Collection.

WHITEHAVEN
Public Library:
Minute Books of the Town and Harbour Trustees, 1829-84.
MS 30 John Spedding's Token Book 1707-14.
MS 44 List of employees at Howgill Colliery, 8 May 1802.
MS 48 Workington Colliery Book 1799-1853.
MS 160 Whitehaven Colliery Book 1828-35.

Cumberland Miners' Association Offices, 35 Queen Street:
Minute Books.
Correspondence and miscellaneous records.

Theses

Bainbridge, T. H., 'Geographical Factors Controlling Industrial Occupations in Cumbria with particular reference to the Coal Measure and Iron Ore Areas'. University of Durham M.Sc., 1934.
Beckett, J. V., 'Landownership in Cumbria 1680-1750'. University of Lancaster Ph.D., 1975.
Eaglesham, A., 'The Growth and Influence of the West Cumberland Shipping Industry, 1660-1800'. University of Lancaster Ph.D., 1977.
Makey, W. H., 'The Place of Whitehaven in the Irish Coal Trade 1600-1750'. University of London M.A., 1952.
Phillips, C. B., 'The Gentry in Cumberland and Westmorland 1600-65'. University of Lancaster Ph.D., 1973.
Williams, J. E., 'The Growth and Decline of the Port of Whitehaven, Cumberland, 1650-1900'. University of Leeds M.A., 1951.
Wood, O., 'The Development of the Coal, Iron and Shipbuilding Industries of West Cumberland, 1750-1914'. University of London Ph.D., 1952.

Articles

Allen, J. S., 'The 1715 and other Newcomen Engines at Whitehaven, Cumberland', *Transactions of the Newcomen Society*, xlv (1972-3).
Bainbridge, T. H., 'The West Cumberland Coalfield', *Economic Geography*, xii (1936).
Bainbridge, T. H., 'Cumberland Population Movements 1871-81', *Geographical Journal* (1946).
Bainbridge, T. H., 'Population Changes over the West Cumberland Coalfield', *Economic Geography* (1949).
Ferguson, R. S., 'The Colliery, Harbour, Lime and Iron Tokens of West Cumberland', CW1 (1898).
Fisher, J., 'Observations and Inquiries made upon and concerning the coal works at Whitehaven, in the county of Cumberland, in the year 1793', *Annual Register*, xxxvi (1794).
Fletcher, I., 'The Archaeology of the West Cumberland Coal Trade', CW1 (1878).
Ford, P., 'Tobacco and Coal: A Note on the Economic History of Whitehaven', *Economica*, ix (1929).
Fox, P. H., 'Cumberland Ports and Shipping in the reign of Elizabeth', CW2 (1921).
Hudleston, C. R., 'An 18th Century Squire's Possessions', CW2 (1957).
Hughes, E., 'The Founding of Maryport', CW2 (1964).
Jarvis, R. C., 'The Appointment of Ports in Cumberland, Westmorland and Lancashire north-of-the-sands', CW2 (1947).
Moore, R. W., 'Coal Mining', Victoria County History, *Cumberland*, ii (1905).
Namier, L. B., 'Anthony Bacon, M.P. An Eighteenth Century Merchant', *Journal of Economic and Business History*, ii (1929).
Prevost, W. A. J., ' "A Trip to Whitehaven to visit the coal works there in 1739" by Sir John Clerk', CW2 (1965).
Steel, J. P., 'Genealogical Gleanings relating to Cumberland', CW2 (1923).
Wood, O., 'A Cumberland Colliery During the Napoleonic War', *Economica*, xxi (1954).
Wood, O., 'The Collieries of J. C. Curwen', CW2 (1971).
Wood, O., 'A Colliery Payroll in 1802', CW2 (1972).
Wood, O., 'The Chapel Bank Colliery Disaster, 1837', CW2 (1973).

Transactions etc.

Cumberland and Westmorland Antiquarian and Archaeological Society, *Transactions*, 1875 to date.

Cumberland and Westmorland Association for the Advancement of Literature and Science, *Transactions*, 1875-92.

North of England Institute of Mining Engineers, *Transactions*, 1852-89.

Federated Institute of Mining Engineers, *Transactions*, 1889 to date.

Whitehaven Scientific Association, *Annual Journal*, 1898-1913.

Newcomen Society, *Transactions*, 1921 to date.

The Scallop Shell (St. James's Church Magazine), 8 vols., 1936-43. (Contains a series of articles on the history of Whitehaven by Mr W. Watson.)

Local Newspapers

The largest collection of local newspapers in Cumberland is in the Carlisle Public Library. A valuable guide to the location and extent of files of Cumberland newspapers is F. Barnes and J. L. Hobbs, *Handlist of Newspapers Published in Cumberland, Westmorland and North Lancashire.* (Kendal, 1951).

Official Publications

Calendar of State Papers Domestic, 1547-1703.

Calendar of State Papers Ireland, 1600-62.

Calendar of Treasury Papers, 1557-1728.

Calendar of Treasury Books, 1660-1718.

Calendar of Treasury Books and Papers, 1729-45.

Historical Manuscripts Commission, *Thirteenth Report, The Manuscripts of the Earl of Lonsdale.* 1893.

Censuses of Population, 1801-1971.

Annual Publications:

Board of Trade, *Statistical Abstract for the United Kingdom*, 1854-1940.

Central Statistical Office, *Annual Abstract of Statistics*, 1948 to date.

Memoirs of the Geological Survey of Great Britain, *Mineral Statistics of the United Kingdom*, 1854-81.

Mines Department, *Mineral Statistics of the United Kingdom*, 1882-96.

Mines Department, *Mines and Quarries General Report with Statistics*, 1897-1920.

Report of Inspectors of Mines, 1851-1973.

Mines Department, Report of the Secretary for Mines, 1921-38.

Mines Department, *Report of the Chief Inspector of Mines*, 1939-74.

Mines Department, *List of Mines and Quarries*, 1897-1950.

Board of Trade, *Abstract of Labour Statistics of the United Kingdom*, 1893-1936.

Board of Trade, *Statistical Tables and Report on Trade Unions*, 1887-1913.

Reports of the Commissioner for the Special Areas in England and Wales, 1935-38.

National Coal Board, *Report and Accounts*, 1947 to date.

Ministry of Fuel and Power, *Statistical Digest*, 1944-67.

Department of Energy, *Digest of United Kingdom Energy Statistics*, 1968 to date.

Annual Statement of the Trade and Navigation of the United Kingdom, 1853 to date.

Reports:

Select Committee on the Coal Trade, 1800.

Select Committee on the Combination Laws, 1825.

Lords' Select Committee on the State of the Coal Trade, 1829.
Lords' Select Committee on the State of the Coal Trade, 1830.
Select Committee on Manufactures, Commerce and Shipping, 1833.
Select Committee on Accidents in Mines, 1835.
Select Committee on the Payment of Wages, 1842.
Children's Employment Commission (Mines), 1842.
General Board of Health, *Report to the General Board of Health on a preliminary inquiry into the sewerage, drainage and supply of water and the sanitary condition of the inhabitants of the town of Whitehaven*, by Robert Rawlinson, 1849.
Commission on the State of the Population in the Mining Districts, 1850-9.
Select Committee on Accidents in Coal Mines, 1853.
Royal Commission on the Condition of all Mines in Great Britain (Health and Safety of Miners), 1862-4.
Committee on Operation of Acts for the Regulation and Inspection of Mines, 1866-7.
Royal Commission on Coal in the United Kingdom, 1871.
Royal Commission on Accidents in Coal Mines, 1879-86. C 4699 (1886).
Royal Commission on the Depression of Trade and Industry, 1886.
Committee on Days per Week and Hours per Shift worked in Coal Mines, 1890.
Royal Commission on Mining Royalties, 1889-93.
Royal Commission on Labour, 1891-2.
General Report on Wages of Manual Labour Classes in the United Kingdom, 1893.
Report on Standard Piece Rates of Wages and Sliding Scales, 1900.
Royal Commission on Coal Resources, 1903-5.
Departmental Committee on the Probable Economic Effects of a Limit of Eight Hours to the Working Day of Coal Miners, 1907. Cmd. 3505.
Report on Collective Agreements, 1910.
Royal Commison on Mines, 1907-11.
Royal Commission on the Coal Industry, 1919.
Royal Commission on the Coal Industry, 1925. Cmd. 2600 (1926).
Royal Commission on Safety in Coal Mines, 1938. Cmd. 5890 (1938).
Ministry of Labour, *Reports of Investigations into Industrial Conditions in certain Depressed Areas*. Cmd. 4728 (1934).
Ministry of Fuel and Power, *Northumberland and Cumberland Coalfields: Regional Survey Report (Northern 'A' Region)*, 1945.
Financial Position of the Coal Mining Industry: Coal Charges Account, Cmd. 6617 (1945).
Pay Board Special Report, *Relative Pay of Mineworkers*. Cmd. 5567 (1974).

Reports on Pit Accidents:
Wellington Pit, Whitehaven, 11 May 1910. Cmd. 5524 (1911).
Haig Pit, Whitehaven, 5 September 1922. Cmd. 1796. (1923).
Haig Pit, Whitehaven, 29 January 1931. Cmd. 3858 (1931).
William Pit, Whitehaven, 3 June 1941. Cmd. 6367 (1942).
Harrington No. 10 Colliery, 9 December 1946. Cmd. 7222 (1947).
William Pit, Whitehaven, 15 August 1947. Cmd. 7410 (1948).
Risehow Colliery, 26 November 1957. Cmd. 396 (1958).

Cumberland Coal Trade:
Cumberland Coal Commission, *Minutes of Proceedings*, 1892 and 1899.
Board of Conciliation, *Minutes of Proceedings*, 1902-7.

Books

The place of publication is London unless otherwise stated.

Aberconway, Lord, *The Basic Industries of Great Britain* (1927).

Abercrombie, P. and Kelly, S. A., *Cumbrian Regional Report* (Liverpool, 1932).

Andrews, P. W. S. and Brunner, S. A., *Capital Development in Steel* (Oxford, 1951).

Arnot, R. P., *The Miners: A History of the Miners' Federation of Great Britain, 1889-1910* (1949).

Arnot, R. P., *The Miners: Years of Struggle* (1953).

Arnot, R. P., *The Miners in Crisis and War* (1961).

Ashton, T. S. and Sykes, J., *The Coal Industry of the Eighteenth Century* (Manchester, 1929).

Ayton, R. and Daniell, W., *A Voyage Round Great Britain*, 2 vols. (1814-15).

Bailey, J. B., *Notes on the History of Maryport, Ancient and Modern* (Preston, 1883).

Beckett, J. C., *The Making of Modern Ireland 1603-1923* (1966).

Beckett, J. V., *Coal and Tobacco: The Lowthers and the Economic Development of West Cumberland, 1660-1760* (Cambridge, 1981).

Berkovitch, I., *Coal on the Switchback: The Coal Industry since Nationalisation* (1977).

Bouch, C. M. L., *Prelates and People of the Lake Counties* (Kendal, 1948).

Boyd., R. N., *Coal Pits and Pitmen* (1892).

Bulman, H. F., *Coal Mining and the Coal Miner* (1920).

Bulmer, T., *History, Topography and Directory of West Cumberland* (Manchester, 1883).

Burn, D. (ed.), *The Structure of British Industry*, 2 vols. (Cambridge, 1958).

Buxton, N. K., *The Economic Development of the British Coal Industry: From the Industrial Revolution to the Present Day* (1978).

Caine, C., *Cleator and Cleator Moor: Past and Present* (Kendal, 1916).

Carvel, J. L., *The Coltness Iron Company* (Privately printed, 1948).

Central Statistical Office, *Statistical Digest of the War* (1951).

Challinor, R. and Ripley, B., *The Miners' Association* (1968).

Chester, Sir Norman, *The Nationalisation of British Industry 1945-51* (1975).

Clapham, J. H., *Economic History of Modern Britain*, 3 vols. (Cambridge, 1931-8).

Cole, G. D. H., *Labour in the Coal-Mining Industry, 1914-21* (Oxford, 1923).

Collingwood, W. G., *Elizabethan Keswick* (Kendal, 1912).

Collingwood, W. G., *The Lake Counties* (1932).

Connell, K. H., *The Population of Ireland 1750-1845* (Oxford, 1950).

Court, W. H. B., *History of the Second World War: Coal* (1951).

Cullen, L. M., *Anglo-Irish Trade 1660-1800* (Manchester, 1968).

Cumberland Development Council Ltd., *Annual Reports* (Whitehaven, 1936-72).

Davis, R., *The Rise of the English Shipping Industry* (1962).

Daysh, G. H. J., *West Cumberland (with Alston): A Survey of Industrial Facilities* (Whitehaven, 1938).

Daysh, G. H. J. and Watson, E. M., *Cumberland: A Survey of Industrial Facilities* (Whitehaven, 1951).

Defoe, D. *A Tour through England and Wales* (Everyman ed.).

Department of Employment, *British Labour Statistics: Historical Abstract 1886-1968* (1971).

Devereux, W. C., *An Industrial Plan for West Cumberland* (Whitehaven, 1944).

Dickie, J. P., *The Coal Problem, A Survey: 1910-36* (1936).

Dixon, J., *The Literary Life of William Brownrigg, M.D., F.R.S., to which are added An Account of the Coal Mines near Whitehaven and Observations on the Means of Preventing Epidemic Fevers* (Whitehaven, 1801).

Dock and Harbour Authorities Association, *Port Statistics for the Foreign Trade of the United Kingdom* (Annual 1961-70).

Donald M. B., *Elizabethan Copper: The History of the Company of Mines Royal* (1955).
Dunn, M., *The Winning and Working of Collieries* (Newcastle, 1848).
Dunning, J. H. and Thomas, C. J., *British Industry* (1963).
Dupin, Baron, *The Commercial Power of Great Britain*, 2 vols. (1825).
Eastwood, T., *The Geology of the Maryport District* (1930).
Eastwood, T., *The Geology of the Whitehaven and Workington District* (1931).
Eden, Sir F. M., *The State of the Poor*, 3 vols. (1797).
Ferguson, R. S., *A History of Cumberland* (1890).
Forsyth, D. J. C. and Kelly, D. M., *Studies in the British Coal Industry* (1969).
Freeman, T. W., *Ireland: Its Physical, Historical, Social and Economic Geography* (1950).
Galloway, R. L., *Annals of Coal Mining and the Coal Trade*, 2 vols. (1898-1904).
Gibson, F. A., *Statistics of the Coal Mining Industry of the United Kingdom* (Cardiff, 1922).
Griffin, A. R., *The British Coalmining Industry* (Buxton, 1977).
Hainsworth, D. R., *Commercial Papers of Sir Christopher Lowther 1611-44*, Surtees Society, vol. clxxxix (1977).
Hay, D., *Whitehaven: An Illustrated History* (Whitehaven, 1979).
Haynes, W. W., *Nationalization in Practice: The British Coal Industry* (1953).
Heinemann, M., *Britain's Coal* (1944).
Hughes, E., *North Country Life in the Eighteenth Century: Cumberland and Westmorland 1700-1830* (Oxford, 1965).
Hull, E., *The Coal Fields of Great Britain* (1905).
Hutchinson, W., *History of the County of Cumberland*, 2 vols. (Carlisle, 1794).
Jackson, W., *Papers and Pedigrees mainly relating to Cumberland and Westmorland*, 2 vols.(Kendal, 1892).
Jars, G., *Voyages Métallurgiques*, 3 vols. (Lyons and Paris, 1774-81).
Jefferson, S., *History and Antiquities of Leath Ward* (Carlisle, 1840).
Jefferson, S., *History and Antiquities of Allerdale Ward above Derwent* (Carlisle, 1842).
Jevons, H. S., *The British Coal Trade* (1915).
Jewkes, J. and Winterbottom, A., *An Industrial Survey of Cumberland and Furness* (Manchester, 1933).
Jollie, F., *Cumberland Guide and Directory* (Carlisle, 1811).
Jones, G. P. and Pool, A. G., *A Hundred Years of Economic Development* (1940).
Jones, J. H., Cartwright, G. and Guenault, P. H., *The Coal-Mining Industry* (1939).
Kirby, M. W., *The British Coalmining Industry 1870-1946: A Political and Economic History* (1977).
Knight, E. M., *Men Leaving Mining: West Cumberland 1966-7. Report to the Ministry of Labour* (Newcastle upon Tyne, 1968).
Lancaster, J. Y. and Wattleworth, D. R., *The Iron and Steel Industry of West Cumberland* (Workington, 1977).
Lee, W. A., *Thirty Years in Coal 1917-47* (1954).
Lewis, B., *Coal Mining in the Eighteenth and Nineteenth Centuries* (1971).
Lewis, M. J. T., *Early Wooden Railways* (1970).
Lysons, D. and S., *Magna Britannia*, vol. iv (1816).
Mannix and Whellan, *History, Directory and Gazetteer of Cumberland* (Beverley, 1847).
Marshall, J. D. and Davies-Shiel, M., *The Industrial Archaeology of the Lake Counties* (1969).
Meade, R., *The Coal and Iron Industries of the United Kingdom* (1882).
Mining Association of Great Britain, *Statistical Review of the Coal Industry* (1925 and 1933-4).
Mitchell, B. R. and Deane, P., *Abstract of British Historical Statistics* (Cambridge, 1962).
Mitchell, B. R. and Jones, H. G., *Second Abstract of British Historical Statistics* (Cambridge, 1971).
National Ports Council, *Digest of Port Statistics* (Annual 1967-77).

Nef, J. U., *The Rise of the British Coal Industry*, 2 vols. (1932).
Neuman, A., *Economic Organisation of the British Coal Industry* (1934).
Nicolson, J. and Burn, R., *The History and Antiquities of the Counties of Westmorland and Cumberland*, 2 vols. (1777).
Northern Economic Planning Council, *Regional Ports Survey* (1969).
Nott-Bower, Sir G. and Walkerdine, R. H., *National Coal Board: The First Ten Years* (1957).
O'Brien, G., *The Economic History of Ireland in the Seventeenth Century* (Dublin, 1919).
O'Brien, G., *The Economic History of Ireland in the Eighteenth Century* (Dublin, 1918).
Page, T., *Report on the Proposed Harbour at Maryport* (1865).
Parson, W. and White, W., *History, Directory and Gazetteer of the Counties of Cumberland and Westmorland* (Leeds, 1829).
Pennant, T., *A Tour in Scotland and Voyage to the Hebrides* (Chester, 1774).
Platt, J., *British Coal* (1968).
Political and Economic Planning, *Report on the British Coal Industry* (1936).
Pollard, S., *The Development of the British Economy, 1914-1967*, 2nd edn. (1969).
Porter, G. R., *The Progress of the Nation* (ed. F. W. Hirst) (1912).
Postlethwaite, J., *Mines and Mining in the Lake District*, 3rd. edn. (Whitehaven, 1913).
Redford, A., *Labour Migration in England 1800-50* (Manchester, 1926).
Redmayne, Sir R. A. S., *The British Coal Mining Industry During the War* (Oxford, 1923).
Redmayne, Sir R. A. S., *Men, Mines and Memories* (1942).
Reid, G. L., Allen, K. and Harris, D. J., *The Nationalized Fuel Industries* (1973).
Robens, Lord, *Ten Year Stint* (1972).
Rowe, J. W. K., *Wages in the Coal Industry* (1923).
Simmons, J., *The Maryport and Carlisle Railway* (1947).
Simpson, E. S., *Coal and the Power Industries in Postwar Britain* (1966).
Sugden, E. H., *A History of the Parish of Arlecdon and Frizington* (Whitehaven, 1897).
Sweezy, P. M., *Monopoly and Competition in the English Coal Trade, 1550-1850* (Harvard, 1938).
Townshend-Rose, H., *The British Coal Industry* (1951).
Victoria County History, *Cumberland*, 2 vols. (1905).
Ward, J. T. and Wilson, R. G. (eds.), *Land and Industry: The Landed Estate and the Industrial Revolution* (Newton Abbot, 1971).
Whellan, W., *The History and Topography of the Counties of Cumberland and Westmorland* (Pontefract, 1860).
Willan, T. S., *The English Coasting Trade, 1660-1750* (Manchester 1938).
Wilson, H., *New Deal for Coal* (1945).
Youngson, A. J., *The British Economy 1920-57* (1960).

INDEX

Compiled by Joyce Cherry